D1799810

METHODS IN MOLECULAR BIOLOGY

Series Editor
John M. Walker
School of Life and Medical Sciences
University of Hertfordshire
Hatfield, Hertfordshire, AL10 9AB, UK

For further volumes:
http://www.springer.com/series/7651

Peptide Microarrays

Methods and Protocols

Second Edition

Editors

Marina Cretich

Consiglio Nazionale delle Ricerche, Milano, Italy

Marcella Chiari

Consiglio Nazionale delle Ricerche, Milano, Italy

 Humana Press

Editors
Marina Cretich
Consiglio Nazionale delle Ricerche
Milano, Italy

Marcella Chiari
Consiglio Nazionale delle Ricerche
Milano, Italy

ISSN 1064-3745   ISSN 1940-6029   (electronic)
Methods in Molecular Biology
ISBN 978-1-4939-3036-4   ISBN 978-1-4939-3037-1   (eBook)
DOI 10.1007/978-1-4939-3037-1

Library of Congress Control Number: 2015950855

Springer New York Heidelberg Dordrecht London

© Springer Science+Business Media New York 2016
This work is subject to copyright. All rights are reserved by the Publisher, whether the whole or part of the material is concerned, specifically the rights of translation, reprinting, reuse of illustrations, recitation, broadcasting, reproduction on microfilms or in any other physical way, and transmission or information storage and retrieval, electronic adaptation, computer software, or by similar or dissimilar methodology now known or hereafter developed.
The use of general descriptive names, registered names, trademarks, service marks, etc. in this publication does not imply, even in the absence of a specific statement, that such names are exempt from the relevant protective laws and regulations and therefore free for general use.
The publisher, the authors and the editors are safe to assume that the advice and information in this book are believed to be true and accurate at the date of publication. Neither the publisher nor the authors or the editors give a warranty, express or implied, with respect to the material contained herein or for any errors or omissions that may have been made.

Printed on acid-free paper

Humana Press is a brand of Springer
Springer Science+Business Media LLC New York is part of Springer Science+Business Media (www.springer.com)

Preface

Peptide microarrays, with their unlimited content flexibility, are formidable tools in many areas of biochemistry and medicine. For example, they enable the design of enzyme inhibitors and enhancers or advance structural and functional information on protein interactions. When employed in epitope mapping, they serve to identify antibody binding domains that can be implemented in vaccine and new therapeutics development.

Peptide microarrays can also diagnose diseases. By the design of peptide arrays incorporating epitope collections, the huge amount of immunological information hidden in the plasma of an individual can be revealed, thus profiling his or her personalized immunoresponse to infection, vaccination, allergens, and autoimmunity and providing hints on new biomarkers. Peptides in a miniaturized and multiplexed format test can simultaneously screen for dozens to thousands of biomarkers in a single assay with easier and less expensive protocols than most DNA or protein microarrays.

Advancements made in recent years in peptide library synthesis, immobilization chemistry, and array production have created a ground from which different new applications are derived, extending the ways in which peptide microarray technology is applied every day. The maturity of the technique is now proven by the many clinical applications published and by the several commercial providers offering support and services.

Peptide Microarrays: Methods and Protocols, Second Edition, is an updated outlook on peptide array technology. The 21 chapters in this volume provide insights into the technological fundamentals together with a comprehensive overview of the potentialities of this technology in basic research and clinical assays. Not all techniques could be described fully, but we have tried to match timely and useful new developments with consolidated approaches for both the experienced researcher and the novice to the field.

The book is divided into three parts each containing an introductory review chapter and a collection of protocols:

Part I, *Peptide arrays: cutting-edge methods and technology*, provides an overview of methodological aspects and is focused on general applications of peptide arrays such as affinity study, enzyme activity screening, epitope mapping, and secondary structure determination.

Part II, *Chemoselective strategies to peptide immobilization*, is devoted to smart methods of peptide binding and oriented display on surfaces.

Part III, *Peptide microarrays for medical applications*, comprises examples of clinical applications in the fields of allergy and autoimmunity.

Taken together, these chapters provide a set of invaluable tools to advance and extend research in peptide and protein chemistry, and we hope that our readers will take advantage of the unique insights and novel solutions offered by the authors of each chapter of this book. It is our hope that this volume will encourage scientists to apply current peptide array protocols to the study of interesting new biochemical and medical questions and will assist researchers aiming at developing new methods to further develop the peptide microarray technology.

We are grateful to all the authors for sharing their valuable knowledge when contributing this volume. We thank Prof. John Walker and the editorial staff of Springer for continuous encouragement and assistance. Finally, we wish to express gratitude to the members of our group and to all those who supported and continue to do so.

Milan, Italy   *Marina Cretich*
Milan, Italy   *Marcella Chiari*

Contents

Contributors

GEORGE P. ANDERSON • *US Naval Research Laboratory, Washington, DC, USA*

FRUZSINA BABOS • *MTA-ELTE Research Group of Peptide Chemistry, Hungarian Academy of Sciences, Budapest, Hungary*

SARA BENEDÉ • *Pediatric Allergy & Immunology, Icahn School of Medicine at Mount Sinai, New York, NY, USA; Institute of Food Science Research (CIAL) CSIC-UAM, Madrid, Spain*

ANDREA BRIONES • *School of Biological Sciences, Centre for Biomedical Sciences, Royal Holloway University of London, Egham, Surrey, UK*

MARCELLA CHIARI • *Istituto di Chimica del Riconoscimento Molecolare (ICRM), Consiglio Nazionale delle Ricerche (CNR), Milan, Italy*

MARINA CRETICH • *Istituto di Chimica del Riconoscimento Molecolare (ICRM), Consiglio Nazionale delle Ricerche (CNR), Milan, Italy*

YANG DENG • *Department of Biomedical Engineering, Yale University, New Haven, CT, USA*

SALVATORE G. DE-SIMONE • *Centro de Desenvolvimento Tecnológico em Saúde (CDTS)/ Instituto Nacional de Ciência e Tecnologia de Inovação em Doenças Negligenciadas (INCT-IDN), FIOCRUZ, Rio de Janeiro, Brazil*

THATIANE G. DE-SIMONE • *FIOCRUZ, Center of Technological Development in Health (CDTS) / National Institute of Science and Technology for Innovation on Neglected Diseases (INCT-IDN) FIOCRUZ, Rio de Janeiro, Brazil*

MAREN ECKEY • *JPT Peptide Technologies GmbH, Berlin, Germany*

GEORG ERHARDT • *Department of Animal Breeding and Genetics, Justus-Liebig University Gießen, Gießen, Germany*

QIAOJUN FANG • *CAS Key Laboratory for Biomedical Effects of Nanomaterials & Nanosafety, National Center for Nanoscience and Technology of China, Beijing, China*

CHIARA FINETTI • *Istituto di Chimica del Riconoscimento Molecolare (ICRM), Consiglio Nazionale delle Ricerche (CNR), Milan, Italy*

GUSTAVO GIMENEZ • *Pediatric Allergy & Immunology, Icahn School of Medicine at Mount Sinai, New York, NY, USA*

ALESSANDRO GORI • *Istituto di Chimica del Riconoscimento Molecolare (ICRM), Consiglio Nazionale delle Ricerche (CNR), Milan, Italy*

RAPHAEL GOTTARDO • *Vaccine and Infectious Disease Division, Fred Hutchinson Cancer Research Center, Seattle, WA, USA*

CARSTEN GRÖTZINGER • *Department of Gastroenterology and Molecular Cancer Research Center (MKFZ), Charité – Universitätsmedizin Berlin, Berlin, Germany*

STEVEN R. HEAD • *DNA Array Core, Department of Immunology & Microbial Science, The Scripps Research Institute, La Jolla, CA, USA*

BELÉN DE LA HOZ • *Servicio de Alergología, Hospital Universitario Ramón y Cajal-IRYCIS, Madrid, Spain*

ZHIYUAN HU • *CAS Key Laboratory for Biomedical Effects of Nanomaterials & Nanosafety, National Center for Nanoscience and Technology of China, Beijing, China*

KRISZTINA HUBER • *Department of Immunology, Eötvös Loránd University, Budapest, Hungary*

FERENC HUDECZ • *Department of Organic Chemistry, Eötvös Loránd University, Budapest, Hungary; MTA-ELTE Research Group of Peptide Chemistry, Hungarian Academy of Sciences, Budapest, Hungary*

GREGORY IMHOLTE • *Vaccine and Infectious Disease Division, Fred Hutchinson Cancer Research Center, Seattle, WA, USA*

JANINA JANSONG • *JPT Peptide Technologies GmbH, Berlin, Germany*

KOTARO KAJIKAWA • *Department of Electronics and Applied Physics, Interdisciplinary Graduate School of Science and Engineering, Tokyo Institute of Technology, Yokohama, Japan*

TOBIAS KNAUTE • *JPT Peptide Technologies GmbH, Berlin, Germany*

MANSUN LAW • *DNA Array Core, Department of Immunology & Microbial Science, The Scripps Research Institute, La Jolla, CA, USA*

TAO LI • *State Key Laboratory of Electroanalytical Chemistry, Changchun Institute of Applied Chemistry, Chinese Academy of Sciences, Changchun, China*

JING LIN • *Pediatric Allergy & Immunology, Icahn School of Medicine at Mount Sinai, New York, NY, USA*

MARIA LISSON • *Department of Animal Breeding and Genetics, Justus-Liebig University Gießen, Gießen, Germany*

XIA LIU • *State Key Laboratory of Electroanalytical Chemistry, Changchun Institute of Applied Chemistry, Chinese Academy of Sciences, Changchun, China*

DIANJUN LIU • *State Key Laboratory of Electroanalytical Chemistry, Changchun Institute of Applied Chemistry, Chinese Academy of Sciences, Changchun, China*

RENATO LONGHI • *Istituto di Chimica del Riconoscimento Molecolare (ICRM), Consiglio Nazionale delle Ricerche (CNR), Milan, Italy*

ANNA MAGYAR • *MTA-ELTE Research Group of Peptide Chemistry, Hungarian Academy of Sciences, Budapest, Hungary*

JAVIER MARTÍNEZ-BOTAS • *Servicio de Bioquímica-Investigación, Hospital Universitario Ramón y Cajal-IRYCIS and CIBER de Fisiopatología de la Obesidad y Nutrición (CIBEROBN), Instituto de Salud Carlos III, Madrid, Spain*

RYAN MCBRIDE • *DNA Array Core, Department of Immunology & Microbial Science, The Scripps Research Institute, La Jolla, CA, USA*

HISAKAZU MIHARA • *Department of Bioengineering, Graduate School of Bioscience and Biotechnology, Tokyo Institute of Technology, Yokohama, Japan*

PALOMA NAPOLEÃO-PÊGO • *Department of Cellular and Molecular Biology, Federal Fluminense University, Biology Institute, Rio de Janeiro, Brazil*

STELLA H. NORTH • *US Naval Research Laboratory, Washington, DC, USA; Latham & Watkins LLP, Washington, DC, USA*

PHILLIP ORDOUKHANIAN • *Protein & Nucleic Acid Research Core, Department of Immunology & Microbial Science, The Scripps Research Institute, La Jolla, CA, USA*

NIKOLAUS PAWLOWSKI • *JPT Peptide Technologies GmbH, Berlin, Germany*

MARC E. PFEIFER • *Institute of Life Technologies, University of Applied Sciences and Arts Western Switzerland, Sion, Switzerland*

JUDIT POZSGAY • *Department of Immunology, Eötvös Loránd University, Budapest, Hungary*

DENIS PRIM • *Institute of Life Technologies, University of Applied Sciences and Arts Western Switzerland, Sion, Switzerland*

ULF REIMER • *JPT Peptide Technologies GmbH, Berlin, Germany*

GABRIELLA SARMAY • *Department of Immunology, Eötvös Loránd University, Budapest, Hungary*

RENAN SAUTERAUD • *Vaccine and Infectious Disease Division, Fred Hutchinson Cancer Research Center, Seattle, WA, USA*

LISA C. SHRIVER-LAKE • *US Naval Research Laboratory, Washington, DC, USA*

LAURA SOLA • *Istituto di Chimica del Riconoscimento Molecolare (ICRM), Consiglio Nazionale delle Ricerche (CNR), Milan, Italy*

MIKHAIL SOLOVIEV • *School of Biological Sciences, Centre for Biomedical Sciences, Royal Holloway University of London, Egham, Surrey, UK*

AMIR SYAHIR • *Department of Biochemistry, Faculty of Biotechnology and Biomolecular Sciences, Universiti Putra Malaysia, Serdang, Malaysia*

ESZTER SZARKA • *Department of Immunology, Eötvös Loránd University, Budapest, Hungary*

CHRIS R. TAITT • *US Naval Research Laboratory, Washington, DC, USA*

VÍCTOR TAPIA MANCILLA • *Institute of Medical Immunology, Charité-Universitätsmedizin zu Berlin, Berlin, Germany*

CHRISTOPH TERSCH • *JPT Peptide Technologies GmbH, Berlin, Germany*

KIN-YA TOMIZAKI • *Department of Materials Chemistry, Ryukoku University, Otsu, Japan*

BENJAMIN E. TURK • *Department of Pharmacology, Yale University School of Medicine, New Haven, CT, USA*

KENJI USUI • *FIRST (Faculty of Frontiers of Innovative Research in Science and Technology), Konan University, Kobe, Japan*

RUDOLF VOLKMER • *Institute of Medical Immunology, Charité-Universitätsmedizin zu Berlin, Berlin, Germany*

WEIZHI WANG • *Laboratory for Biomedical Effects of Nanomaterials & Nanosafety, National Center for Nanoscience and Technology of China, Beijing, China*

ZHENXIN WANG • *CAS Key Laboratory for Biomedical Effects of Nanomaterials and Nanosafety, Chinese Academy of Sciences, Changchun, China*

JOHANNES ZERWECK • *JPT Peptide Technologies GmbH, Berlin, Germany*

FAN ZHANG • *School of Biological Sciences, Centre for Biomedical Sciences, Royal Holloway University of London, Egham, Surrey, UK*

CATERINA ZILIO • *Istituto di Chimica del Riconoscimento Molecolare (ICRM), Consiglio Nazionale delle Ricerche (CNR), Milan, Italy*

Part I

Peptide Arrays: Cutting-Edge Methods and Technology

Chapter 1

Peptide Arrays on Planar Supports

Víctor Tapia Mancilla and Rudolf Volkmer

Abstract

On a past volume of this monograph we have reviewed general aspects of the varied technologies available to generate peptide arrays. Hallmarks in the development of the technology and a main sketch of preparative steps and applications in binding assays were used to walk the reader through details of peptide arrays. In this occasion, we resume from that work and bring in some considerations on quantitative evaluation of measurements as well as on selected reports applying the technology.

Key words Peptide, Analyte, Capture assay, Microarray, Spot synthesis, Immobilization

1 Introduction

Peptide arrays are high-throughput devices for binding assays. The array format refers to the spatially addressable presentation of peptide probes immobilized to discrete areas of a support surface. This principle has been extensively and fruitfully applied in RNA hybridization experiments using cDNA microarray devices [1–3]. Development in this area has led to the establishment of advanced technical aspects respecting immobilization chemistries, the industrialization of dedicated equipment, and tailor-made bioinformatic tools. Current aims to globally analyze cellular behavior relying on cDNA capture molecules include mutation and polymorphism analysis, the determination of clinically predictive genes, as well as the continuous efforts to relate specific expression profiles to a particular cellular state. However, distribution of cellular labor between DNA and protein structures may render measurements of genetic change insufficient to explain emergent phenotypes [4] involved in normal or pathological cellular behavior. Situations in which this point cannot be overseen are the intricate balance between promiscuity and selectivity of protein interactions as well as in the regulation of enzyme activity. Here, assays at the protein level are required to point out gene function at the level of the phenotype. Moreover, in pragmatic terms, RNA levels do not

Marina Cretich and Marcella Chiari (eds.), *Peptide Microarrays: Methods and Protocols*, Methods in Molecular Biology, vol. 1352, DOI 10.1007/978-1-4939-3037-1_1, © Springer Science+Business Media New York 2016

correlate with protein abundance [5–7] and proteins, rather than genes, are the immediate causative agents of cellular function.

Several technologies aim to bridge this causative gap and are coined under the research field of proteomics. Rudely stated, their object of study is the proteome, originally defined as "the entire complement of proteins expressed by a genome, cell, tissue or organism" [8]. Advanced protein science is characteristic of the field and it is dedicated to satisfy the original expectations of functional genomics to globally analyze cellular organization. Just as genomics has reached an endpoint in gene discovery, namely the sequencing of the human genome, an equivalent ambition to discover the human proteome is borne by protein science [9, 10].

Protein arrays are predestined to support proteomic research. Due to their structural and functional complexity, proteins are highly informative systems and well suited to investigate protein interactions occurring over extensive complementary surfaces. However, several critical factors such as stability, native folding, or activity of the immobilized proteins pose stringent demands on production, storage, and experimental conditions. Due to that it is not amazing that the first reports on protein arrays came just in time for the new millennium [11–13].

Peptides, on the contrary, are chemically quite resistant and have more modest dynamics to achieve their active conformations. In general, peptides retain partial aspects of protein function and, thus, peptide arrays are suitable to support proteomic research in a more reductionist approach. Not surprisingly, peptides paved the way for array technologies: in 1991 two techniques for the preparation of peptide arrays were published.

Firstly, Frank et al. presented the SPOT technique—a stepwise synthesis of peptides on filter paper [14]. Secondly, Fodor et al. reported the concept of light-directed spatially addressable chemical synthesis [15]. Both techniques are milestones in the advancement of the peptide array technology but the former has found wider spread as preparative source of peptides for varied biological applications and for binding assays carried out in situ. The assay format and work flow are similar to ELISA applications and consist in the capture of analyte in a biological sample by a collection of peptide probes found attached to discrete areas of a supporting solid phase.

Figure 1 illustrates the four main procedures involved in a representative peptide array-based assay: (a) biological sample collection, (b) peptide array preparation, (c) analyte capture assay, and (d) evaluation of measured responses. Biological samples or specimens may be collected by diverse techniques like venipuncture, plasma fractionation, biopsy, and urination. Repositories of standardized biosamples used to train supervised learning machine models for data mining may also be available. Procedures for their collection and preparation are highly tailor-made for individual applications, escaping the frame of this review.

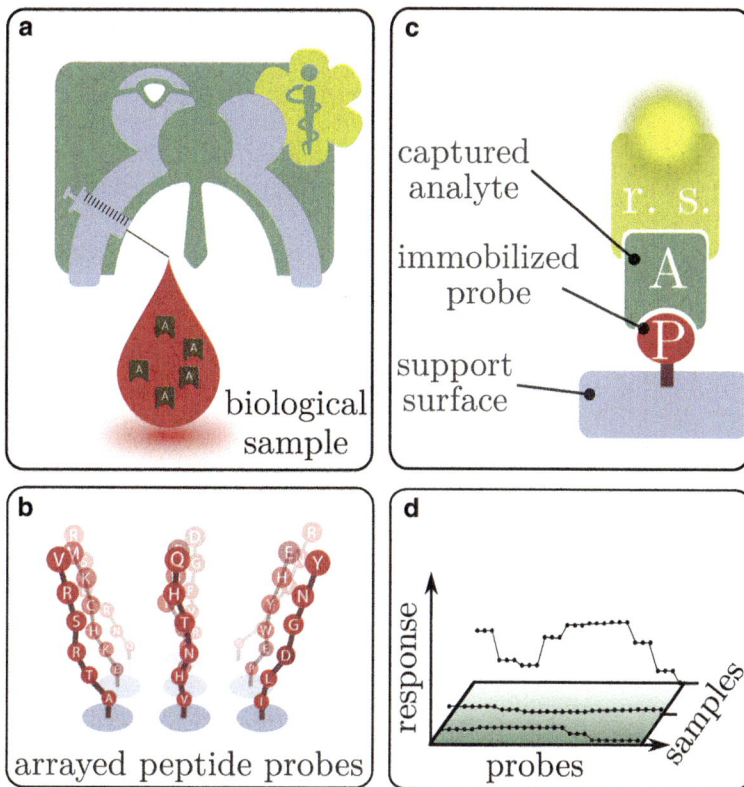

Fig. 1 Analyte capture assay using arrayed peptide probes: general work flow and terminology. (**a**) Sample collection or preparation. Sample refers to the solution containing the analytes under study. (**b**) Preparation of peptide probes. A collection of peptides may be synthesized in situ (e.g., spot synthesis) or independently with subsequent immobilization (e.g., microarrays on glass chips). (**c**) The main features of a capture assay are specific binding of diffusing analyte to immobilized probes and detection of captured analyte via direct or indirect labels (response source, "r. s."). (**d**) Evaluation of capture assays may be qualitative or quantitative depending on the potential of the response and control of adjustable experimental parameters

In the following we succinctly review hallmarks in the development of preparative techniques and stretch out some concepts concerning the quantitative evaluation of measurements with mechanistic models.

2 Advancement of Peptide Arrays

The technological advancement of peptide arrays can roughly be reviewed by focusing (at least) on three milestones of stepwise peptide synthesis. The development of solid-phase peptide synthesis (SPPS) by Bruce Merrifield [16, 17] and adaptions of this procedure [18] set the chemical ground for innovative technologies to follow. Secondly, the development of the "Pin" method by H. Geysen [19] introduces the array format to peptide synthesis. Definitive establishment of peptide arrays came along with the development of the SPOT synthesis by Roland Frank [20, 21] which simplified chemical synthesis of peptide arrays to the addressable deposition of reagents on a cellulose sheet. Moreover, chemical synthesis allows incorporation of nonnatural building blocks, preparation of branched and cyclic structures, and labeling with chromophores. Modern peptide synthesis approaches and molecular biology make peptides accessible in a high degree of structural diversity. The two greatest drawbacks of synthetic peptide arrays are peptide length, with a quality threshold between 30 and 50 amino acids, as well as the restriction to linear motives, since the mimicry of nonlinear motives with linear peptide constructs is still under development [22].

Peptide arrays also represent a milestone in the advancement of analytical binding assay systems. Since the 1990s a major aspect of development to achieve the required sensitivities to analyze biological samples has been the miniaturization of analytical devices [23]. It is important to note that miniaturization is not only a matter of high throughput and economy. Miniaturization is an essential factor that should provide saturation of binding sites under low analyte concentrations without significantly altering its bulk concentration upon capturing. In this sense, the first application of a peptide microarray device in 1991, anticipating even the application of cDNA arrays, achieved already the impressive feature density of about 1024 peptides in 1.6 cm^2 by means of in situ light-directed parallel synthesis [15]. The several methods available to generate peptide arrays on planar solid surfaces offer a range between 16 peptides per cm^2, in the case of SPOT macroarrays [24, 25], and 2000–4000 peptides in 1.5 cm^2, in the case of microarrays generated by digital photolithography [26–29].

In order to support synthesis, planar materials have to fulfill several requirements including stability towards solvent and reagent deposition. The functional groups on the surface must also be biochemically accessible for chemical derivatization. Furthermore, upon solid-phase binding assay, generated peptides must be functionally displayed to allow molecular recognition with a binding partner in the solution phase. In particular nonspecific interactions should be ruled out. Flexible porous supports such as cellulose [30, 31], cotton [32, 33], or membranes [34–36] are preferentially used for peptide array generation. Rigid, nonporous materials such as glass [37], gold films [38–40], or silicon [15, 29] have also been used for in situ synthesis, but are much more technically demanding. On the other side, rigid materials have a number of advantages over porous supports for functional display of molecules. Impermeability and smooth two dimensionality of the material do not limit diffusion of the binding partner and lead to more accurate kinetics of recognition events. Finally the flatness and transparency of glass improve image acquisition and simplify the use of fluorescence dyes for the readout process. In some cases, assembled 3D structure on a nonporous surface could be a fruitful approach.

Three types of peptide arrays are illustrated in Fig. 2. These selected examples are centered around the spot synthesis as preparative technique for large peptide collections. The core steps of the technique consist in an initial cellulose functionalization (step 1) and subsequent C- to N-term elongation cycles (steps 2 and 3). It offers an adjustable Fmoc synthesis strategy from low scale for in situ assays (steps 1–4) to a higher scale for a preparative approach (steps 1–3). Adjustment is achieved by, respectively, setting reactant amounts on the same equipment. Perhaps it may clear confusion to mention that compared to resin-supported classical peptide synthesis [41], cellulose-supported peptide synthesis is by all means of adjustment a low-scale synthesis. For more details on the spot technique refer to [21, 42] and the literature therein. Additionally, for a restricted sample of applications refer to [43–47].

A low-scale synthesis for in situ spot assays may be the method of choice whenever the peptide collection is either small in number or large but with only few replicas. On the contrary, if the array

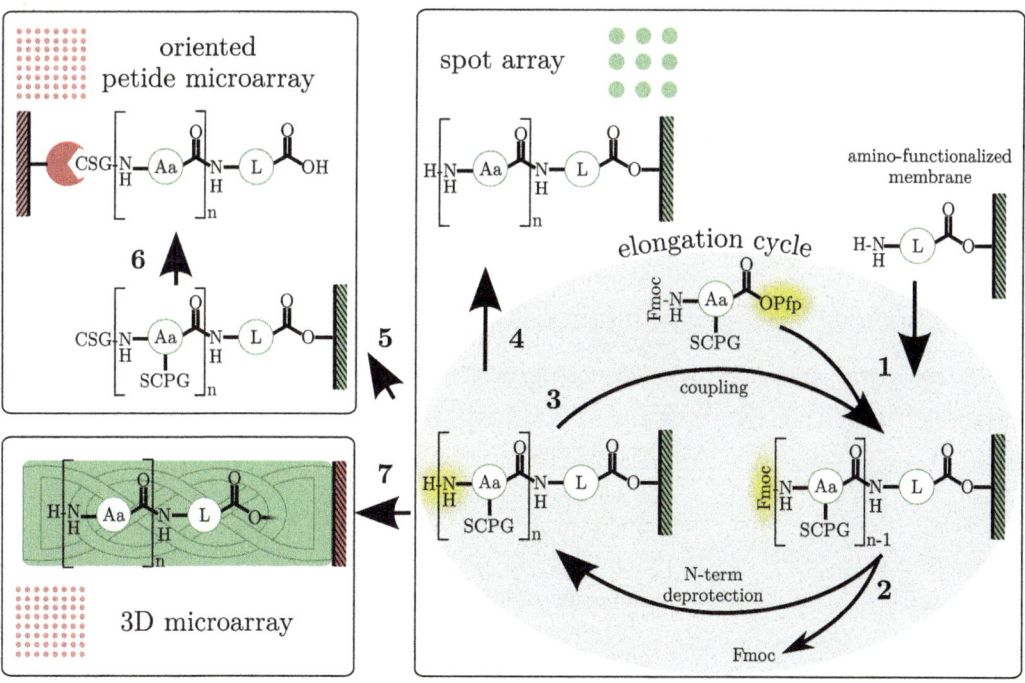

Fig. 2 Preparation of peptide microarrays. Spot arrays: steps *1–4*. Oriented peptide microarray: steps *1–3* and *5–6*. 3D microarrays or SC2 miniarrays: steps *1–3* and *7*

design contains a large number of probes with repeats, or you expect to require several assay batches, then it is advisable to run a high-scale synthesis once and distribute individual probes to array clones in or between batches. Continuing the later preparative strategy via step 5 in Fig. 2 allows to choose between several N-term modifications of peptide probes that couple selectively to surface functionalities (step 6). This choice, which involves peptide cleavage from the preparative surface, may turn rewarding in terms of the possibility to rigorously analyze individual probe solutions before immobilization, exclusion of truncated probe versions, and miniaturization of device size with concomitant more discrete analyte consumption.

Miniaturization not only means the generation of microarray devices with more than 200 spots per cm² on glass [24]; it may be crucial for high-throughput quantitative approaches as pointed out below. The major advantage of immobilizing per-synthesized peptides derives from the fact that the elongation cycles interlaced with amino acetylation steps allow exclusive N-term modification of full-length peptides. This translates into selective immobilization of full-length peptides and their homogeneous oriented display by covalent attachment via N-term. Some recommended strategies for selective immobilization are shown in Table 1 along with respective references to procedure details.

Table 1

Overview of selected peptide immobilization strategies for oriented display

Surface function	N-terminal modification	Selected references
Aldehyde	Amino-oxy-acetyl	Panse et al. [48]; Rychlewski et al. 49][a]; Schutkowski et al. [25]; Shigaki et al. [50]; see also pioneer work by Falsey et al. [37][a] and MacBeath et al. [11]
Epoxide[b]	Amine	Shreffler et al. [51]; Cerecedo et al. [52]; Lin et al. [53]. See also De Vos et al. [54] for alternative hetero-bifunctional PEG loading of silane-epoxide surfaces
Maleimide	Thiol	MacBeath et al. [55]; Han et al. [56]; Inamori et al. [57, 58]; see also references in Mori et al. [59] for some alternatives
Capture protein	Tag[c]	Lessaicherre et al. [60]; Uttamchandani et al. [61]; Andresen et al. [62, 63]
C-intein	N-intein	Lessaicherre et al. [64]; Camarero et al. [65]; Kwon et al. [66]; Sun et al. [67]; Xu et al. [68]; Shah et al. [69]. See also Mootz [70] and Shah and Muir [71] for review on the biotechnological potential of split inteins

[a]1-Amino-4,7,10-trioxa-13-tridecanamine succinimic acid (Ttds-OH) was alternatively used as N-term modification
[b]The pH in the printing buffer is essential to reach optimal conditions to preferentially target primary amines
[c]Affinity attachment

Finally, a third case of microarrays generated by preparative spot synthesis is designated SC^2 microarrays and illustrated in Fig. 2, step 7. The main feature of this approach is a variation of the side-chain deprotection step carried out on isolated but still cellulose-bounded peptide probes to generate a peptide-cellulose cocktail and then, in the words of his inventor, "spotting compound-support conjugates" [72]. As a result the cocktail is glued to the nonfunctionalized glass surface to yield 3D spots of about 750 μm of horizontal diameter and densities of about 25 spots per cm². This has led some authors to refer to these devices as miniarrays. We classify them among a longer tradition of 3D microarrays and expect them to meet the promise of greater sensitivity harbored along the diverse variants of this technique [58, 73–80].

3 Accurate Affinity Determination by Miniaturization

Cellular function is usually explained in biology and medicine through models, in which a redundant or degenerate input is transformed into a meaningful pattern. A key concept of such explanations is the specificity of molecular recognition. In fact, the most often driving force behind the technological developments observed in high-throughput strategies to analyze molecular interactions is to describe recognition specificity in terms of probability or affinity.

Despite the ubiquitous use of the concept, any attempt to define it has been as controversial as a definition of living systems [81]. Nevertheless, scientific debate about specificity, e.g., in cellular signal transduction, has led to some delightful ideas around the apparent paradox between promiscuous binding events and functional outputs.

Instead of giving an absolute definition, which to our best knowledge is not satisfactorily given in the literature, focus is given here to a paradigm shift pointed out early by Ladbury & Arold [82] and others. The most widely used model to explain the functional outputs of cellular signal transduction is a correction of the most simplistic linear chain of binding events through differential expression, subcellular compartmentalization, additive effects of multiple binding sites between interaction partners, and cooperative allosteric effects upon assembly of molecular complexes [83]. While the linear-chain model relies on an affinity-driven concept of inherent specificity, corrections of it give account for the general observation that affinity for natural ligands may be comparable to that for competing ligands. The corrected linear-chain model relies on the concept of an effective specificity driven by the factors mentioned above.

Beyond corrections, a different model is mostly used in the analysis of molecular interaction networks. Therein, inherent specificity of pairwise interactions is replaced by a network of interactions with dynamic equilibrium for assembly to functional molecular complexes. The gain in information content that can be assigned to individual interactions in such a probabilistic view and its potential to explain the evolutionary plasticity of protein interactions [84] is, in fact, very seductive.

Most of the work dealing with binding specificity to peptide probes takes a pragmatic position in this debate. The basic strategy consists in defining an in vitro context of extensive peptide probes to define specificity as a profile of biochemical binding potential based on semiquantitative data (inherent specificity) and validating selected potential interactors via pull-down and co-localization experiments (effective specificity). The modular analysis of specificity, i.e., focusing on interactions mediated by peptide recognition modules (PRMs), is a further common feature.

A peptide array may also characterize the specificity of an analyte such as proteins, antibodies, and PRMs by using a tailor-made collection of peptide probes and a mechanistic mathematical model to determine the equilibrium constant of the binding reaction. The quantification of specificity in terms of affinity is a key factor in the development of reagents with the potential to identify, quantify, and purify substances of biological and medical importance, e.g., receptor-modulating ligands, therapeutic monoclonal antibodies [85], and biomarker capture agents.

Approaches that can determine, or more modestly estimate the binding affinities of an analyte to a collection of peptides, are referred to as *functional binding assays* [10]. Peptide arrays are replicated and each replica is incubated with different analyte concentrations. The subsequent evaluation of data is similar to other, more established methods for determination of affinity: biosensors, calorimetry, and, most nearly, radioimmunoassays.

In more detail, datasets of measurements (captured analyte with values of response intensity) of each peptide probe are nonlinearly regressed to the values of total or applied analyte concentration (here treated as independent variable). The regression or fitting curve is defined by a model derived from the mass-action principle and a Levenberg–Marquardt algorithm to find the least square distances to data points [86–88].

We have previously explored the quality of quantitative measurements using peptide microarrays and two mechanistic models [89]. One equation is a straightforward derivation of the mass-action principle (Eq. 1) and the other one is a simplified version (Eq. 2). The latter is acquired by assuming that the total or applied analyte concentration is greater than the apparent peptide concentration. This is compatible with the ambient analyte theory, which was a critical driving force in the earliest development of microarray techniques [23, 90–94].

The cited equations are written for a functional binding assay as follows:

$$y = \frac{(KD_i + P_0 + x)}{2} - \sqrt{\frac{(KD_i + P_0 + x)^2}{4} - P_0 \cdot x} \qquad (1)$$

$$y = P_0 \cdot \frac{x}{KD_i + x} \qquad (2)$$

$$R = m \cdot y + n \qquad (3)$$

where y (dependent variable) is the concentration at equilibrium of captured analyte, x (independent variable) is the total analyte concentration, KD is the equilibrium dissociation constant treated here as adjustable parameter with index i for each value, P_0 is the apparent peptide concentration, and R is the measured response intensity assumed to be linearly related to y.

Peptide arrays generated by methods using presynthesized peptide probes as illustrated in Fig. 2 are expected to generate devices with constant P_0 across all peptide probes even if measurements at different analyte concentrations are made using separate support devices. The main challenge to this assumption is the use of different synthesis batches to generate the collection of peptide probes or their replicas. Consequently, this condition allows to search for a fixed value of P_0 across all datasets, an essential requirement for a robust application of Eq. 1.

The simplicity of Eq. 2 can be appreciated by the fact that the optimization algorithm can be set to search globally for a value of the product $m*P_0 = R_{max}$, thus reducing the number of parameters. This gain in robustness of the regression analysis occurs at the underestimated price of loosing knowledge of the apparent P_0. The value of apparent P_0 is a measure of miniaturization and, as such, it has a rarely explored potential for experimental design and the quality of the regression analysis.

The usefulness of P_0 for the evaluation process can be demonstrated in formal terms by comparing simulation curves drawn by both equations. Equation 1 is usually too complex to estimate reliable parameter values when fitting real data, but it should most accurately simulate the binding reaction under the mass-action principle; hence we use it in Fig. 3 to plot the idealized scattering of data points. On the contrary, Eq. 1 is by large preferable for nonlinear regression of real data and has been notably applied at the MacBeath Lab [95]; hence, we use it in Fig. 3 to identify discrepancy from the curve traces of Eq. 2 with the same input KDi values.

Panels a and b are used to find orientation in the negative-logarithmical plot of the independent variable, basically resulting

in curve traces with weak binding affinities shown to the left and
those with strong binding affinities shown to the right of the plot.
The simulations shown in panel c are defined by a constant value
of $P_0 = 1$ μM and the following adjustments to the affinity parameter: $KD_1 = 10^*P_0$ (representing weak affinities), $KD_2 = P_0$, and
$KD_3 = P_0/10$ (representing strong affinities). While estimations of
weak affinities from Eq. 2 (continuous line) are observed to coincide well with the ideal data scattering from Eq. 1 (circles), the

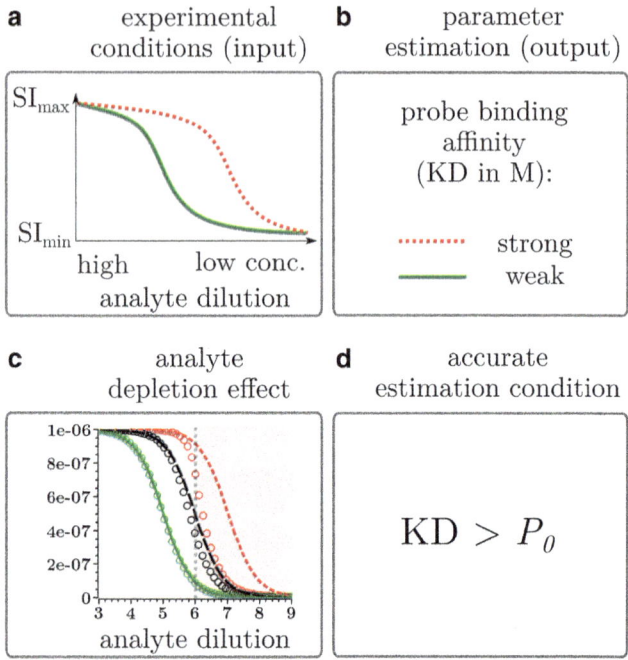

Fig. 3 Accurate parameter estimation on a functional assay with mechanistic
models. Panel **a** shows a sketch of the trace for a weak affinity (i.e., high **KD**
values, *green continuous line*) and a strong affinity (i.e., low **KD** values, *red dotted line*). The independent variable contains values of total analyte concentrations applied to different array replicas. Panel **b** shows the legend to the curve
traces in **a**. Panel **c** shows the measures of captured analyte resulting from
changes in the total analyte concentrations in M plotted as neg-log values. The
gray dotted vertical line in panel **c** shows the input value of the peptide concentration (6 corresponding to 1 μM). The curve traces of different **KD**s shown with
circles represent the ideal scattering of data (Eq. *1*), while curve traces with *continuous lines* represent the regression lines obtained by the simplified model
usually applied to estimate affinity (Eq. *2*). The input **KD** values are **KD1 = 10*P0**
(*green*), **KD2 = P0** (*black*), and **KD3 = P0/10** (*red*). The *red region* of the plot
shows a collapse of data points and a stubborn even horizontal distribution of
traces of the simplified model. Panel **d** makes the observation that affinity is only
accurately estimated when affinity is weaker than the apparent peptide concentration. The peptide concentration is an effective parameter fixed at a global
value across all datasets

right half of the plot affecting strong affinities shows a severe discrepancy between both equations.

Measurements at low total analyte concentrations (to the right of P_0) collapse ever closer to each other with growing affinity strength. Strong affinity datasets, that heavily depend on these measurements, may disregard the initial condition to derive Eq. 2. As a consequence, the forced even distribution of traces with changing KD in Eq. 2 may strongly underestimate the parameter value of strong affinities and, most probably, will not be able to distinguish data mainly distributed to the right of the plot. Panel d puts it all in a nutshell: accurate affinity determination with Eq. 2 is possible only if $KD > P_0$.

Miniaturization is a great advantage since lowering the effective peptide concentration to the sensitivity limits of the assay may expand the range of affinity which can be measured, specially improving the accuracy of strong affinity estimations. A well-proven good practice to optimize the accuracy of parameter estimation and the sensitivity of captured analyte detection is preparing spots of minimal size and highest density of peptide probes.

Acknowledgements

Victor Tapia is supported by grants from the German Research Association (DFG VO 885/8-1).

References

1. Bier FF, von Nickisch-Rosenegk M, Ehrentreich-Forster E, Reiss E, Henkel J, Strehlow R, Andresen D (2008) DNA microarrays. Adv Biochem Eng Biotechnol 109:433–453

2. Southern EM (2001) DNA microarrays. History and overview. Methods Mol Biol 170:1–15

3. DeRisi J, Penland L, Brown PO, Bittner ML, Meltzer PS, Ray M, Chen Y, Su YA, Trent JM (1996) Use of a cDNA microarray to analyse gene expression patterns in human cancer. Nat Genet 14:457–460

4. Szathmary E, Smith JM (1995) The major evolutionary transitions. Nature 374:227–232

5. Anderson L, Seilhamer J (1997) A comparison of selected mRNA and protein abundances in human liver. Electrophoresis 18:533–537

6. Chen G, Gharib TG, Huang CC, Thomas DG, Shedden KA, Taylor JM, Kardia SL, Misek DE, Giordano TJ, Iannettoni MD, Orringer MB, Hanash SM, Beer DG (2002) Proteomic analysis of lung adenocarcinoma: identification of a highly expressed set of proteins in tumors. Clin Cancer Res 8:2298–2305

7. Gygi SP, Rochon Y, Franza BR, Aebersold R (1999) Correlation between protein and mRNA abundance in yeast. Mol Cell Biol 19:1720–1730

8. Wilkins MR, Pasquali C, Appel RD, Ou K, Golaz O, Sanchez JC, Yan JX, Gooley AA, Hughes G, Humphery-Smith I, Williams KL, Hochstrasser DF (1996) From proteins to proteomes: large scale protein identification by two-dimensional electrophoresis and amino acid analysis. Biotechnology (N Y) 14:61–65

9. Aebersold R (2003) Constellations in a cellular universe. Nature 422:115–116

10. Phizicky E, Bastiaens PIH, Zhu H, Snyder M, Fields S (2003) Protein analysis on a proteomic scale. Nature 422:208–215

11. MacBeath G, Schreiber SL (2000) Printing proteins as microarrays for high-throughput function determination. Science 289:1760–1763

12. Zhu H, Bilgin M, Bangham R, Hall D, Casamayor A, Bertone P, Lan N, Jansen R, Bidlingmaier S, Houfek T et al (2001) Global analysis of protein activities using proteome chips. Science 293:2101–2105

13. Zhu H, Klemic JF, Chang S, Bertone P, Casamayor A, Klemic KG, Smith D, Gerstein M, Reed MA, Snyder M (2000) Analysis of yeast protein kinases using protein chips. Nat Genet 26:283–289

14. Frank R, Güler S, Krause S, Lindenmaier W (1991) Facile and rapid spot-synthesis of large numbers of peptides on membrane sheets. In: Giralt E, Andreu D (eds) Peptides. ESCOM Science Publishers B. V, Leiden, pp 151–152

15. Fodor SP, Read JL, Pirrung MC, Stryer L, Lu AT, Solas D (1991) Light-directed, spatially addressable parallel chemical synthesis. Science 251:767–773

16. Gutte B, Merrifield RB (1969) The total synthesis of an enzyme with ribonuclease A activity. J Am Chem Soc 91:501–502

17. Merrifield RB (1965) Automated synthesis of peptides. Science 150:178–185

18. Fields GB, Noble RL (1990) Solid phase peptide synthesis utilizing 9-fluorenylmethoxycarbonyl amino acids. Int J Pept Protein Res 35:161–214

19. Geysen HM, Meloen RH, Barteling SJ (1984) Use of peptide synthesis to probe viral antigens for epitopes to a resolution of a single amino acid. Proc Natl Acad Sci U S A 81:3998–4002

20. Frank R (1992) Spot-synthesis: an easy technique for the positionally addressable, parallel chemical synthesis on a membrane support. Tetrahedron 48:9217–9232

21. Frank R (2002) The SPOT-synthesis technique: synthetic peptide arrays on membrane supports—principles and applications. J Immunol Methods 267:13–26

22. Goede A, Jaeger IS, Preissner R (2005) SUPERFICIAL—surface mapping of proteins via structure-based peptide library design. BMC Bioinformatics 6:223

23. Ekins RP (1998) Ligand assays: from electrophoresis to miniaturized microarrays. Clin Chem 44:2015–2030

24. Reimer U, Reineke U, Schneider-Mergener J (2002) Peptide arrays: from macro to micro. Curr Opin Biotechnol 13:315–320

25. Schutkowski M, Reimer U, Panse S, Dong L, Lizcano JM, Alessi DR, Schneider-Mergener J (2004) High-content peptide microarrays for deciphering kinase specificity and biology. Angew Chem Int Ed Engl 116:2725–2728

26. El Khoury G, Laurenceau E, Dugas V, Chevolot Y, Merieux Y, Duclos MC, Souteyrand E, Rigal D, Wallach J, Cloarec JP (2007) Acid deprotection of covalently immobilized peptide probes on glass slides for peptide microarrays. Conf Proc IEEE Eng Med Biol Soc 2007:2242–2246

27. Gao X, Pellois JP, Na Y, Kim Y, Gulari E, Zhou X (2004) High density peptide microarrays. In situ synthesis and applications. Mol Divers 8:177–187

28. Pellois JP, Wang W, Gao X (2000) Peptide synthesis based on t-Boc chemistry and solution photogenerated acids. J Comb Chem 2:355–360

29. Pellois JP, Zhou X, Srivannavit O, Zhou T, Gulari E, Gao X (2002) Individually addressable parallel peptide synthesis on microchips. Nat Biotechnol 20:922–926

30. Eichler J, Beyermann M, Bienert M (1989) Application of cellulose paper as support in simultaneous solid phase peptide synthesis. Collect Czechoslov Chem Commun 54:1746–1752

31. Frank R, Döring R (1988) Simultaneous multiple peptide synthesis under continuous flow conditions on cellulose paper disks as segmental solid supports. Tetrahedron 44:6031–6040

32. Eichler J, Bienert M, Stierandova A, Lebl M (1991) Evaluation of cotton as a carrier for solid-phase peptide synthesis. Pept Res 4:296–307

33. Schmidt M, Eichler J (1993) Multiple peptide synthesis using cellulose-based carriers: synthesis of substance P—diastereoisomers and their histamine-releasing activity. Bioorg Med Chem Lett 3:441–446

34. Daniels SB, Bernatowicz MS, Coull JM, Köster H (1989) Membranes as solid supports for peptide synthesis. Tetrahedron Lett 30:4345–4348

35. Wang Z, Laursen RA (1992) Multiple peptide synthesis on polypropylene membranes for rapid screening of bioactive peptides. Pept Res 5:275–280

36. Wenschuh H, Volkmer-Engert R, Schmidt M, Schulz M, Schneider-Mergener J, Reineke U (2000) Coherent membrane supports for parallel microsynthesis and screening of bioactive peptides. Biopolymers 55:188–206

37. Falsey JR, Renil M, Park S, Li S, Lam KS (2001) Peptide and small molecule microarray for high throughput cell adhesion and functional assays. Bioconjug Chem 12:346–353

38. Houseman BT, Huh JH, Kron SJ, Mrksich M (2002) Peptide chips for the quantitative evaluation of protein kinase activity. Nat Biotechnol 20:270–274

39. Jonsson U, Fagerstam L, Ivarsson B, Johnsson B, Karlsson R, Lundh K, Lofas S, Persson B, Roos H, Ronnberg I et al (1991) Real-time biospecific interaction analysis using surface

plasmon resonance and a sensor chip technology. Biotechniques 11:620–627

40. Malmqvist M (1993) Biospecific interaction analysis using biosensor technology. Nature 361:186–187

41. Coin I, Beyermann M, Bienert M (2007) Solid-phase peptide synthesis: from standard procedures to the synthesis of difficult sequences. Nat Protoc 2:3247–3256

42. Hilpert K, Winkler DF, Hancock RE (2007) Peptide arrays on cellulose support: SPOT synthesis, a time and cost efficient method for synthesis of large numbers of peptides in a parallel and addressable fashion. Nat Protoc 2:1333–1349

43. Rüdiger S, Schneider-Mergener J, Bukau B (2001) Its substrate specificity characterizes the DnaJ co-chaperone as a scanning factor for the DnaK chaperone. EMBO J 20:1042–1050

44. Reineke U, Ivascu C, Schlief M, Landgraf C, Gericke S, Zahn G, Herzel H, Volkmer-Engert R, Schneider-Mergener J (2002) Identification of distinct antibody epitopes and mimotopes from a peptide array of 5520 randomly generated sequences. J Immunol Methods 267:37–51

45. Nady N, Min J, Kareta MS, Chédin F, Arrowsmith CH (2008) A SPOT on the chromatin landscape? Histone peptide arrays as a tool for epigenetic research. Trends Biochem Sci 33:305–313

46. Nady N, Krichevsky L, Zhong N, Duan S, Tempel W, Amaya MF, Ravichandran M, Arrowsmith CH (2012) Histone recognition by human malignant brain tumor domains. J Mol Biol 423:702–718

47. Ulbricht A, Eppler FJ, Tapia VE, van der Ven PFM, Hampe N, Hersch N, Vakeel P, Stadel D, Haas A, Saftig P et al (2013) Cellular mechanotransduction relies on tension-induced and chaperone-assisted autophagy. Curr Biol 23:430–435

48. Panse S, Dong L, Burian A, Carus R, Schutkowski M, Reimer U, Schneider-Mergener J (2004) Profiling of generic antiphosphopeptide antibodies and kinases with peptide microarrays using radioactive and fluorescence-based assays. Mol Divers 8:291–299

49. Rychlewski L, Kschischo M, Dong L, Schutkowski M, Reimer U (2004) Target specificity analysis of the Abl kinase using peptide microarray data. J Mol Biol 336:307–311

50. Shigaki S, Yamaji T, Han X, Yamanouchi G, Sonoda T, Okitsu O, Mori T, Niidome T, Katayama Y (2007) A peptide microarray for the detection of protein kinase activity in cell lysate. Anal Sci 23:271–275

51. Shreffler WG, Lencer DA, Bardina L, Sampson HA (2005) IgE and IgG4 epitope mapping by microarray immunoassay reveals the diversity of immune response to the peanut allergen, Ara h 2. J Allergy Clin Immunol 116:893–899

52. Cerecedo I, Zamora J, Shreffler WG, Lin J, Bardina L, Dieguez MC, Wang J, Muriel A, de la Hoz B, Sampson HA (2008) Mapping of the IgE and IgG4 sequential epitopes of milk allergens with a peptide microarray-based immunoassay. J Allergy Clin Immunol 122:589–594

53. Lin J, Bardina L, Shreffler WG, Andreae DA, Ge Y, Wang J, Bruni FM, Fu Z, Han Y, Sampson HA (2009) Development of a novel peptide microarray for large scale epitope mapping of food allergens. J Allergy Clin Immunol 124:315–322

54. De Vos K, Girones J, Popelka S, Schacht E, Baets R, Bienstman P (2009) SOI optical microring resonator with poly(ethylene glycol) polymer brush for label-free biosensor applications. Biosens Bioelectron 24:2528–2533

55. MacBeath G, Koehler AN, Schreiber SL (1999) Printing small molecules as microarrays and detecting protein–ligand interactions en masse. J Am Chem Soc 121:7967–7968

56. Han A, Sonoda T, Kang JH, Murata M, NIidome T, Katayam Y (2006) Development of a fluorescence peptide chip for the detection of caspase activity. Comb Chem High Throughput Screen 9:21–25

57. Inamori K, Kyo M, Nishiya Y, Inoue Y, Sonoda T, Kinoshita E, Koike T, Katayama Y (2005) Detection and quantification of on-chip phosphorylated peptides by surface plasmon resonance imaging techniques using a phosphate capture molecule. Anal Chem 77:3979–3985

58. Inamori K, Kyo M, Matsukawa K, Inoue Y, Sonoda T, Tatematsu K, Tanizawa K, Mori T, Katayama Y (2008) Optimal surface chemistry for peptide immobilization in on-chip phosphorylation analysis. Anal Chem 80:643–650

59. Mori T, Inamori K, Inoue Y, Han X, Yamanouchi G, Niidome T, Katayama Y (2008) Evaluation of protein kinase activities of cell lysates using peptide microarrays based on surface plasmon resonance imaging. Anal Biochem 375:223–231

60. Lesaicherre ML, Uttamchandani M, Chen GY, Yao SQ (2002) Developing site-specific immobilization strategies of peptides in a microarray. Bioorg Med Chem Lett 12:2079–2083

61. Uttamchandani M, Chen GY, Lesaicherre ML, Yao SQ (2004) Site-specific peptide immobilization strategies for the rapid detection of kinase activity on microarrays. Methods Mol Biol 264:191–204

62. Andresen H, Grotzinger C, Zarse K, Kreuzer OJ, Ehrentreich-Forster E, Bier FF (2006) Functional peptide microarrays for specific and sensitive antibody diagnostics. Proteomics 6:1376–1384

63. Andresen H, Zarse K, Grotzinger C, Hollidt JM, Ehrentreich-Forster E, Bier FF, Kreuzer OJ (2006) Development of peptide microarrays for epitope mapping of antibodies against the human TSH receptor. J Immunol Methods 315:11–18

64. Lesaicherre M-L, Lue RYP, Chen GYJ, Zhu Q, Yao SQ (2002) Intein-mediated biotinylation of proteins and its application in a protein microarray. J Am Chem Soc 124:8768–8769

65. Camarero JA, Kwon Y, Coleman MA (2004) Chemoselective attachment of biologically active proteins to surfaces by expressed protein ligation and its application for "protein chip" fabrication. J Am Chem Soc 126:14730–14731

66. Kwon Y, Coleman MA, Camarero JA (2006) Selective immobilization of proteins onto solid supports through split-intein-mediated protein trans-splicing. Angew Chem Int Ed 45:1726–1729

67. Sun L, Rush J, Ghosh I, Maunus JR, Xu MQ (2004) Producing peptide arrays for epitope mapping by intein-mediated protein ligation. Biotechniques 37:430

68. Xu M-Q, Ghosh I, Kochinyan S, Sun L (2007) Intein-mediated peptide arrays for epitope mapping and kinase/phosphatase assays. In: Rampal JB (ed) Microarrays. Humana Press, Totowa, NJ, pp 313–338

69. Shah NH, Vila-Perelló M, Muir TW (2011) Kinetic control of one-pot trans-splicing reactions by using a wild-type and designed split intein. Angew Chem Int Ed Engl 50:6511–6515

70. Mootz HD (2009) Split inteins as versatile tools for protein semisynthesis. Chembiochem 10:2579–2589

71. Shah NH, Muir TW (2013) Inteins: nature's gift to protein chemists. Chem Sci 5:446–461

72. Dikmans A, Beutling U, Schmeisser E, Thiele S, Frank R (2006) SC2: a novel process for manufacturing multipurpose high-density chemical microarrays. QSAR Comb Sci 25:1069–1080

73. Zubtsov DA, Savvateeva EN, Rubina AY, Pan'kov SV, Konovalova EV, Moiseeva OV, Chechetkin VR, Zasedatelev AS (2007) Comparison of surface and hydrogel-based protein microchips. Anal Biochem 368:205–213

74. Mori T, Yamanouchi G, Han X, Inoue Y, Shigaki S, Yamaji T, Sonoda T, Yasui K, Hayashi H, Niidome T, Katayama Y (2009) Signal-to-noise ratio improvement of peptide microarrays by using hyperbranched-polymer materials. J Appl Phys 105:102020

75. Angenendt P, Glokler J, Murphy D, Lehrach H, Cahill DJ (2002) Toward optimized antibody microarrays: a comparison of current microarray support materials. Anal Biochem 309:253–260

76. Seurynck-Servoss SL, Baird CL, Miller KD, Pefaur NB, Gonzalez RM, Apiyo DO, Engelmann HE, Srivastava S, Kagan J, Rodland KD, Zangar RC (2008) Immobilization strategies for single-chain antibody microarrays. Proteomics 8:2199–2210

77. Angenendt P, Glokler J, Sobek J, Lehrach H, Cahill DJ (2003) Next generation of protein microarray support materials: evaluation for protein and antibody microarray applications. J Chromatogr A 1009:97–104

78. Seurynck-Servoss SL, Baird CL, Rodland KD, Zangar RC (2007) Surface chemistries for antibody microarrays. Front Biosci 12:3956–3964

79. Sobek J, Aquino C, Schlapbach R (2007) Quality considerations and selection of surface chemistry for glass-based DNA, peptide, antibody, carbohydrate, and small molecule microarrays. Methods Mol Biol 382:17–31

80. Engelmann BW, Kim Y, Wang M, Peters B, Rock RS, Nash PD (2014) The development and application of a quantitative peptide microarray based approach to protein interaction domain specificity space. Mol Cell Proteomics 13:3647–3662

81. Pawson T, Nash P (2000) Protein–protein interactions define specificity in signal transduction. Genes Dev 14:1027–1047

82. Mayer BJ (2001) SH3 domains: complexity in moderation. J Cell Sci 114:1253–1263

83. Ladbury JE, Arold S (2000) Searching for specificity in SH domains. Chem Biol 7:R3–R8

84. Beltrao P, Serrano L (2007) Specificity and evolvability in eukaryotic protein interaction networks. PLoS Comput Biol 3(3):e70

85. Bee C, Abdiche YN, Pons J, Rajpal A (2013) Determining the binding affinity of therapeutic monoclonal antibodies towards their native unpurified antigens in human serum. PLoS One 8:e80501

86. Levenberg K (1944) A method for the solution of certain non-linear problems in least squares. Q J Appl Math II 2:164–168

87. Marquardt DW (1963) An algorithm for least-squares estimation of nonlinear parameters. J Soc Ind Appl Math 11:431–441

88. Press WH (1992) Numerical recipes in C: the art of scientific computing, 2nd edn. Cambridge University Press, Cambridge

89. Tapia V, Bongartz J, Schutkowski M, Bruni N, Weiser A, Ay B, Volkmer R, Or-Guil M (2007) Affinity profiling using the peptide microarray technology: a case study. Anal Biochem 363:108–118

90. Ekins R, Chu F, Biggart E (1990) Multispot, multianalyte, immunoassay. Ann Biol Clin (Paris) 48:655–666

91. Ekins RP (1989) Multi-analyte immunoassay. J Pharm Biomed Anal 7:155–168

92. Joos TO, Stoll D, Templin MF (2002) Miniaturised multiplexed immunoassays. Curr Opin Chem Biol 6:76–80

93. Templin MF, Stoll D, Bachmann J, Joos TO (2004) Protein microarrays and multiplexed sandwich immunoassays: what beats the beads? Comb Chem High Throughput Screen 7:223–229

94. Hartmann M, Toegl A, Kirchner R, Templin MF, Joos TO (2006) Increasing robustness and sensitivity of protein microarrays through microagitation and automation. Anal Chim Acta 564:66–73

95. Jones RB, Gordus A, Krall JA, MacBeath G (2006) A quantitative protein interaction network for the ErbB receptors using protein microarrays. Nature 439:168–174

Chapter 2

High-Throughput Microarray Incubations Using Multi-Well Chambers

Johannes Zerweck, Ulf Reimer, Janina Jansong, Nikolaus Pawlowski, Christoph Tersch, Maren Eckey, and Tobias Knaute

Abstract

Peptide microarrays are ideal tools for a variety of applications ranging from epitope mapping to immune monitoring. Here we present a method for high-throughput screening of biological samples using only standard microtiter plate equipment. Parallel incubation of a large number of samples with a small library of peptides is enabled by printing multiple identical mini-arrays on one microarray slide and further combining four slides to yield an incubation frame possessing the dimensions of a 96-well microtiter plate. Applying conventional lab equipment such as ELISA washers, hundreds of samples can be processed in 1 day yielding approx. 200 data points in triplicates per sample.

Key words Peptide microarray, High-throughput screening, Biomarker verification, Multi-well incubation chamber

1 Introduction

High-density peptide microarrays are efficient tools for mapping the immune response at the epitope level down to single-residue resolution [1–4]. The assay requires only minor amounts of antibody/serum or other patient samples (1 µl per assay) to perform a complete binding study on large numbers of overlapping peptides representing multiple antigen epitopes.

However, in many cases only a limited number of peptides are of interest and need to be probed with a high number of samples. To keep this process cost and time efficient we developed an assay protocol enabling screening of a limited number of peptides with a large number of individual samples in parallel.

The peptides are immobilized in a chemo-selective and directed fashion onto functionalized glass slides [5]. The schematic multi-well layout is shown in Fig. 2 (image 1). One microarray slide displays 21 mini-arrays, each with three complete sets of peptides (63 copies of library per microarray slide). Four microarray slides

Marina Cretich and Marcella Chiari (eds.), *Peptide Microarrays: Methods and Protocols*, Methods in Molecular Biology, vol. 1352, DOI 10.1007/978-1-4939-3037-1_2, © Springer Science+Business Media New York 2016

can be combined in an incubation frame exhibiting the dimension of 96-well plate. Thus, the incubation chamber is fully compatible with standard microtiter plate processing devices, e.g., ELISA equipment such as plate washers or pipetting robots.

For the assay, each mini-array is incubated with a biological sample (antibody, protein, blood serum/plasma, saliva, or other body fluids) individually. Antibodies from the sample bind specifically to the peptides and are detected by a fluorescently labeled secondary antibody. Eighty-four parallel reactions (four slides, each containing 21 mini-arrays) are applied either in high-throughput screening setups or for sophisticated assay development allowing testing of different buffer conditions, detection antibodies, concentrations, or other parameters to the individual wells.

For demonstrating the high-throughput capacity of the platform described above, microarrays displaying 158 peptides derived from EBNA1 (EBV) were generated in multi-well format. Each peptide was deposited 63 times generating 21 mini-arrays per microarray slide. Four microarray slides were combined in an incubation frame and probed with samples. Mini-arrays of each microarray slide were incubated with 18 human sera donated from healthy volunteers and two commercially available sera. Simultaneously, one mini-array was incubated with the detection antibody only to identify false-positive responses. Each of the four microarray slides was probed with a different detection antibody. In total, 84 mini-array incubations were performed. The determination of signal intensities was performed using a spot recognition software. The MMC2 value (*see* Subheading 3.4) for each peptide was calculated and visualized.

For the investigated sera, diverse and sample-specific responses to the EBNA1-derived peptides were observed (Fig. 1) indicating a wide range of immune states of the donors with respect to the Epstein-Barr virus. The different detection antibodies generally show comparable binding patterns for individual serum samples. Only minor differences were observed for the four used detection antibodies. Here, deviations were seen primarily in signal intensities but the signal pattern was largely unaffected. Antibody 1 shows a somewhat weaker signal compared to others whereas antibody 4 usually yields the strongest intensities. The most pronounced binding is observed on peptides covering amino acids 69–87:

```
    GVRRPQKRPSCIGCK              EBNA1_069
        PQKRPSCIGCKGTHG          EBNA1_073
```

as well as on peptides covering amino acids 393–415:

```
    SPPRRPPPGRRPFFH              EBNA1_393
        RPPPGRRPFFHPVGE          EBNA1_397
            GRRPFFHPVGEADYF      EBNA1_401
```

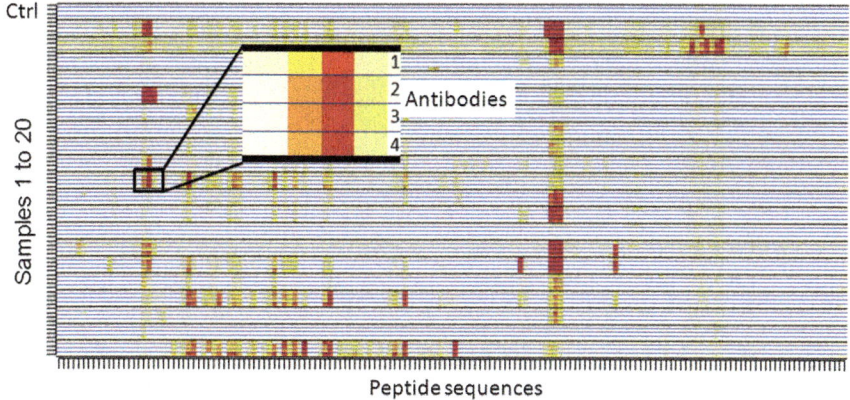

Fig. 1 Results of EBNA1 multi-well incubation: Epitopes recognized by antibodies contained in the individual samples are shown. The MMC2 values are displayed as colors: color coding is from *white* (low intensity) over *yellow* (middle intensity) to *red* (high intensity); *x*-axis: EBNA1 protein sequence represented by overlapping peptides; *y*-axis: one line represents one individual samples. Each incubation is visualized in one row. Each sample was probed with four different detection antibodies (*1–4*)

For several samples only weak signals are observed indicating no significant presence of EBV-targeted antibodies. For the two commercially acquired sera a raised background was observed mirroring the compounded nature of these normalized serum pools.

The approach of parallel microarray incubation applying antibody/patient samples shown here allows for high-throughput screening of a limited number of peptides with a high number of samples. The experiment demonstrated in this example is performed in an incubation chamber featuring a standard microtiter plate footprint. Parallel application and handling of several chambers are easily done since all pipetting and washing steps are fully compatible with multichannel pipettes or liquid-handling robots and ELISA washers. Screening of several hundreds of samples on peptide libraries is feasible in the time frame of a few hours. This throughput enables a wide range of applications like verification of biomarkers from screening campaigns; parallelized elaborated assay development, especially for protein-protein interactions; epitope mapping for small antigens; antibody characterization; immune monitoring during vaccine development; or time-resolved tracking of immune status.

2 Materials

2.1 Hardware

1. Eightfold Multipette adjustable from 20 to 250 μl with corresponding plastic tips.

2. Single pipette adjustable from 1 to 10 μl with corresponding plastic tips.

3. Centrifuge for serum vials.

4. Standard 96-well plate.

5. Shaking system.

6. Multi-well incubation chamber (The Gel Company, San Francisco, CA).

7. Microplate washing system.

8. Microarray high-speed centrifuge.

9. Microarray fluorescence scanner or imaging system which is able to perform microarray fluorescence scans with at least 10 μm/pixel resolution.

2.2 Buffers and Reagents

1. TBS buffer (Tris-buffered saline): 5.7 g Tris-HCl (Tris (hydroxymethyl)-aminomethane), 8.0 g NaCl, and 200 mg KCl dissolved in 1 l double-distilled water, pH adjusted to 8.0 with HCl (*see* **Note 1**).

2. TBS-based sample dilution buffer (Pierce, Rockford, USA).

3. Biological samples from anonymous healthy volunteers and commercially available human sera.

4. Fluorescently labeled secondary antibodies (all antibodies were applied in a concentration of 1 μg/ml) (*see* **Note 2**).

   • Cy5-anti-human IgG.

   • DyLight649-anti-human IgG.

   • Cy5-anti-human IgG.

   • Alexa Fluor 647-anti-human.

5. Peptide microarrays (JPT Peptide Technologies GmbH, Berlin, Germany) printed in multi-well format.

2.3 Software

1. Spot-recognition software GenePix Pro 7.2 (Molecular Devices, USA).

2. Software package R (http://www.r-project.org/).

3 Methods

3.1 Preparation of Multi-Well Incubation Chamber

All handling steps are shown in Fig. 2.

1. Insert the slides into the lower carrier tray of the chamber and secure the slide with the side thumb screw.

2. Make sure that the surface the peptides are printed on is facing up. Do not touch the printed surface under any circumstances.

3. Insert the gasket into the upper part of the chamber. The side displaying the narrow rims of the incubation wells should be facing up.

   Make sure that the gasket is lying flat and tight inside the upper structure plate.

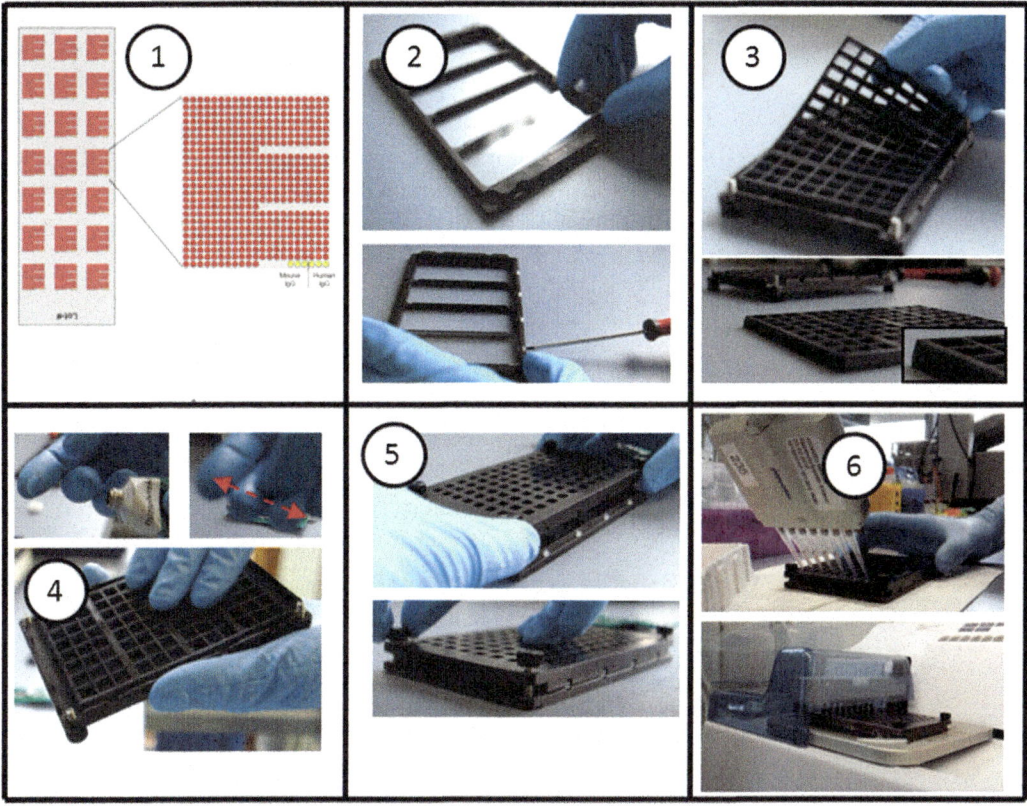

Fig. 2 Assembly, preparation, and use of multi-well incubation chamber: (*1*) schematic layout of multi-well array; (*2*) embedding of slides in lower incubation frame; (*3*) setting the seal in top frame; (*4*) preparation of seal; (*5*) assembling of incubation chamber; (*6*) pipetting of solutions and washing of incubation chamber

4. Put a very thin layer of silicone grease on the gasket (*see* **Note 3**).

   (a) Take a tiny amount of silicone grease.

   (b) Spread it between your fingers to create a thin layer.

   (c) Carefully, apply the silicone grease on the gasket to create a uniform film on all surfaces touching the microarrays.

5. Close and fix the chamber by putting the lower and upper parts together and fix the device by tightening the four thumb screws in the corners. The gasket should be pressed up against the glass slide with the narrow-rimmed sides of the incubation wells.

6. Pipette samples and solutions prepared as described in Subheading 3.2 into individual wells of incubation chamber. Make sure that the total incubation time of the samples on the microarray slide is nearly identical. Use an ELISA washer for efficient and automated washing of individual wells (*see* **Note 4**).

3.2 Sample Preparation

1. In case of patient samples it is advisable to spin down the sample containers to avoid pipetting any precipitation (samples described here were diluted 1:200 in assay dilution buffer).

2. Prepare individual sample dilutions in individual wells of a separate 96-well microplate (*see* **Note 5**). The microarray slides used have an engraved batch number in lower row; therefore row H of microplate was ignored resulting in a total of 84 sample preparations.

3.3 Assay and Washing Procedure (See Note 6)

1. Incubation of sample:

   (a) Add 150 μl sample diluted in sample dilution buffer per well.

   (b) Incubate sample on peptide microarray for 1 h at 30 °C. Orbital shaking of the incubation chamber improves homogeneity throughout the incubated well (a setting of 300 rpm is recommended) (*see* **Note 7**).

2. Washing procedure of peptide microarray 2.1. Wash each well with 4× 300 μl TBS using an ELISA washer (*see* **Notes 8** and **9**).

3. Incubation of secondary antibody:

   (a) Add 150 μl secondary antibody diluted in sample dilution buffer per well (*see* **Note 10**).

   (b) Incubate secondary antibody on peptide microarray for 45 min at 30 °C. Orbital shaking of the incubation chamber improves homogeneity throughout the incubated well (a setting of 300 rpm is recommended) (*see* **Note 11**).

4. Final washing procedure of peptide microarray:

   (a) Wash each well with 4× 300 μl TBS using an ELISA washer.

   (b) Wash each well with 2× 300 μl ddH_2O using an ELISA washer.

5. Disassemble incubation chamber and dry peptide microarrays by centrifugation (*see* **Note 12**).

3.4 Readout and Evaluation

1. Scan the dried microarray using a microarray scanner with appropriate laser settings (*see* **Notes 13** and **14**).

2. Quantify the scanned images using the provided .gal files with spot-recognition software. Save the resulting values for the signal intensities and background.

3. Generate MMC2 values using appropriate scripts. The MMC2 equals the mean value of all three replicates on the microarray if the coefficient of variation (CV)—standard deviation divided by the mean value—is smaller than 0.5. If the variation is larger the mean of the two closest values (MC2) is assigned to MMC2.

4. Visualize MMC2 values using heatmap diagrams.

4 Notes

1. The main cause for artifacts or scratches is dust particles in the incubation or washing solutions. To avoid this, we recommend to filter all solutions and buffers used in the experiment.

2. JPT strongly recommends using dyes corresponding to red laser wavelengths such as Dylight 649, or Cy5. Dyes sensitive to shorter wavelengths are more affected by sample composition, aberrations in slide layer thickness, and variation in washing conditions.

3. In case the artifact occurs at the borders of the incubation area, the main cause could be excess silicone grease on the seals of the incubation chamber. Reduce the amount of grease applied before assembling the chamber.

4. The settings of ELISA (microplate) washer should be adjusted in advance using standard glass slides.

5. Increased concentrations of sample/second antibody may yield high background signals caused by unspecific binding to the slide surface. In such a case we recommend to perform dilution series to identify the optimal concentration range for each analyte.

6. The major causes for high or irregular background within the incubation areas are the composition or nature of the sample, insufficient washing procedures, or non-optimal assay conditions. Directly fluorescently labeled proteins tend to induce background signals via unspecific binding to the slide surface. Changing of buffer conditions in the incubation step can reduce background signals very efficiently. Additional washing steps after the incubation is finished can also reduce unwanted background. Alternatively, a variation of blocking buffers can also reduce nonspecific binding of the samples. For avoiding the sample to stick to the peptide array, we recommend a variation of blocking buffers as well as additional washing steps after the primary and/or the secondary incubation steps.

7. To improve the signal-to-noise ratio the incubation time can be increased. In some cases, overnight incubations at 4 °C can vastly improve S/N ratio.

8. Make sure that the washing head does not touch the microarray surface. Any contact of the washing tip to the surface will result in an artifact. In this case, the artifact usually occurs in the center of the incubation area.

9. Alternatively, the 96-well chamber may be washed manually like a conventional ELISA plate.

10. Bleaching effects during the incubation with fluorescently labeled secondary antibody or prior to scanning can dramatically affect the performance of the fluorescence dye. Make sure to always protect the incubated microarrays from light and to keep the waiting time between incubation and scanning to an absolute minimum! Also, some dyes are sensitive towards ozone. Make sure to keep the slides in an ozone-free environment prior to scanning.

11. The secondary antibody might induce false-positive signals due to unspecific interaction. Control incubations using labeled second antibody alone should always be performed in parallel to the actual experiment to ensure that identified signals are not caused by nonspecific binding of the second antibody to the immobilized peptide.

12. Alternatively to using a slide centrifuge the slides may be dried by a gentle stream of nitrogen.

13. Major causes for signal intensities above the scanner saturation level are inappropriate scanner settings or non-optimal concentrations for the fluorescently labeled secondary antibodies. Adjust the scanner settings to reduce the overall signal intensities. Usually the scan power or the PMT gain can be adjusted. Perform a dilution series to identify the optimal concentration range for the fluorescently labeled secondary antibodies.

14. Most often insufficient signal intensities are caused by non-optimal scanning/storage or assay conditions not reflecting the expected binding affinity range. In case the overall signal is too low, adjust the scanner conditions in order to increase the total signal gain.

References

1. Ruprecht K, Scheibenbogen C (2014) Multiple sclerosis: the elevated antibody response to Epstein-Barr virus primarily targets, but is not confined to, the glycine-alanine repeat of Epstein-Barr nuclear antigen-1. J Neuroimmunol 272:56–61

2. Hoppe S, Bier FF, von Nickisch-Rosenegk M (2013) Rapid identification of novel immunodominant proteins and characterization of a specific linear epitope of Campylobacter jejuni. PLoS One 8(5):e65837

3. Zolla-Pazner S, deCamp AC, Cardozo T, Karasavvas N, Gottardo R et al (2013) Analysis of V2 antibody responses induced in vaccinees in the ALVAC/AIDSVAX HIV-1 vaccine efficacy trial. PLoS One 8(1):e53629

4. Stuyver LJ (2013) An antibody response to human polyomavirus 15-mer peptides is highly abundant in healthy human subjects. Virol J 10:192

5. Funkner A et al (2013) Peptide binding by catalytic domains of the protein disulfide isomerase-related protein ERp46. J Mol Biol 425(8):1340–1362

Chapter 3

Analysis of Protein Tyrosine Kinase Specificity Using Positional Scanning Peptide Microarrays

Yang Deng and Benjamin E. Turk

Abstract

Protein tyrosine kinases phosphorylate their substrates within the context of specific consensus sequences surrounding the site of modification. We describe a peptide microarray approach to rapidly determine tyrosine kinase phosphorylation site motifs. This method uses a peptide library that systematically substitutes each of the amino acid residues at multiple positions surrounding a central tyrosine residue. Peptide substrates are synthesized as biotin conjugates for immobilization on avidin-coated slides. Following incubation of the slide with protein kinase and radiolabeled ATP, the relative extent of phosphorylation of each of the peptides is quantified by phosphor imaging. This method allows small quantities of kinase to be analyzed rapidly in parallel, facilitating analysis of large numbers of kinases.

Key words Protein tyrosine kinases, Peptide microarrays, Enzyme specificity, Substrate profiling, Kinase assay, Signal transduction

1   Introduction

Protein tyrosine kinases (PTKs) are key enzymes in the initiation and propagation of cellular signaling pathways in multicellular eukaryotes [1]. Because PTKs often relay cell growth and survival signals, their deregulation frequently contributes to the development of human cancer [2]. Accordingly, PTKs have emerged as important drug targets in oncology [3]. PTKs function by catalyzing the phosphorylation of specific tyrosine residues in protein substrates. The rate of phosphorylation by a PTK is influenced by the amino acid sequence surrounding the phosphorylated Tyr residue [4]. Each PTK has a specific preferred consensus phosphorylation site sequence, or phosphorylation motif. Knowing which sequences are preferred by a kinase facilitates the production of model peptide substrates that can be used in small-molecule screening in vitro or in cultured cells to identify new kinase inhibitors [5, 6]. In addition, bioinformatics tools can be used to identify

Marina Cretich and Marcella Chiari (eds.), *Peptide Microarrays: Methods and Protocols*, Methods in Molecular Biology, vol. 1352, DOI 10.1007/978-1-4939-3037-1_3, © Springer Science+Business Media New York 2016

potential PTK substrates by scanning protein sequence databases for sites conforming to the phosphorylation site motif [7, 8].

Approaches to determine consensus motifs phosphorylated by protein kinases typically involve using peptide libraries, in which phosphorylated sequences are selected from a large pool of potential peptide substrates. One approach has been to phosphorylate a complex peptide mixture in solution with the kinase of interest, isolate the phosphopeptides from the pool, and sequence the phosphorylated products using either Edman degradation or mass spectrometry [4, 9]. Alternatively, one can examine in parallel the relative phosphorylation rates of a series of peptides either in solution or immobilized on a solid support [10, 11]. Peptide microarrays offer the capacity to analyze large numbers of sequences and require small quantities of enzyme, thus facilitating analysis of multiple kinases [12, 13]. Here we describe analysis of a positional scanning peptide library (PSPL) in a microarray format [14]. Microarrays utilizing PSPLs are unusual in that each spot corresponds to a peptide mixture rather than an individual peptide sequence. The PSPL we describe consists of 189 peptide mixtures of identical length, each having a single Tyr residue at a fixed position. Multiple positions flanking the central Tyr residue are composed of equimolar mixtures of 18 of the 20 amino acid residues. Each microarray spot corresponds to a distinct peptide mixture in which a single residue is fixed at one of the nine positions surrounding the Tyr residue (Fig. 1). Microarrays are incubated with the PTK of interest using radiolabeled ATP as a substrate. After washing to remove unincorporated radiolabel, phosphorylation of each peptide mixture on the array is quantified by phosphor imaging. The relative rate of phosphorylation of each component of the PSPL reveals which amino acid residues are preferred or deselected by the PTK at each position within the peptide.

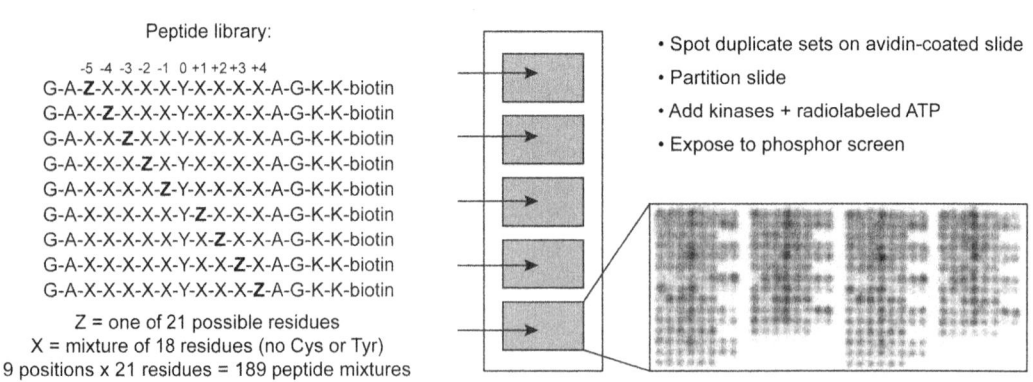

Fig. 1 Overview of procedure for PTK analysis with peptide microarrays. The sequence of peptide mixtures in the PSPL is shown at *left*. Positions marked X are composed of identical degenerate mixtures of amino acid residues. Positions marked Z consist of a single residue, either 1 of 19 unmodified amino acids (excluding Cys), pThr or pTyr. Representative data for the PTK EphA3 is shown at the *bottom right*

2 Materials

2.1 Microarray Components

1. Chicken avidin (*see* **Note 1**).

2. Binding buffer: 10 mM Tris–HCl, pH 7.5, 140 mM NaCl, 10 % glycerol, 0.1 % Triton X-100.

3. TBST: 10 mM Tris–HCl, pH 7.5, 140 mM NaCl, 0.1 % Triton X-100.

4. Humidified chamber: A sealed box with a piece of water-saturated paper towel at the bottom.

5. High-density biotin-coated glass slides (2.5×7.5 cm, Microsurfaces, Inc.).

6. Distilled, deionized water (ddH$_2$O).

7. Nitrogen gas.

8. The PSPL (Fig. 1, synthesized by Anaspec, Inc.) is a set of 189 peptide mixtures having the general sequence G-A-X-X-X-X-X-Y-X-X-X-X-G-A-K-K(biotin) in which X represents an equimolar mixture of the 18 amino acids (excluding Tyr and Cys) and K(biotin) is ε-(*N*-biotin-6-aminohexanoyl)-lysine. In each peptide mixture, one of the X positions is replaced with one of the 19 unmodified amino acids (excluding Cys), phosphothreonine (pThr), or phosphotyrosine (pTyr) (*see* **Note 2**).

9. Dimethyl sulfoxide (DMSO).

10. Spotting buffer: 10 mM Tris–HCl, pH 7.5, 140 mM NaCl, 15 % glycerol, 0.01 % Triton X-100 (*see* **Note 3**).

11. Tabletop centrifuge with microplate rotor.

12. Contact printing microarray robot (for example GeneMachines Omnigrid robotic printer) equipped with low-volume delivery pins (*see* **Note 4**).

13. TBST-biotin buffer: To 1.22 g biotin and 100 ml ddH$_2$O, add 1 M NaOH dropwise with stirring until biotin dissolves completely. Add 700 ml ddH$_2$O water and 100 ml 10× TBST (100 mM Tris–HCl, pH 7.5, 1.4 M NaCl, 1 % Triton X-100). Adjust the pH to 8.0 and add ddH$_2$O to a final volume of 1 l (*see* **Note 5**).

2.2 Kinase Assay Components

1. Purified kinase(s) to be assayed (*see* **Note 6**).

2. Kinase reaction buffer: If the optimal buffer for the kinase is not known, the following buffer works for most protein kinases: 50 mM HEPES, pH 7.5, 150 mM NaCl, 10 mM MgCl$_2$, 1 mM DTT, 0.5 % Triton X-100, and 0.05 μCi/μl [γ-^{33}P]ATP (3000 Ci/mmol) (*see* **Note 7**).

3. Quenching buffer: 10 mM Tris–HCl, pH 7.5, 140 mM NaCl, 0.1 % SDS.

4. Humidified chamber: We use a P1000 pipet tip box containing 50–100 ml water with the insert turned upside-down so that slides can be placed a few centimeters above the surface of the water.

5. Standing incubator, set to 30 °C.

6. PAP pen (hydrophobic marker) for marking slides (Abcam).

7. Phosphor storage screen and cassette (GE Healthcare).

8. Phosphor imager instrument (for example Bio Rad Molecular Imager FX Pro Plus).

3   Methods

3.1   Preparation of Avidin-Coated Slides for Microarray Printing

1. Warm biotin-derivatized slides to room temperature (*see* **Note 8**).

2. Dilute avidin in binding buffer to 50 μg/ml and keep on ice.

3. Apply approximately 200 μl of the diluted avidin solution to each slide to cover the entire surface.

4. Place slides on a piece of parafilm in the humidified chamber. Close the box and incubate for 30 min at room temperature.

5. Wash slides thoroughly with washing buffer, and then rinse with ddH$_2$O.

6. Blow-dry slides with nitrogen gas. Slides may be stored at 4 °C for up to 4 weeks until printing.

3.2   Preparation of Peptide Stock Solutions

1. Dissolve each peptide in DMSO to achieve a concentration of ~65 mg/ml.

2. Dilute an aliquot of each peptide stock 500-fold in 20 mM HEPES, pH 7.4, and measure the absorbance at 280 nm of duplicate samples.

3. Based on the peptide molar absorptivity, calculate the true peptide concentration in the original DMSO solution (*see* **Note 9**).

4. Add the appropriate volume of DMSO to each stock to bring the peptide concentration to 10 mM (*see* **Note 10**).

5. Dilute peptides from 10 mM DMSO stock solutions to a final concentration of 1 mM in spotting buffer and aliquot to a round-bottom 384-well plate (10 μl/well) (*see* **Note 11**).

3.3   Microarray Printing

1. Remove the aqueous peptide stock plate from the freezer and allow to warm to room temperature. Mix the plate well by manual shaking, and spin the plate in a centrifuge at 2000×*g* for 5 min.

2. Mount the peptide stock plate and avidin-coated slides into the plate holder and the slide holder of the microarrayer, respectively.

3. Spot peptides in the humidified chamber of the microarrayer. Peptides should be spotted in duplicate with an inter-spot distance of 0.5 mm, so that each set of 396 (2×189) peptides occupies an 18 mm $\times$ 8 mm area. Up to four duplicate sets of the PSPL can be printed on each slide.

4. Transfer slides to a 50 ml conical tube. Seal the tube and incubate for 1 h.

5. Remove excess peptide by washing once with 50 ml TBST-biotin and then three times with 50 ml TBST, 5 min per wash.

6. Rinse slides briefly with ddH$_2$O and then blow-dry under a stream of nitrogen. Printed slides should be stored at 4 °C and used within a few weeks.

3.4 Peptide Microarray Analysis of Protein Tyrosine Kinases

1. Equilibrate the humidified chamber to 30 °C by placing in the standing incubator for 2 h.

2. Partition the slide by circumscribing each duplicate set of peptides with a PAP pen.

3. Add purified kinase (1–10 ng) in 20 μl reaction buffer to each duplicate peptide set, and place the slide in the pre-warmed humidified chamber. Seal the chamber with parafilm and incubate slides at 30 °C for 30 min.

4. Wash the slide three times with 50 ml quenching buffer and two times with 50 ml 2 M NaCl, 10 min per wash.

5. Give the slide a final brief rinse with ddH$_2$O and dry under a stream of nitrogen. Cover the slide with plastic wrap and expose to a storage phosphor screen.

6. Scan the phosphor screen on a phosphor imager with a resolution of no larger than 50 μm. Quantify the extent of radiolabel incorporation into each spot using software accompanying the instrument. These raw spot intensity values should be exported into a spreadsheet for subsequent data analysis.

7. Average the data for duplicate spots on the array that correspond to the same peptide. Normalize the data by dividing each signal by the average value for all peptides with the same fixed position. In this way, the average value at each position is 1, with positively selected residues having a value greater than 1 and negatively selected residues having a value less than 1.

4   Notes

1. To avoid potential phosphorylation of avidin itself during the microarray procedure, we have used a bacterially expressed mutant (Y33F) in which the sole Tyr residue is changed to Phe [15]. However, because the Tyr residue is buried within the

avidin structure, it is unlikely to be accessible to kinases. Wild-type avidin, either native or recombinant, should be sufficient for use in this procedure.

2. An advantage of using the PSPL is that it provides an indication of the relative rate of phosphorylation of peptides having every possible amino acid at each position within the peptide. However, the PSPL is costly to synthesize, and because it consists of random peptide mixtures some highly specific PTKs may provide an insufficiently low signal. An alternative is to use sets of peptides with defined sequence. Several companies offer either premade sets of biotinylated PTK substrates (JPT Peptide Technologies) or economically parallel synthesis of custom peptides (Sigma-Aldrich, Thermo Scientific). We have found that the presence of Cys residues in peptides results in spuriously high signals on the microarray [14], perhaps due to immobilization of kinase to the array surface by disulfide formation. We therefore recommend excluding Cys if using a set of custom peptides.

3. Glycerol (15 %) is included in the spotting buffer to maintain hydration of the peptide spots during the incubation period after arraying. A "ring effect" is a common problem in protein microarrays that can hamper accurate quantitative analysis. This problem can be eliminated by adding detergent (0.01 % Triton X-100) to the spotting buffer [16].

4. The avidin-coated microarray surface used in this protocol is extremely hydrophilic relative to standard slide use for DNA microarray fabrication. As a result, the use of standard microarray pins will cause spots to merge together due to excess spreading. It is therefore important to use low-volume delivery pins. We used Stealth 3 Micro Spotting Pins (SMP3, ArrayIt), which are rated to deliver 0.7 nl fluid.

5. TBST-biotin buffer can be stored at 4 °C for up to 2 months and reused at least 20 times.

6. A major advantage of the microarray method over other methods for characterizing PTK substrate specificity is that very small quantities of kinase are required for analysis (typically 10 ng or less). This facilitates the use of low-yield mammalian expression systems that have a generally high success rate for producing active kinases compared with bacterial expression.

We have typically expressed kinases as either GST or FLAG epitope tag fusion proteins by transient transfection in HEK293T cells and purified by one-step affinity chromatography [14, 17]. While mammalian and insect cell expression systems are useful for producing kinases in their active form, a potential problem is that preparations may be contaminated with endogenous kinases that can phosphorylate peptides on the microarray. For this reason, we highly recommend that an inactive mutant of the kinase be generated and analyzed alongside the WT kinase. If the preparation is free of contaminating kinase activity, the mutant protein should produce no signal on the microarray.

7. It is best to use a buffer known to provide optimal activity for a particular PTK. The most common buffer components influencing kinase activity are monovalent and divalent salts. Some kinases are inhibited by physiological concentrations of monovalent cations (Na^+ or K^+), so their salts are often omitted from kinase assay buffers. Kinases require a divalent metal ion cofactor [18]. While the physiological cofactor is Mg^{2+}, some kinases (including many PTKs) are more active in the presence of Mn^{2+}. If the suggested buffer does not provide sufficient activity, substituting $MnCl_2$ for $MgCl_2$ is recommended.

8. Slides are generally stored at –80 °C as per manufacturers' instructions. Condensation forming on slides during warming to room temperature does not affect subsequent steps.

9. Because "dry" peptides can have substantial and variable residual solvent content, weight does not provide a faithful measure of the amount of peptide present. The best way to conveniently determine true peptide concentration is to measure UV absorbance arising from Trp and Tyr residues in the peptide sequence ($\varepsilon_{280} = 5690$ M^{-1} cm^{-1} for Trp and 1280 M^{-1} cm^{-1} for Tyr) [19]. Assuming that the mixture positions are 5.5 % Tyr and 5.5 % Trp, the molar absorptivity is 4350 M^{-1} cm^{-1} for most peptide mixtures in the PSPL, 5550 M^{-1} cm^{-1} for peptides containing Tyr at the fixed positions, and $10,040$ M^{-1} cm^{-1} for peptides containing Trp at the fixed positions.

10. Peptide DMSO stocks can be stored at –20 °C for several years.

11. Diluted aqueous peptide stocks can be sealed and stored at –20 °C for at least 8 months.

References

1. Manning G, Whyte DB, Martinez R, Hunter T, Sudarsanam S (2002) The protein kinase complement of the human genome. Science 298:1912–1934

2. Brognard J, Hunter T (2011) Protein kinase signaling networks in cancer. Curr Opin Genet Dev 21:4–11

3. Zhang J, Yang PL, Gray NS (2009) Targeting cancer with small molecule kinase inhibitors. Nat Rev Cancer 9:28–39

4. Songyang Z, Carraway KL III, Eck MJ, Harrison SC, Feldman RA, Mohammadi M et al (1995) Catalytic specificity of protein-tyrosine kinases is critical for selective signalling. Nature 373:536–539

5. Allen MD, DiPilato LM, Rahdar M, Ren YR, Chong C, Liu JO et al (2006) Reading dynamic kinase activity in living cells for high-throughput screening. ACS Chem Biol 1:371–376

6. Zaman GJ, Garritsen A, de Boer T, van Boeckel CA (2003) Fluorescence assays for high-throughput screening of protein kinases. Comb Chem High Throughput Screen 6:313–320

7. Yaffe MB, Leparc GG, Lai J, Obata T, Volinia S, Cantley LC (2001) A motif-based profile scanning approach for genome-wide prediction of signaling pathways. Nat Biotechnol 19:348–353

8. Linding R, Juhl Jensen L, Pasculescu A, Olhovsky M, Colwill K, Bork P et al (2008) NetworKIN: a resource for exploring cellular phosphorylation networks. Nucleic Acids Res 36:D695–D699

9. Kettenbach AN, Wang T, Faherty BK, Madden DR, Knapp S, Bailey-Kellogg C et al (2012) Rapid determination of multiple linear kinase substrate motifs by mass spectrometry. Chem Biol 19:608–618

10. Hutti JE, Jarrell ET, Chang JD, Abbott DW, Storz P, Toker A et al (2004) A rapid method for determining protein kinase phosphorylation specificity. Nat Methods 1:27–29

11. Tegge WJ, Frank R (1998) Analysis of protein kinase substrate specificity by the use of peptide libraries on cellulose paper (SPOT-method). Methods Mol Biol 87:99–106

12. Schutkowski M, Reimer U, Panse S, Dong L, Lizcano JM, Alessi DR et al (2004) High content peptide microarrays for deciphering kinase specificity and biology. Angew Chem Int Ed 43:2761–2674

13. Uttamchandani M, Chan EW, Chen GY, Yao SQ (2003) Combinatorial peptide microarrays for the rapid determination of kinase specificity. Bioorg Med Chem Lett 13:2997–3000

14. Deng Y, Alicea-Velazquez NL, Bannwarth L, Lehtonen SI, Boggon TJ, Cheng HC et al (2014) Global analysis of human nonreceptor tyrosine kinase specificity using high-density Peptide microarrays. J Proteome Res 13:4339–4346

15. Marttila AT, Hytonen VP, Laitinen OH, Bayer EA, Wilchek M, Kulomaa MS (2003) Mutation of the important Tyr-33 residue of chicken avidin: functional and structural consequences. Biochem J 369:249–254

16. Deng Y, Zhu XY, Kienlen T, Guo A (2006) Transport at the air/water interface is the reason for rings in protein microarrays. J Am Chem Soc 128:2768–2769

17. Deng Y, Couch BA, Koleske AJ, Turk BE (2012) A peptide photoaffinity probe specific for the active conformation of the Abl tyrosine kinase. Chembiochem 13:2510–2512

18. Adams JA (2001) Kinetic and catalytic mechanisms of protein kinases. Chem Rev 101:2271–2290

19. Edelhoch H (1967) Spectroscopic determination of tryptophan and tyrosine in proteins. Biochemistry 6:1948–1954

Chapter 4

Secondary Structure Determination of Peptides and Proteins After Immobilization

Stella H. North and Chris R. Taitt

Abstract

The presentation of immobilized peptides and other small biomolecules attached to surfaces can be greatly affected by the attachment chemistry and linking moieties, resulting in altered activity and specificity. For this reason, it is critical to understand how the various aspects of surface immobilization—underlying substrate properties, tether structure, and site of linkage—affect the secondary and quaternary structures of the immobilized species. Here, we present methods for attaching cysteine-containing peptides to quartz surfaces and determining the secondary structure of surface-immobilized peptides. We specifically show that, even when covalently immobilized, changes in peptide conformation can still occur, with measurement occurring in real time.

Key words Circular dichroism, CD spectroscopy, Antimicrobial peptide, Bioimmobilization, Peptide structure, Secondary structure, Quaternary structure

1 Introduction

Peptide microarrays can be used for myriad applications, including determination of protein binding and enzymatic specificities [1, 2], epitope mapping [3, 4], seromarker discovery [5, 6], biosensing [7–9], and development of diagnostic tools and therapeutics [4]. Most are produced either by in situ synthesis or, more commonly, by spotting onto pre-activated surfaces. Although attachment chemistry can be varied when using bead-based arrays, use of planar surfaces for production of peptide microarrays typically requires that all peptides are immobilized on the surface using the same chemistry. Use of such "universal" attachment chemistry for peptides differing in structure and function may result in suboptimal (or nonnative) presentation of the immobilized species. Indeed, we and others have found that presentation of immobilized peptides and other small biomolecules can be greatly affected by the attachment chemistry and linking moieties, resulting in altered binding affinity and specificity [10–15]. For this reason, it is critical

Marina Cretich and Marcella Chiari (eds.), *Peptide Microarrays: Methods and Protocols*, Methods in Molecular Biology, vol. 1352, DOI 10.1007/978-1-4939-3037-1_4, © Springer Science+Business Media New York 2016

to understand how the various aspects of surface immobilization—underlying substrate properties, tether structure, and site of linkage, to name a few—affect the secondary and quaternary structures of the immobilized species. Once these effects are known, we can begin to predict, design, and engineer new biomaterials with novel, unique, and desirable properties.

Circular dichroism (CD) is a powerful tool for determining secondary structure of proteins and peptides in solution. Differential absorption of the right- and left-handed circularly polarized components of plane-polarized light by a protein or peptide gives rise to elliptical polarization. This dichroism can be measured as a function of wavelength (CD spectrum), with the various resulting spectral bands assigned to distinct structural features of the molecule measured (for excellent reviews, see [16, 17]). To apply this technology to surface-immobilized species, micro- and nanoparticles have been used as substrates [18–20] to increase the effective peptide or protein concentration and achieve sufficient light absorption and CD signals. The same approach can be used by stacking silica substrates onto which peptides or proteins have been immobilized [21–24].

CD has an important role in measuring the structural determinants of proteins and peptides. However, the real power of CD is in its ability to analyze secondary structural changes upon perturbation or to compare the structures of engineered peptides/proteins to their native counterparts, from which interesting peptide/protein candidates can be selected for more detailed analysis or testing. As part of ongoing research on using antimicrobial peptides (AMPs) for broad-based screening of bacterial, fungal, and viral targets [7, 9, 25], we wish to obtain structural information on immobilized AMPs to help determine the best options for semiselective capture and detection.

Here, we first describe the methods and materials for covalent immobilization of a model α-helical cationic AMP, cecropin A, to silica slides. The silica substrates are first silanized, then activated with a heterobifunctional cross-linker, and finally incubated with cecropin A peptide for covalent linkage (Fig. 1). To achieve oriented, site-specific immobilization, a unique cysteine was appended to the C-terminus of the cecropin A peptide. We then describe structural characterization of the immobilized AMPs and demonstrate the ability of this surface-tethered AMP to undergo conformational changes in response to its interaction with a detergent mimic of its natural membrane targets. Although the immobilization techniques described here are used for CD analysis of a single immobilized peptide, the attachment methods are amenable to any thiol-containing peptide, protein, or other biomolecule arrayed on a surface by various methods [26–29].

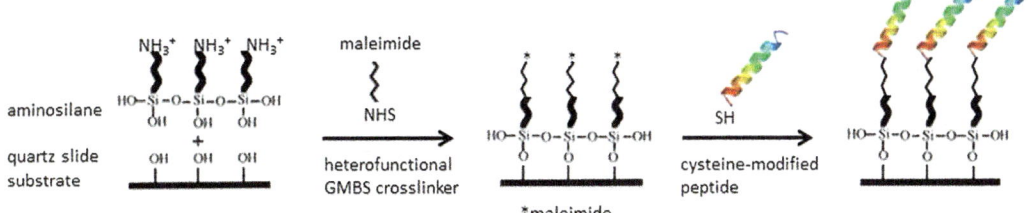

Fig. 1 Schematic showing site-directed attachment of cysteine-modified peptide covalently immobilized to maleimide-functionalized quartz slides

2 Materials

2.1 Peptide Immobilization

1. Quartz slides: Bare quartz slides (30 mm × 9.5 mm × 0.63 mm) were custom designed from Chemglass Life Sciences (Vineland, NJ, USA) to fit in standard, open-top quartz or disposable cuvettes.

2. Racks for quartz slides: Alumina cover glass staining racks from Thomas Scientific (Swedesboro, NJ, USA; catalog no. 8542E40) (*see* **Note 1**). The alumina rack is in the form of a 90° V-trough with an 8 mm open slot at the bottom. The sides have 12 equispaced grooves for supporting the quartz slides.

3. Staining dish or beakers (for cleaning, silanization, cross-linker addition).

4. Spacers: U-shaped polyetheretherketone (PEEK) spacers with the following dimensions (Fig. 2): 30 mm (height), 9.5 mm (base width), 2 mm (arm width), and 0.1 (thickness).

5. Chamber for peptide immobilization: Standard disposable 1.5 mL capacity cuvettes are used as incubation chambers for the slide/spacer assemblies during peptide deposition (*see* **Note 2**).

6. Piranha cleaning solution: Add seven parts concentrated H_2SO_4 to three parts H_2O_2 (*see* **Note 3**).

7. RCA cleaning solution[1]: Add one part H_2O_2 and one part NH_4OH to five parts ultrapure water (*see* **Note 3**).

8. Amino silane solution: Mix 2 mL (3-aminopropyl)triethoxysilane with 98 mL anhydrous toluene. Prepare this solution immediately before use within a glove bag (*see* **Notes 4** and **5**).

9. Toluene (*see* **Note 5**).

---

[1]The RCA cleaning technique was developed during the 1960s by Werner Kern at RCA Laboratories (hence, the moniker of this solution) and has become a gold standard method for removing surface impurities from silicon semiconductors and glass substrates.

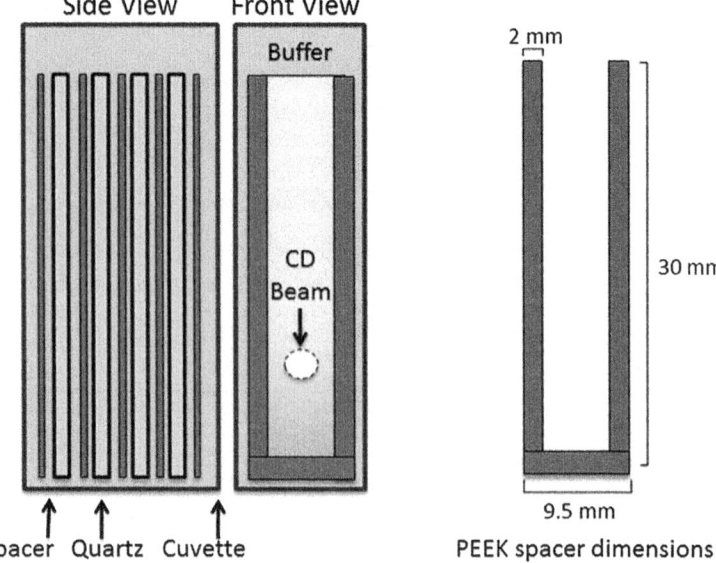

Fig. 2 Diagram of the slide-spacer assembly for attachment of peptide and subsequent CD analysis. Stacks of four slides separated by three U-shaped PEEK spacers (0.1 mm thick; dimensions shown), with one spacer on each side, are assembled in a 0.5 cm cuvette. A standard disposable cuvette can be used for peptide immobilization, while a quartz cuvette is required for CD analysis (modified from [21])

10. Cross-linker solution: Dissolve 25 mg N-(γ-maleimidobutyryloxy) succinimide ester (GMBS; Pierce, Rockland, IL) in 0.25 mL anhydrous dimethyl sulfoxide (DMSO). Add the GMBS/DMSO to 86 mL 100 mM phosphate buffer (PB), pH 7.5. Prepare this solution immediately before use (*see* **Note 6**).

11. Phosphate-buffered saline (PBS), pH 7.3.

12. Peptides: Peptide cecropin A was custom synthesized with a terminal cysteine (cecA-C38: KWKLFKKIEKVGQNIRDG IIKAGPAVAVVGQATQIAKC) and purified to 90 % purity by Biosynthesis, Inc. (Lewisville, TX, USA). Lyophilized cecA-C38 was dissolved in ultrapure water to 1 mg/mL (*see* **Note 7**). Before use, cecA-C38 is treated with tris(2-carboxyethyl)phosphine) (TCEP, 5 μM final concentration) for 30 min at room temperature prior to use (*see* **Note 8**).

2.2 Structural Determination

1. Quartz slides with immobilized peptides: from Subheading 3.1.4.

2. Peptide solutions: CecA-C38 or its base peptide cecropin A (cecA: KWKLFKKIEKVGQNIRDGIIKAGPAVAVVGQA TQIAK; Biosynthesis, Inc., Lewisville, TX, USA) dissolved to 1 mg/mL in ultrapure water.

3. Spacers: U-shaped polyetheretherketone (PEEK) spacers (*see* **Note 9**).

4. Quartz spectrophotometer cell (0.5 cm) from Starna Cells, Inc. (Atascadero, CA, USA). Cuvettes for CD measurements should be clean and dry. For standard measurements, a 0.5 cm cuvette supports stacks of four individual quartz slides and five PEEK spacers.

5. Ultrapure deionized water (*see* **Note 10**).

6. 1 M sodium dodecyl sulfate (SDS): Dissolve 28.84 g SDS in 80 mL of ultrapure water by stirring. Add water to a final volume of 100 mL. Store at room temperature.

7. Instrumentation: All measurements were carried out on a Jasco J-815 CD spectropolarimeter (Eaton, MD, USA). Spectra were recorded at 20 °C in quartz cuvettes of 5 mm path length over the wavelength range of 190–250 nm. CD spectra were acquired at a scan speed of 20 nm/min, and results were obtained by averaging three scans.

3   Methods

3.1   Immobilization of Peptides onto Quartz Slide Surfaces

The chemoselective attachment of cysteine-modified antimicrobial peptides onto quartz slides is illustrated in Fig. 1. The overall process entails four basic steps: (1) cleaning of the surface; (2) functionalization of the surface with an amine-terminated silane; (3) addition of a heterobifunctional cross-linker, resulting in a maleimide-decorated surface; and (4) addition and covalent linking of a cysteine-containing peptide.

3.1.1   Cleaning Slides

Any contaminants present on slide surfaces will interfere with uniform functionalization of the slides and will also interfere with optical measurements. For this reason, the slides are subjected to an initial acidic piranha cleaning step, followed by a subsequent alkali-based cleaning step (RCA clean).

1. Rinse each slide under a stream of distilled water, rubbing both sides to remove particulates.

2. Dry each slide under an air stream.

3. Place the slides in the alumina rack within a glass beaker or staining dish. Move all to a chemical hood.

4. Cover the slides and rack with piranha solution and incubate for 30 min (*see* **Note 3**).

5. Carefully remove the slides and rack (they can be held temporarily in a clean, dry beaker).

6. Dispose of used piranha solution within the staining dish as hazardous waste according to government and institutional regulations (*see* **Note 11**).

7. Place slides and rack back into staining dish and move to a shielded sink.

8. Fill the staining dish containing the slides to overflowing with copious amounts of deionized water to effectively dilute the remaining cleaning solution before removing and rinsing the slides. Rinse exhaustively with deionized water.

9. Using forceps to handle the slides, rinse each slide individually with deionized water and follow by drying with flowing nitrogen gas or air.

10. Place slides into a clean, dry alumina rack. Place rack within a new, clean glass beaker or staining dish and move to a chemical hood.

11. Cover the slides and rack with RCA cleaning solution and incubate for 30 min (*see* **Note 3**).

12. Carefully remove the slides and rack (they can be held temporarily in a clean, dry beaker).

13. Dispose of used RCA cleaning solution within the staining dish as hazardous waste according to government and institutional regulations (*see* **Note 11**).

14. Place slides and rack back into staining dish and move to a shielded sink.

15. Fill the staining dish containing the slides to overflowing with copious amounts of deionized water to effectively dilute the remaining cleaning solution before removing and rinsing the slides. Rinse exhaustively with deionized water.

16. Using forceps to handle the slides, rinse each slide individually with deionized water and follow by drying with flowing nitrogen gas or air.

17. Place slides into a clean, dry alumina rack within a clean glass beaker with cover.

3.1.2 Silanization

Treatment of the cleaned slides with an amine silane provides a reproducible, uniform surface functionalized with pendant amines to which a cross-linker can be attached.

1. Transfer the beaker containing the slides and rack to a nitrogen-filled glove bag within a chemical fume hood.

2. Within the glove bag, prepare the amino silane solution and immediately pour over the quartz slides. Cover the beaker to minimize evaporation and incubate for 30 min within the glove bag (*see* **Notes 4** and **5**).

3. Remove the beaker from the glove bag.

4. Using forceps to handle the slides, rinse each slide in toluene three times by swishing the slide three to five times sequentially in three 100 mL beakers filled with fresh toluene.

5. Dry each slide with flowing nitrogen gas or air. Ensure no solvent remains on the sides of each slide.

6. Return silanized quartz slides to a clean, dry alumina rack. Silanized slides may be kept dried under nitrogen for up to 1 week.

3.1.3 Attachment of Cross-Linker

The heterobifunctional cross-linker used here, GMBS, has an *N*-hydroxysuccinimidyl terminus for attachment to pendant amines on the silanized surface. In subsequent steps, the maleimide terminus will be used to attach cysteine-containing peptides, via their unique thiol moiety (*see* **Note 12**).

1. Place silanized slides (within a clean alumina rack, from Subheading 3.1.2) into a clean glass beaker.

2. Pour freshly prepared cross-linker solution over the rack and slides (*see* **Note 6**). Place a lid on the glass beaker and incubate for 30 min at room temperature.

3. Using forceps to handle the slides, rinse each slide three times by swishing the slide three to five times sequentially in three 100 mL beakers filled with deionized water.

4. Immediately after rinsing, dry each slide with a flowing nitrogen gas or air.

5. Proceed immediately to peptide immobilization (Subheading 3.1.4).

3.1.4 Chemoselective Immobilization of Cysteine-Modified Antimicrobial Peptides

Peptides containing a unique cysteine are incubated with the functionalized slide surface; pendant maleimides from the surface-attached GMBS cross-linker covalently link to the cysteine's thiol moiety. To process multiple slides simultaneously, slides are assembled into a stack, with spaces placed between the slides and at the top and bottom of the stack. The assembled stack is then loaded into a standard disposable cuvette for incubation with peptide (*see* **Note 2**).

1. Assemble four GMBS-treated slides, separated by PEEK spacers, in a standard 1.5 mL capacity disposable cuvette. A slide-spacer assembly comprised of four slides and five spacers is required for the analysis of each peptide (*see* Fig. 2).

2. Fill the cuvette chamber with 100 μg/mL cysteine-modified peptide (e.g., cecA-C38) in 10 mM phosphate buffer, pH 6.0; for the 1.5 mL capacity, 1 mL of peptide solution is required to cover the entire slide-spacer assembly.

3. Cover with parafilm to minimize evaporation and incubate overnight at 4 °C.

4. After overnight incubation at 4 °C, remove each slide from the stack using forceps and rinse by swishing the slide three to five

times sequentially in three 100 mL beakers filled with deionized water to remove reversibly bound peptides.

5. Coated slides are stored in a small beaker or cuvette filled with water (*see* **Note 13**).

6. The peptide-coated slides and spacers are then reassembled and transferred into a clean 0.5 cm quartz cuvette filled with 1 mL ultrapure water, ready for CD analysis (*see* **Note 14**).

3.2 Circular Dichroism Measurements

Circular dichroism measurements are obtained for both solution-phase peptides and peptide immobilized onto slides. In each case, SDS is added to the cuvettes (solution-phase peptide, immobilized peptide) to mimic the peptide's natural membrane target. In solution, upon addition of SDS, the peptide should undergo a change in secondary structure from a disordered, random orientation to a conformation with high helical content [30, 31].

3.2.1 CD Spectrum Measurement of Peptides in Solution

The solution-phase measurements are performed with the cysteine-modified and native peptides to ensure that incorporation of the cysteine does not affect its native structure and ability to undergo conformational changes. These data should be collected two to three times to ensure the sample has reached equilibrium. Once it has been determined that the signal is not changing over time, data can be averaged and blanks (no peptide) subtracted out (*see* **Note 15**).

1. Settings for the Jasco J-815 circular dichroism spectrometer are as follows: wavelength scan from 190 to 250 nm in a thermally controlled (20 °C) quartz cell having a 0.5 cm path length. The data were gathered with data pitch of 1 nm, D.I.T. of 8 s, bandwidth of 1 nm, and a scan speed of 50 nm/min.

2. Determine the spectrum of the blank. Fill the cell with 1 mL water and determine the CD and absorbance spectra. Collect data three times and average the data.

3. Rinse cell with ultrapure deionized water.

4. Replace water with 1 mL of peptide solution (>30 µg/mL) and collect the CD and absorbance spectra.

5. Collect data three times for a total of three replicates with each peptide concentration (±SDS). Average the data.

6. Repeat **steps 3–5** for the next peptide sample.

3.2.2 CD Spectrum Measurement of Slide-Immobilized Peptides

Determining percentages of various secondary structures generally requires knowing the exact protein concentrations and path length of the cuvette. When analyzing surface-immobilized peptides, measurement of concentration is impractical and often unreliable. However, if the same sample and cell are used for both CD and absorbance measurements, concentration and path length determinations are not necessary (*see* **Note 16**).

1. Settings for the Jasco J-815 circular dichroism spectrometer are as follows: wavelength scan from 190 to 250 nm in a thermally controlled (20 °C) quartz cell having a 0.5 cm path length. The data were gathered with data pitch of 1 nm, D.I.T. of 8 s, bandwidth of 1 nm, and a scan speed of 50 nm/min.

7. Determine the spectrum of the blank. Fill the cell with 1 mL water. Assemble the slides that have been treated with only silane and cross-linker (no peptides) with the PEEK spacers in the cell. Be careful to avoid the formation of air bubbles. Determine the CD and absorbance spectra. Collect data three times and average the data.

2. Remove the slide-spacer assembly and store in ultrapure deionized water.

3. Determine the spectrum of the blank in the presence of SDS. Rinse the cell with ultrapure deionized water. Fill the cell with 1 mL of SDS solution (20 mM). Reassemble the blank slide-spacer stack in the SDS solution and collect the CD and absorbance spectra. Collect data three times and average the data.

4. Rinse cell with ultrapure deionized water.

5. Fill the cell with 1 mL ultrapure deionized water. Assemble peptide-immobilized slides with PEEK spacers, and collect the CD and absorbance spectra. Collect data three times and average the data.

6. Remove the slide-spacer assembly and store in ultrapure deionized water.

7. Replace water with 1 mL of SDS solution (20 mM) and reassemble the peptide-treated slide-spacer stack in the SDS solution and collect the CD and absorbance spectra. Collect data three times and average the data.

8. Repeat **steps 4–6** for the next set of peptide-immobilized slides.

3.3 Data Analysis

The raw spectrophotometric data can be used to extract information about secondary and quaternary structures through data deconvolution. Appropriate controls and replicates are therefore absolute requirements for meaningful interpretation of structural data.

While the spectra themselves can be used to determine secondary structures, the ratios of mean residue molar ellipticities (for peptides in solution) or g-factor values (for immobilized peptides) at 222 and 208 nm can be used as an indication of the presence of coiled coils or other quaternary assemblies of helical peptides. A 222 nm/208 nm ratio higher than 1.0 is generally indicative of a coiled coil formation; ratios of less than 0.9 generally indicate the presence of isolated, monomeric helices.

3.3.1 Peptides in Solution

When comparing the CD spectra of different solution peptide constructs, or the same construct at different concentrations, it is important to normalize the instrument units of millidegree ellipticity by conversion to mean residue molar ellipticity, Θ, using the following equation:

$$\Theta\left(\mathrm{deg} \times \mathrm{cm}^2 / \mathrm{dmol}\right) = \left(\mathrm{ellipticity}\left(\mathrm{mdeg}\right) \times 10^6\right) / \mathrm{Path\ length\,(mm)} \times [\mathrm{peptide}](\mu M) \times n \quad (1)$$

where n is the number of peptide bonds and *ellipticity* is the raw data from the instrument.

1. Subtract baseline from the spectrum of the sample. The ellipticity for most peptides should be close to zero at 250 nm.

2. Convert the data to mean residue ellipticity using Eq. 1. *See* Fig. 3 for an example.

3. Analyze the data using the CDPro package or equivalent online software (DichroWeb) to evaluate secondary structural content (α-helix, β-strand, turns, etc.) using data collected between 190 and 240 nm.

4. Calculate 222 nm/208 nm ratios for each spectrum.

5. Calculate the mean and 95 % confidence interval (CI) for all of the sets of experimental data collected. Perform appropriate statistical analysis (e.g., Student's unpaired t test) to determine whether differences in content are statistically significant.

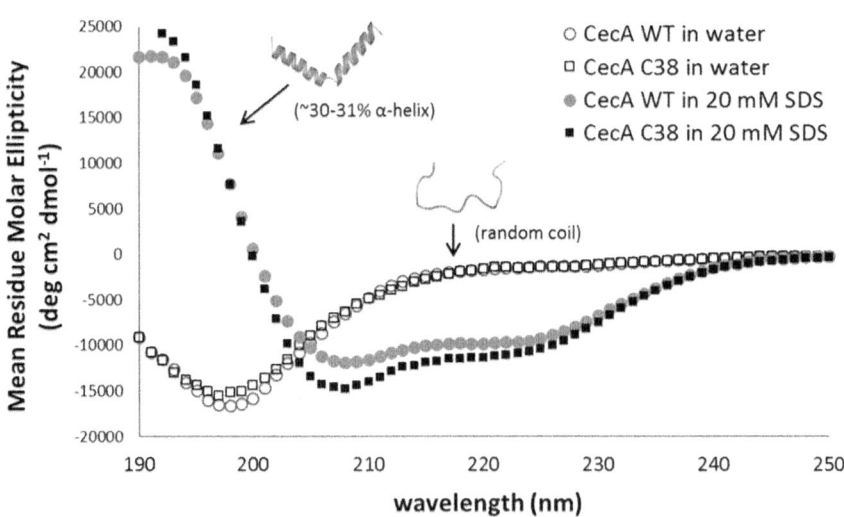

Fig. 3 CD spectra of native cecropin A and C-terminally modified cecropin A. Even without surface immobilization, it is important to determine if the incorporation of a thiol group on the C-terminus affects the α-helical folding behavior of cecA. For this reason, we investigated—in solution—the SDS-induced, secondary structure transition of cysteine-modified cecA (cecA C38) compared to the wild-type peptide (cecA WT). The presence of a cysteine at the C-terminus did not change the native structure in water or the transition from unstructured to α-helix in the presence of SDS

3.3.2 Immobilized Peptides

Both the shape of the CD spectrum (peak positions) *and* its absolute magnitude influence calculations of structural content (e.g., α-helices, parallel and antiparallel β-sheet, turns, and "other" structures). Although obtaining accurate concentrations of immobilized peptides and proteins can be difficult, reliable structural content calculations can be performed without knowing the exact concentrations of immobilized peptide by *g*-factor analysis [32]. The dimensionless *g*-factor is independent of path length, concentration, amino acid content, and molecular weight and is calculated when the same sample and cell are used for both CD and absorption measurements. Secondary structure can then be deduced from the *g*-factor spectra of samples with known structure. A *g-factor* spectrum is calculated by dividing the differential absorbance of left- and right-handed circularly polarized light (A_l and A_r, respectively) by the absorbance at each wavelength (A):

$$g = (A_l - A_r) / A \qquad (2)$$

Once the *g*-factor spectrum is determined, the data are deconvoluted using CDPro software to obtain a quantitative assessment of the secondary structural content of the peptide. Information on quaternary structure (inter-peptide interactions) can be extracted from the 222 nm/208 nm ratio. The 222 nm/208 nm ratio ≥1.0 is indicative of coiled coils, while ≤0.9 is indicative of α-helix monomers [33].

1. Subtract slide-spacer baseline from the spectrum of the sample. The ellipticity for most peptides should be close to zero at 250 nm.

2. Calculate *g*-factor spectra based on Eq. 2 for the peptide-treated samples. *See* Fig. 4 (left panel) for an example.

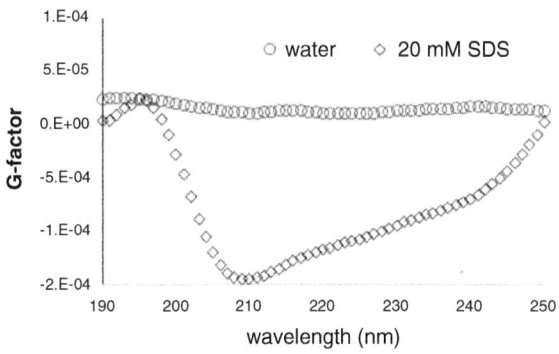

%	water	SDS
α-Helix	3	8
β-Strand	42	39
Turns	21	23
Remainder ("other")	34	30

Fig. 4 Representative *g*-factor spectra of surface-immobilized cecropin A in the presence or absence of SDS (20 mM) and calculated structural contents. Upon addition of SDS (*left panel*), there is an apparent change in conformation of the immobilized peptide. The table shown (*right*) indicates the relative percentages of various secondary structural motifs ± SDS, based on spectral data deconvoluted using CDPro. Although the SDS-induced change in α-helical content is small, this change is statistically significant ($p < 0.05$)

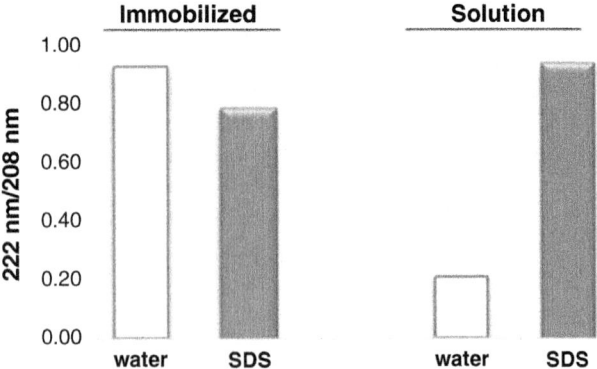

Fig. 5 The ratio of 222 nm/208 nm can be used to assess quaternary structure. A ratio of 222 nm/208 nm greater than 1.0 is indicative of fully folded coiled coils. A ratio of 222 nm/208 nm less than 0.9 is indicative of fully folded single-stranded α-helices. In solution, CecA C38 exhibits a 222 nm/208 nm ratio of 0.2 in water (random coil), which increases to 0.9 in the presence of SDS, indicative of monomeric α-helices. In contrast, when immobilized to the surface, cecA C38 has a 222 nm/208 nm ratio of 0.8–0.9, suggesting the fully helical yet single-stranded state both in the presence or absence of SDS

3. Analyze the data using the CDPro package or equivalent online software (DichroWeb) to evaluate secondary structural content (α-helix, β-strand, turns, etc.) using data collected between 190 and 240 nm (*see* Fig. 4, right panel).

4. The ratio of 222 nm/208 nm is calculated by dividing the observed *g*-factor value at 222 nm by the observed *g*-factor value at 208 nm for each spectrum and is used to discern quaternary structure (*see* Fig. 5).

5. Calculate the mean and 95 % confidence interval (CI) for all sets of experimental data collected. Perform appropriate statistical analysis (e.g., Student's unpaired *t* test) to determine whether differences in content are statistically significant.

4   Notes

1. Metal and plastic racks may be incompatible with the solutions for cleaning and silanization of the quartz slides.

2. As peptides and proteins are expensive, we use standard disposable spectrophotometer cuvettes as incubation chambers during peptide addition to decrease the volume of peptide solution needed. If using less expensive biologicals, a staining dish or beaker may be used instead. Use of slide/spacer assemblies allows multiple slides to be modified on both sides at the same time.

3. All work with the piranha solution and RCA cleaning solution should be performed in a chemical hood by personnel wearing appropriate personal protective equipment (lab coat, gloves, and face shield) who have been trained in the use and handling of oxidizers. Due to the self-decomposition of hydrogen peroxide, piranha solution should be used freshly prepared. In piranha, H_2O_2 reacts with H_2SO_4 exothermically and will start to bubble and heat up. Piranha solution can be explosive near or in contact with acetone, propanol, or any organic solvents. Also, explosions may occur if the peroxide solution concentration is more than 50 %. Used piranha solutions must be allowed to cool and off-gas in an open container left inside of a chemical fume hood for 24 h after use. After the initial 24 h has passed, the cooled piranha solution may be transferred to a piranha waste bottle—glass bottle with a vented cap.

4. Silane is sensitive to humidity. To avoid hydrolysis and subsequent cross-linking, open the silane bottle and perform all subsequent steps in an inert atmosphere, such as a nitrogen-filled glove bag.

5. Both silane and toluene are hazardous to human health; toluene is a reproductive hazard and hepatotoxin. All manipulations of solutions containing toluene should be performed in a chemical hood by personnel equipped with appropriate personal protective equipment.

6. The GMBS cross-linker targets both amines and sulfhydryls. The amine-targeting moiety of the cross-linker, *N-hydroxysuccinimidyl* ester, hydrolyzes rapidly in aqueous buffers, especially at basic pHs. To avoid unnecessary hydrolysis, store the cross-linker in a dessicator at 4 °C and warm to room temperature before opening both the dessicator and vial. *Prepare solutions immediately before use* and avoid using any amine-containing buffer, such as Tris.

7. Lyophilized peptides may contain a significant % bound water and salts by weight. Therefore, it is critical to determine peptide concentration prior to use. Protein concentrations may be verified by UV spectrophotometry at 280 nm or by a bicinchoninic acid (BCA) assay (Pierce, Rockford, IL, USA).

8. TCEP is a reducing agent that is added to prevent disulfide bond formation among individual peptide molecules. Reduction of existing disulfides ensures that a pendant thiol moiety is available for covalent immobilization to the maleimide groups displayed on the slide surface.

9. Circular dichroism analysis on a planar surface is limited by the low surface area of the substrates and the finite surface cover-

age of the immobilized peptides. To overcome these limitations, the quartz slides are stacked, separated by 0.1 mm thick polyetheretherketone (PEEK) spacers, to increase the effective peptide concentration in the optical path [34, 35].

10. Solutions for CD spectroscopy should not contain any materials that are optically active and should be as transparent as possible. Peptides dissolved in water alone have the highest transparency. However, many samples denature in the absence of salt. Buffers with low concentrations of salts and are amenable for use in CD analysis include 10 mM potassium phosphate, 100 mM KCl; 10 mM potassium phosphate, 100 mM NaCl; Dulbecco's phosphate-buffered saline; and 2 mM HEPES, 50 mM NaCl, 2 mM EDTA, and 1 mM dithiothreitol. It is important to note that different buffers will demonstrate different lower wavelength limits, below which the signal to noise is poor and the ellipticity is no longer a linear function of the path length of the cell [16].

11. The beaker/dish used for temporary storage of piranha- or RCA-cleaned slides should also be rinsed with copious amounts of water.

12. Although we have found that GMBS provides the optimal length and tether characteristics for immobilizing our peptides of interest, other linkers with similar amine- and thiol-targeting termini can be used. Care should be exercised when using other linkers, as we have observed that some linkers may interact directly with peptide domains not involved in chemical linking. For example, the long-chain hydrophobic linker, KMUS (Pierce), causes misfolding of hydrophobic peptides in solution.

13. Although we typically use our coated slides on the day of preparation, they can be used within 2 days with no decrease in performance.

14. A slide-spacer assembly without any immobilized peptide serves as a blank for background spectra. This control should be measured before any peptide-functionalized slides are measured.

15. Although many CD instruments can normalize for background values, we prefer to do this post-analysis data processing to maintain statistical relevance.

16. The Jasco CD spectrometer is capable of measuring both absorbance and CD spectra simultaneously. Measurements on other instruments incapable of simultaneous measurements will need to be performed in sequence.

Acknowledgments

This work was supported through the Office of Naval Research and the Naval Research Laboratory Core research programs. The views expressed herein are those of the authors and do not represent those of the US Naval Research Laboratory, the US Navy, the US Department of Defense, or the US Government.

References

1. Gosalia DN, Salisbury CM, Maly DJ et al (2005) Profiling serine protease substrate specificity with solution phase fluorogenic peptide microarrays. Proteomics 5:1292–1298

2. Thiele A, Zerweck J, Schutkoweki M (2009) Peptide arrays for enzyme profiling. Methods Mol Biol 570:19–65

3. Buus S, Rockberg J, Forsström B et al (2012) High-resolution mapping of linear antibody epitopes using ultrahigh-density peptide microarrays. Mol Cell Proteomics 11:1790–1800

4. Price JV, Tangsombatvisit S, Xu G et al (2012) On silico peptide microarrays for high-resolution mapping of antibody epitopes and diverse protein-protein interactions. Nat Med 18:1434–1440

5. Carmona SJ, Sanrtor PA, Leguizamon MS et al (2012) Diagnostic peptide discovery: prioritization of pathogen diagnostic markers using multiple features. PLoS One 7, e50748

6. Cooley G, Etheridge RD, Boehlke C et al (2008) High throughput selection of effective serodiagnostics for *Trypanosoma cruzi* infection. PLoS Negl Trop Dis 2, e316

7. Kulagina N, Taitt C, Anderson GP et al (2014) Affinity-based detection of biological targets. US Patent no. 8,945,856 B2 (issued 3 Feb 2015)

8. Kulagina NV, Lassman ME, Ligler FS et al (2005) Antimicrobial peptides for detection of bacteria in biosensor assays. Anal Chem 77:6504–6508

9. Taitt CR, North SH, Kulagina NV (2009) Antimicrobial peptide arrays for detection of inactivated biothreat agents. Methods Mol Biol 570:233–255

10. Han X, Liu Y, Wu F-G et al (2014) Different interfacial behaviors of peptides chemically immobilized on surfaces with different linker lengths and via different termini. J Phys Chem B 118:2904–2912

11. Han X, Uzarski JR, Mello CM et al (2013) Different interfacial behaviors of N- and C-terminus cysteine-modified cecropin P1 chemically immobilized onto polymer surface. Langmuir 29:11705–11712

12. Ngundi MM, Taitt CR, Ligler FS (2007) Crosslinkers modify affinity of immobilized carbohydrates for cholera toxin. Sens Lett 5:621–624

13. North S, Lock E, Walton S et al (2014) Processing microtitre plates for covalent immobilization chemistries. US Patent no. 8,651,158 B2 (issued 18 Feb 2014)

14. North S, Wojciechowski J, Chu V et al (2012) Surface immobilization chemistry influences peptide-based detection of lipopolysaccharide and lipoteichoic acid. J Pept Sci 18:366–372

15. Shriver-Lake LC, North SH, Dean SN et al (2013) Antimicrobial peptides for detection and diagnostic assays. In: Piletsky SA, Whitcomb MJ (eds) Designing receptors for the next generation of biosensors. Springer, Heidelberg, pp 85–104

16. Greenfield NJ (2007) Using circular dichroism spectra to estimate protein secondary structure. Nat Protoc 1:2876–2890

17. Kelly SM, Jess TJ, Price NC (2005) How to study proteins by circular dichroism. Biochim Biophys Acta 1751:119–139

18. Nygren P, Lundqvist M, Broo K et al (2008) Fundamental design principles that guide induction of helix upon formation of stable peptide–nanoparticle complexes. Nano Lett 8:1844–1852

19. Read MJ, Burkett SL (2003) Asymmetric α-helicity loss within a peptide adsorbed onto charged colloidal substrates. J Colloid Interface Sci 261:255–263

20. Stevens MM, Flynnn NT, Wang C et al (2004) Coiled-coil peptide-based assembly of gold nanoparticles. Adv Mater 16:915–918

21. Fears KP, Petrovykh DY, Photiadis SJ et al (2013) Circular dichroism analysis of cyclic β-helical peptides adsorbed on planar fused quartz. Langmuir 29.32 (2013): 10095–10101.

22. Gallardo IF, Webb LJ (2012) Demonstration of α-helical structure of peptides tethered to

gold surfaces using surface infrared and circular dichroic spectroscopies. Langmuir 28:3510–3515

23. Sivaraman B, Fears KP, Latour RA (2009) Investigation of the effects of surface chemistry and solution concentration on the conformation of adsorbed proteins using an improved circular dichroism method. Langmuir 25:3050–3056

24. Vermeer AWP, Norde W (2000) CD spectroscopy of proteins adsorbed at flat hydrophilic quartz and hydrophobic Teflon surfaces. J Colloid Interface Sci 225:394–397

25. Kulagina NV, Shaffer KM, Anderson GP et al (2006) Antimicrobial peptide-based array for *Escherichia coli* and *Salmonella* screening. Anal Chim Acta 575:9–15

26. Fu J, Reinhold J, Woodbury NW (2011) Peptide-modified surfaces for enzyme immobilization. PLoS One 6, e18692

27. Lesaicherre M-L, Uttamchandani M, Chen GYJ et al (2002) Developing site-specific immobilization strategies of peptides in a microarray. Bioorg Med Chem Lett 12:2079–2083

28. Ngundi MM, Taitt CR, Ligler FS (2006) Simultaneous determination of kinetic parameters for the binding of cholera toxin to immobilized sialic acid and monoclonal antibody using an array biosensor. Biosens Bioelectron 22:124–130

29. Ngundi MM, Taitt CR, McMurry SA et al (2006) Detection of bacterial toxins with monosaccharide arrays. Biosens Bioelectron 21:1195–1201

30. Silvestro L, Axelsen PH (2000) Membrane-induced folding of cecropin A. Biophys J 79:1465–1477

31. Lee E, Jeong K-W, Lee J et al (2013) Structure-activity relationships of cecropin-like peptides and their interactions with phospholipid membrane. BMB Rep 46:282–287

32. McPhie P (2001) Circular dichroism studies on proteins in films and in solution: estimation of secondary structure by g-factor analysis. Anal Biochem 293:109–119

33. Becktel WJ, Schellman JA (1987) Protein stability curves. Biopolymers 26:1859–1877

34. Blondelle SE, Ostresh JM, Houghten RA et al (1995) Induced conformational states of amphipathic peptides in aqueous/lipid environments. Biophys J 68:351–359

35. Fears KP, Latour R (2009) Assessing the influence of adsorbed-state conformation on the bioactivity of adsorbed enzyme layers. Langmuir 25:13926–13933

Chapter 5

Peptides and Anti-peptide Antibodies for Small and Medium Scale Peptide and Anti-peptide Affinity Microarrays: Antigenic Peptide Selection, Immobilization, and Processing

Fan Zhang, Andrea Briones, and Mikhail Soloviev

Abstract

This chapter describes the principles of selection of antigenic peptides for the development of anti-peptide antibodies for use in microarray-based multiplex affinity assays and also with mass-spectrometry detection. The methods described here are mostly applicable to small to medium scale arrays. Although the same principles of peptide selection would be suitable for larger scale arrays (with 100+ features) the actual informatics software and printing methods may well be different. Because of the sheer number of proteins/peptides to be processed and analyzed dedicated software capable of processing all the proteins and an enterprise level array robotics may be necessary for larger scale efforts. This report aims to provide practical advice to those who develop or use arrays with up to ~100 different peptide or protein features.

Key words Proteomics, Peptidomics, Peptide array, Antibody array, Affinity assay, Antigenicity, Immunization, Polyclonal antibodies

1 Introduction

Microarray technology is now well established and used widely for simultaneous measurement of the expression level of many genes or proteins. Applications of peptide microarrays include antibody epitope mapping, a multitude of other protein–protein and protein–peptide interactions studies, various diagnostics and functional analyses and proteomics applications. Unlike nucleic acid arrays, protein molecules normally require all assays and treatments to be under non-denaturing conditions, which introduce strong constraints on the allowed signal-to-noise ratios. The latter translates into the need to have a high binding capacity substrates (e.g., porous or 3D array surfaces) and high affinity of interaction between the sample and capture molecules. Unlike DNA arrays, where affinity depends on the degree of sequence

Marina Cretich and Marcella Chiari (eds.), *Peptide Microarrays: Methods and Protocols*, Methods in Molecular Biology, vol. 1352, DOI 10.1007/978-1-4939-3037-1_5, © Springer Science+Business Media New York 2016

complementarity, the length of nucleotide fragments and could be easily predicted and manipulated. Selecting and manipulating affinities of protein–protein or protein–peptide interactions is often impossible or impractical. Protein–peptide interactions are often characterized by low affinity, and this often becomes a bottleneck in devising and using peptide microarrays. Selecting good peptide sequences for generating high affinity antibodies remains largely unresolved problem.

Continuous efforts to predict peptides' and proteins' antigenicity since mid-1970s yielded many useful tools and resources. Advances in recombinant technologies revolutionized protein engineering and the development of recombinant antibody technologies. The explosion of interest in structural proteomics stimulated further research aiming to understand the molecular mechanisms of protein–protein recognition and to reveal the intricate molecular mechanics of macromolecular interactions, including these of antibodies and their antigens. Other major research areas, which are the subject of substantial research effort, are modeling, rational engineering and affinity maturation of protein binding sites. Numerous papers contain data on the role and significance of different amino acids in forming binding sites and molecular docking. The consensus view is that stability and high affinity of protein–protein interactions stem from multiple factors such as precise molecular complementarity, charge complementarity, the presence of multiple hydrogen bonds and van der Waals contacts between the interacting molecules. However, the focus of research in this area has recently moved from predicting antigenic epitopes to improving the affinity of interaction between two given molecules in a given complex. Consequently, the majority of the reported outcomes are computer assisted structure design, mutagenesis and affinity maturation approaches. Predicting of antigenic epitope gave way to high throughput and high cost epitope mapping services, such as provided by PEPperPRINT, GeneScript, or Pepscan, to name a just few suppliers. However, the use of such services becomes prohibitively expensive or impractical if more than one, let alone hundreds of individual antibody–peptide pairs are to be optimized. Limited advice is available on which peptide to use for example for generating anti-peptide antibodies.

The Affinity Peptidomics approach provides a cheaper and simpler alternative to PEPperPRINT peptide array-based method for antibody epitope mapping. The Affinity peptidomics approach to protein arrays also resolved one other major issue of many protein affinity screening applications, including microarrays based multiplex assays, namely protein sample stability and issues related to protein unfolding and denaturing. In the Affinity Peptidomics approaches, samples are first proteolytically digested before the assays, and then anti-peptide antibodies are used to assay the generated protein digests using a variety of formats, including the microarrays [1–3]. The key advantages of this technique are much reduced heterogeneity of the physical properties of the assayed proteins, the reduced dependence of each individual affinity assays on the individual proteins being tested and the increased multiplexing capabilities, reduced costs (peptide antigens and anti-peptide antibodies are easier and cheaper to produce) and compatibility with array based screenings and mass-spectrometry detection [3]. Another indirect advantage is that experimental protein samples do not any longer required careful storage and preservation (of the original intact protein folding), because the assay is not for an intact protein (as would be any traditional affinity-based assays, such as protein microarrays), but for short peptide fragments of that protein (e.g., tryptic fragments). Samples may be proteolytically digested and thus "preserved" right at the moment of being collected or shortly thereafter or at a later date. The effect of protein degradation and misfolding/denaturation during folding, on the assay performance is therefore greatly reduced or void.

The majority of antigenic prediction tools available to date rely on protein structural information or are limited to epitopes based on the protein surface and are not suitable for use with Affinity peptidomics approach. Such tools have only limited usability in the analysis of tryptic peptides for their antigenicity and may miss sequences which are "antigenic" but are not fully solvent exposed. We previously described the preferred formats for such multiplex antibody–peptide affinity assays (on microarrays) and reported the key principles of selecting peptides for antibody detection. There is some limited yet clear correlation between the key physical and chemical properties of tryptic peptides and their ability to yield high titre anti-peptide antibodies capable of capturing proteolytic peptides in a MALDI-TOF-MS assay or a microarray formats. The original parent protein structure, folding and fragment solvent exposure play no role in determining *tryptic* peptides' antigenicities, thus making the majority of existing antigenicity prediction tools useless. The approach detailed below could be useful for ranking and selecting the best tryptic peptide sequences for anti-peptide antibody development in situations when a choice of peptide epitopes from a single protein target is available. We use this approach for selecting antigenic peptides for generation of anti-peptide antibodies for use in Affinity Peptidomics assays or a variety of similar microarray assays. Here we provide a simple practical guide for selecting the best antigenic proteolytic peptide for developing anti-peptide antibodies.

2 Materials

2.1 Selection of Peptides for Anti-peptide Antibody Development

1. Protein or nucleotide databases. There are many; the service currently provided by NCBI appears to offer the most comprehensive search facilities. Use FASTA (text) display option for extracting multiple entries:

   http://www.ncbi.nlm.nih.gov/protein
   http://www.ncbi.nlm.nih.gov/gene

2. *PeptideMass* (see **Note 1**) online service for predicting all proteolytic peptides from specified proteins using a wide range of proteases. This tool is also used to predict potential posttranslational modification sites in these peptides and therefore indicates which peptides might be preferred or avoided when selecting sequences for anti-peptide antibody generation.

   http://web.expasy.org/peptide_mass/

2.2 Affinity Peptidomics: Antibody Microarrays

1. A microarray printer. Whilst we are suing Flexys contact microarray gridding robot (from Genomic Solutions Inc.), many other contact and non-contact microarrayers exist (see **Note 2**).

2. A microarray scanner. Whilst we are suing BioChip microarray Scanner (Packard Bioscience) with 16 bit TIFF readout, many other suitable scanners exist (see **Note 3**).

3. ArrayIt® SuperNylon Microarray Substrate or Biodyne® Positively charged nylon membrane (0.45 μm) or a similar membrane (see **Note 4**).

4. Anti-peptide antibodies and control antibodies (see **Note 5**).

5. Size exclusion chromatography (SEC) setup: Waters 600E pump and system controller (Waters) and Spectroflow 757 Absorbance detector (Applied Biosystems); Sephadex® G-25 column (5 mL bed volume) (see **Note 6**).

6. Microarray reaction cassette with optional compression fit silicone gasket to make multiple wells on any standard glass microarray (see **Note 7**).

7. Proteins: Bovine serum albumin (BSA): 9 % (w/v) in water.

8. Trypsin inhibitor: 10 mM phenylmethanesulfonylfluoride (PMSF) in isopropanol, store at –20 °C.

9. Complete Mini EDTA-free Protease Inhibitor Cocktail Tablets (Roche Applied Science). Prepare 25× stock solution by dissolving one tablet in 400 μL of water, store at –20 °C.

10. Size Exclusion Chromatography (SEC) running buffer (use PBS): 10 mM phosphate buffer, 2.7 mM potassium chloride, 137 mM sodium chloride, pH 7.4.

11. Microarray blocking and assay buffer: 9 % BSA, 0.1 % Tween 20 in PBS (see **Note 8**).

12. Microarray washing buffer: 0.1 % BSA, 0.02 % Tween 20 in PBS (*see* **Note 8**).

13. COOMASSIE® Brilliant Blue G-250 for making colored spots on arrays to facilitate manipulation and identification (*see* **Note 9**).

14. Rhodamine B isothiocyanate (RITC) (*see* **Note 10**).

15. Sequencing grade Trypsin.

3 Methods

3.1 Selection of Peptides for Anti-peptide Antibody Development

Although the common trend is to focus on anti-protein antibodies, custom development of such is often very expensive, takes long time and is not practical or affordable when antibody–antigen pairs are for use in a microarrays format (often 100+ individual features). Furthermore, these are often generated against recombinant expressed fragments, rather than original protein antigens and would require extensive validation efforts. This section aims to describe how to select peptide sequences for use as antigens for antibody production; methods of antibody production are outside the scope of this paper. Many commercial providers are now offering polyclonal antibody services.

1. Enter protein sequence or database accession number of the protein of interest into the *PeptideMass* program (*see* **Notes 11** and **12**).

2. Select "reduced" option for Cysteines, select no acrylamide adducts, no Methionine oxidation, $(M+H)^+$ and monoisotopic masses. Select "Trypsin", choose "no missed cleavages" and select to display all peptides (i.e., larger than 0 Da). Choose to sort peptides by peptide masses (*see* **Note 13**).

3. Choose to display all posttranslational modification, database conflicts, all polymorphisms and splice variants (*see* **Note 14**).

4. Perform the analysis; the *PeptideMass* program will display a list of predicted tryptic peptides, their masses and any information on splice variants, isoforms, and database conflicts. Peptide ranging between 10 and 20 amino acids in length might become useful in developing anti-peptide antibodies. For the ease of use, copy the table and paste for example into EXCEL datasheet.

5. Select a subset of peptides suitable for *chemical synthesis* (*see* **Note 15**). Selection criteria:

    (a) Peptide lengths should be between 5 and 30 amino acids.

    (b) Avoid multiple Prolines, Serines, Aspartic Acids, and Glycines.

Table 1
Amino acid preferences for selecting peptides for immunizations

	Desired amino acids	Allowed amino acids	Make little difference	Should be avoided if possible
Polar	Lys, Arg	His, Asn	Gln, Ser, Thr	Asp, Glu
Nonpolar	Leu, Ile, Val	Met, Gly	Ala, Pro, Trp	Tyr, Phe

(c) Avoid internal Cysteines.

(d) Avoid the following duplets of amino acids: Ser-Ser, Asp-Gly, Asp-Pro.

(e) Avoid the following triplets of amino acids: Gly-Asn-Gly, Gly-Pro-Gly.

(f) Avoid charge clustering and fewer than one in five charged amino acid side chains.

6. Select a subset of peptides suitable for *antibody generation*. Peptides containing 10–15 amino acids make good and economical antipeptide epitopes. Peptide ranking criteria are listed below and summarized in Table 1 (*see* **Notes 16–18**).

(a) Peptides must contain basic amino acids, the total number of these is not limited.

(b) Peptides must contain large aliphatic amino acids, the total number of these is not limited.

(c) Amino acids with acidic side chains should be avoided if at all possible.

(d) Aromatic nonpolar amino acids should be avoided if at all possible.

(e) The presence of polar non-charged amino acids is often necessary to maintain the overall hydrophilicity and solubility of the peptide (Gln, Asn).

(f) Small numbers of these amino acids are allowed (Met, Pro, Gly).

3.2 Affinity Peptidomics: Antibody Microarrays

In a traditional direct binding affinity assay, a labeled antigen or antibody is added to the immobilized antibody or antigen, respectively, and the detected signal is proportional to the concentration of antigen. In a competitive binding assay, where the amount of bound labeled antigen is reduced (displaced) by binding of the unlabeled antigen, the signal detected will generally be inversely proportional to the concentration of the assayed unlabeled antigen. Microarray experiments often employ two-color assay systems (two samples, two different fluorescent dyes, the sample is mixed and the array scanned twice to measure each of the two analytes).

The result is typically a ratio of two signals measured for each spot. Two-color detection is a competitive affinity assay. Affinity pepti-domics microarrays are most suitable for use with competition based detection (either one or two colors).

To simplify affinity microarray experiments, we prefer using single label competitive assays rather than traditional direct binding two-color assays. The justification of the choice can be found here [4, 5]. Briefly, our approach allows to avoid repetitive labeling of the experimental samples and compensates for the heterogeneity of the antibody affinities. Our protocols were originally devised for use with recombinant scFv anti-peptide antibodies developed using Phage display [6], but were later adapted for use with traditional anti-peptide polyclonal antibodies. Such peptide affinity assays are widely applicable to the detection and quantification of the proteo-lytic or naturally occurring peptides. The protocol below exempli-fies a single color detection approach (the simplest); with minor modifications it can be also used for two-color detection.

3.3 Proteolysis and Labeling of Serum Protein Samples

1. Use 100 μL aliquots of each of the serum samples to be tested. Add a few microliters of 1 M K_2HPO_4 or 1 M Tris pH 9 to bring the pH of the sample to pH 8, check pH by spotting the buffered serum onto pH paper (*see* **Note 19**).

2. Make one pooled serum sample by mixing equal volumes from all serum samples being tested (*see* **Note 20**).

3. Add Trypsin to each sample, including the pooled serum sample: use 1 μg per ~20–50 μg of the total serum protein and incubate at 37 °C overnight (*see* **Note 21**).

4. Stop proteolysis by adding 20 μL of 10 mM PMSF (*see* **Note 22**).

5. To fluorescently label the pooled serum sample add 80 μL PBS to a 20 μL aliquot of the digested serum, then add 100 μL of 1 % RITC. Incubate at room temperature for 30–60 min (*see* **Note 23**).

6. Stop labeling reaction by adding 20 μL of 1 M Tris pH 8.5.

7. Purify labeled peptides using size exclusion chromatography (SEC) setup (*see* **Note 24**).

3.4 Microarrays for Fluorescent Detection and Quantification of Peptides (See Note 25)

1. Set up microarray spotting instrument. The Flexys microarray gridding robot allows for three washing buffers to be used for cleaning the pins and the washing program should be set as follows: 1 % Tween 20 wash for 30 s; followed by PBS wash for 10 s, followed by another wash in 1 % Tween 20 for 30 s and PBS wash for 10 s. The final wash is in 0.1 % BSA in PBS with 0.1 % Tween 20 for 30 s (*see* **Note 26**).

2. To check pins quality and to match the pins, perform a trial run by spotting the same fluorescently labeled protein and scan the slides to determine the efficiency of protein transfer for each individual pin (*see* **Notes 27–29**).

3. To measure sample volumes required for spotting, add an even number of identical ~20 μL aliquots of any sample to a microwell plate, and insert it in the robotic spotter. Samples should have the same protein concentration and buffer as that in the antibody samples to be spotted. Choose the wells (or pins) such that half of the samples are transferred to the membrane, and half are not used. Run a number of transfers (e.g., ~100). Remove the plate from the robot and measure the remaining sample volumes, compare volume in the used and unused wells, average the difference and divide by the number of transfers (*see* **Note 30**).

4. Transfer the required volumes of antibodies to microwell plates, insert them into the robot holder and run the spotting program using the parameters specified and tested in previous steps (*see* **Note 31**).

5. Remove slides from the robot and transfer them into a sealed chamber containing a few milliliters of 37 % formaldehyde. Incubate overnight in a fume hood at room temperature (*see* **Note 32**).

6. Block the membranes using large volume of Microarray blocking and assay buffer for at least 2 h.

7. Assemble the assay mixtures as follows (exemplified for 200 μL final volume sample): use ~10 μL of the unlabeled serum digest (or the equivalent amount of the purified proteolytic peptides), add 1 μL of the 25× Protease Inhibitor Cocktail, incubate for 15 min at room temperature. Add 50 μL of the labeled and purified pooled sera digest and 140 μL of fresh Microarray blocking and assay buffer. Assemble an individual assay mixture for each of the tested sera samples (*see* **Notes 33** and **34**).

8. Transfer Microarrays to reaction cassettes with optional compression fit silicone gasket. Add the assay mix and complete the assembly of the cassette. Incubate at room temperature in the dark for at least 2 h. Arrays must not be allowed to dry out (*see* **Note 35**).

9. To wash the arrays transfer them to a flask containing ~50 mL of the Microarray washing buffer for 10 s, change buffer and incubate for 5 min, change buffer again and incubate for 10 min (*see* **Note 36**).

10. Dry the arrays (arrayed side up) in darkness overnight.

11. Scan arrays using a suitable instrument. We use a BioChip microarray Scanner. The scanner settings (focus, laser intensity, and photomultiplier attenuation) should be adjusted to the 3D slides used, but should not be changed between the slides (*see* Fig. 1).

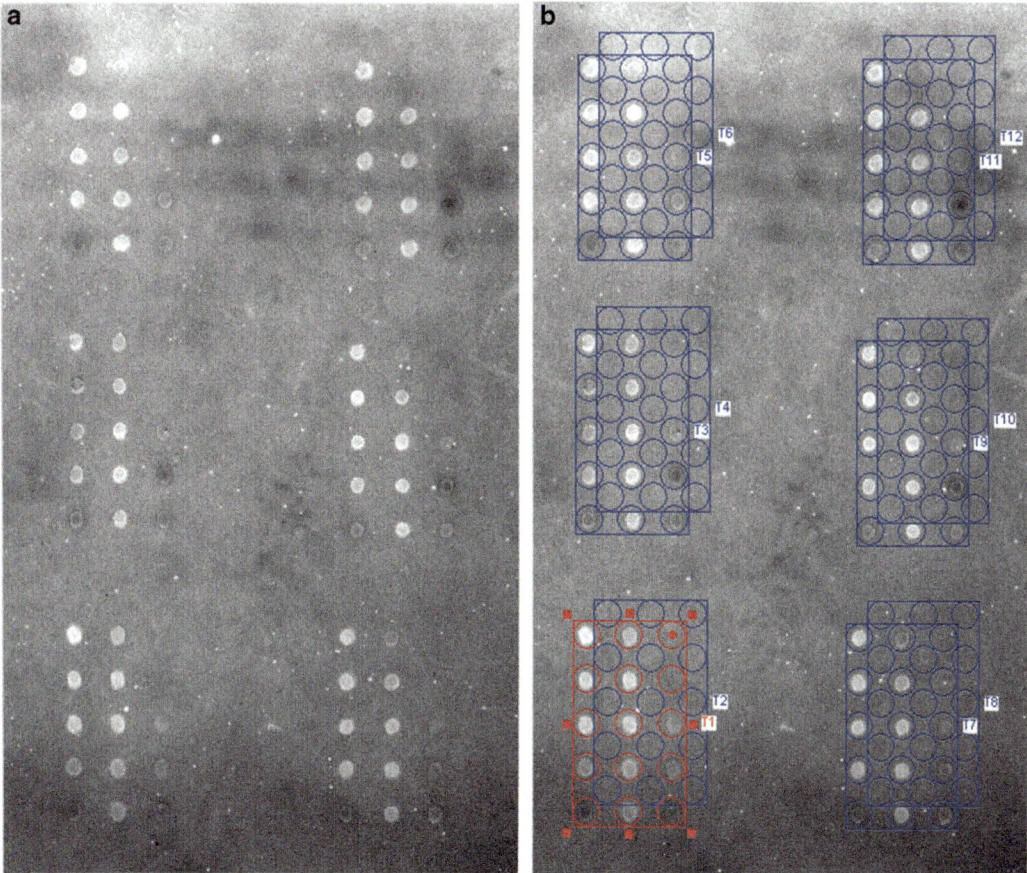

Fig. 1 A typical anti-peptide antibody array. (**a**) Fluorescent readout of a fragment of a microarray following a competitive binding experiment. Six spotted sections are shown with the total of 17 anti-peptide antibodies. Each section is spotted three times for reproducibility. First and the last few spots in each section contain Coomassie dye and are therefore not easily distinguishable on the fluorescent scan (550 nm). (**b**) The same as in panel (**a**) but with the readout grids (T1 to T12) shown. Each section has one grid for reading the fluorescent signal intensities and one identical grid to red background signal for each individual spot

12. Data analysis depends on whether competitive or non-competitive assay was used and also on the set of normalization spots used. In most cases, however, readouts should be normalized pin-to-pin and array-to-array (*see* **Note 37**).

4   Notes

1. Many online tools are available for predicting proteolytic digestion sites, the *PeptideMass* (maintained by Swiss Institute of Bioinformatics and available form SIB ExPASy Bioformatics Resources Portal) provides a wide choice of proteolytic enzymes, including many native digestive enzymes and a good

range of display options generating conveniently formatted data suitable for further processing [7].

2. A range of suitable hardware is very large and extends from a handheld MicroCaster™ Arrayer to enterprise scale NanoPrintTM workstation for printing up to 17,000 microarrays with complete automation. For the methods described here, a smaller scale desktop microarray instruments would be the most suitable. For example SpotBot® 2 Personal Microarrayer from Arrayit corporation which is capable of printing up to 384 samples onto the maximum of 14 slides.

3. A range of suitable hardware is very large. The user may want to consider for example SpotLight™ Personal Two-Color Fluorescence Microarray Scanner or ArrayIt® SpotWare™ Scanner, the latter capable of reading multiple[le slides and enables the use of colorimetric kits based on alkaline phosphatase (AP) and horseradish peroxidase (HRP).

4. 3D porous substrate provide with very high protein binding capacity and are therefore preferred over flat 2D substrates. Ready-made and commercially available membrane substrates such as immobilized Nylon or immobilized Nitrocellulose are available from multiple suppliers, e.g., SuperNylon or SuperNitro from ArrayIt® or FAST™ and CAST™ slides from Schleicher and Schuel. ArrayIt® SuperNylon Microarray Substrates are 25×76 mm ($W \times D$) and have a 150 μm thick immobilized nylon membrane. In addition to their use with proteins or peptides, SuperNylon Microarray Substrates can be used with DNAs, carbohydrates and any other molecules that bind nylon. The binding capacity is 2 μg protein per mm^2. SuperNylon is compatible with most of microarray scanners and can be used with fluorescent, colorimetric, radioactive, and chemiluminescent labels. ArrayIt® SuperNitro Microarray Substrate slides or supported nitrocellulose membrane may provide a good alternative to nylon based supports. These are similar in their performance to the SuperNylon Microarray Substrates, but there might be buffer compatibility issues. Ordinary membranes provide cheaper alternative but are more fiddly to use. These have to be attached to a glass slide during printing, e.g., with a small piece of tape. The advantage is that these can be later treated as soft membranes, rather than rigid and large microscope slides.

5. Antibody sample purity and the protein binding capacity of the microarray substrate material will affect the amount of retained antibodies and therefore the maximum signal obtainable. We typically use total IgG fraction of antisera to spot on the arrays; these require supports with higher protein binding capacities to ensure that sufficient amount of the specific antibody is attached to the membrane. Surfaces with lower binding capacities may be used with purified antibodies.

6. Liquid chromatography setups vary and any suitable equipment and properly sized columns could be used. Gravity flow may also be used for peptide purification, but care should be taken to properly calibrate the elution times of the protein (Trypsin) fraction, the peptides and the unincorporated RITC. The flow rate will vary if gravity flow is used, so calibration should be done by the volume eluted (weigh each tube containing each sample and subtract the weight of the tube), rather than the elution time.

7. Membranes can be assayed in small Petri dishes or sticky gaskets may be used to create small assay/hybridization chambers on the surface of the slides. A small strip of Parafilm or another similar laboratory film may be used to cover a drop of assay buffer on the surface of the slide and might provide sufficient barrier to stop evaporation. Placing array face down into a drop of assay buffer in a small Petri dish works fine too.

8. A 5 % solution of dry fat-free milk powder in 10 mM Na_2CO_3 pH 9 works well for blocking membranes. Alternatively, suitable array buffers are available from Arrayit for a range of proteins, antibodies, and peptide arrays. The Arrayit Protein Microarray Buffer Kit contains the following components: Microarray Activation Buffer, Protein Microarray Reaction Buffer, Protein Microarray Wash Buffer, Protein Microarray Rinse Buffer.

9. Alternatively, Ponceau S may be used. This dye is fully reversible and can be added to printing buffer to all antibodies, proteins, or peptides to facilitate visualizing the spotted array position.

10. A multitude of fluorescent dyes is now available. Amino group-reactive dyes will provide most useful when labeling tryptic digests and thiol group-reactive dyes when labeling synthetic peptides containing cysteine.

11. Entering UniProtKB, Swiss-Prot, or TrEMBL accession numbers is the preferred option, since this would allow to also include in the analysis posttranslational modification, database sequence conflicts, alternative splicing variants and polymorphisms.

12. This tool is convenient for the analysis of individual or small sets of proteins. We created a simple proteolytic digestion tool using EXCEL, which we use for in silico digestion and comparison of individual or groups of proteins. Other existing tools for predicting proteolytic peptides can be used; the choice of the method should not affect the outcome of the predictions. *PeptideCutter* (http://web.expasy.org/peptide_cutter/) is another convenient tool for predicting proteolytic digestion sites. Use the "Table of sites, sorted sequentially by amino acid number" display option.

13. Although mass calculations are not critical at this point, it is worth selecting this and other options, as these would become useful later.

14. Having this information handy will help prevent errors in the subsequent anti-peptide antibody generation program, which may be costly and which may cause very long delays, e.g., if an antibody has to be remade. Posttranslational modifications sites are best being avoided if an anti-peptide antibody is to be generated.

15. The selection of the subset of suitable peptide can be achieved simply by selecting the range of lengths 10–15 amino acids in the EXCEL file, containing the output of the *PeptieMass* program, followed by a quick check for any of the unwanted amino acids. We have entered the above rules into a Visual Basic Macro which is run in Excel, making the selection easy even if multiple proteins are analyzed. Sorting the *PeptieMass* results by mass allows to very easily select a range of peptides of suitable size. We also used truncated tryptic sequences for antibody generation (i.e., just partial tryptic peptide sequence, if the native predicted fragments were too long).

16. Cysteine is not included in the selection criteria because it is often the amino acid which is added to the peptide sequence to provide sulfhydryl for peptide cross-linking and conjugation.

17. Peptides generated using Trypsin, will often contain only one Lysine or Arginine, therefore presence of histidines may become an important selection criterion. Basic peptides are less likely to yield good antibodies and are also usually less suitable for use with positive mode MALDI-MS and thus should be avoided.

18. Much has been published on the prediction of antigenic epitopes from protein sequences [8–16]. Most of the tools however aim to identify linear epitopes in the larger protein sequence. None of these address the affinity of the predicted antibody–antigen pairs. Previously reported tools are based on the amino acid propensity scales, which take into account hydrophilicity, surface accessibility and segmental mobility of amino acids and are not therefore applicable for selecting best peptides for anti-peptide antibody generation. We use our own simple peptide selection and ranking tool (a Macro run within EXCEL).

19. This amount (~100 µL) should be sufficient for more than one assay, but the choice should depend on the volume and the number of assays by the user.

20. The pooled serum is used for fluorescent labeling and as a reference sample in a competitive binding assay. We first make a pooled sample and then proteolytically digest it. Alternatively, individually digested samples cane be pooled after the proteolysis.

21. One may assume that the total serum protein concentration is below 10 %, hence 100 μL of serum should not contain more than 10 mg protein. Hence 0.2–0.5 mg Trypsin should be sufficient.

22. PMSF will inactivate Trypsin irreversibly. PMSF will hydrolyse in water, especially at high pH, and may not work at high salt concentrations, so if in doubt, samples should be diluted and the pH shall be adjusted to pH 7 prior to adding PMSF. Alternatively trypsin may be inactivated by boiling. However, the high total protein concentration in the sample could result in the formation of protein precipitate which will complicate the extraction of peptides.

23. Fluorescence dye NIR-664-iodoacetamide may be used to label peptides through cystine side chains. Such approach would be more suitable for synthetic peptide mixtures where all peptides contain Cysteines.

24. Crude Tryptic digests may be used for affinity assays with or without additional purification (as long as Trypsin is inactivated). Fluorescently labeled peptides must be purified from the unincorporated fluorescent molecules. We use SEC on Sephadex® G-25 to separate the labeled peptides from both Trypsin and the unincorporated RITC. The same procedure can be applied to unlabeled tryptic digests. There is a large choice of commercially available SEC or reverse phase C18 cartridges and purification tools.

25. Irrespective of the type of spotting instrument used (even if using a hand-held "MicroCaster" spotter, Whatman/Schleicher and Schuell), similar key principles have to be followed:

    (a) Spotting should be done at least in triplicate for each individual antibody. The number of replicates is usually not a limiting factor (hundreds or thousands of spots can be made on each array), we found that having six replicates is sufficient in most cases.

    (b) Careful consideration must be given to the array layout: replicates should be spread over the whole array area to minimize staining and scanning artifacts. Our instrument (Flexys robotic spotter) produces blocks of densely arranged spots (grids, having from 5×5 to 12×12 spots each) whilst each grid is well separated from each other. In such a case each grid may contain only a single copy of any antibody, but the patterns should be replicated at least three (better six) times and be spread over the whole array area.

    (c) Relevant negative controls must be included. For example if polyclonal rabbit anti-peptide antibodies are used, pre-immunization sera or just total rabbit IgGs would make a suitable negative control. IgG concentration should be ideally the same as in other (specific) antibody samples and at least the same number of replicates should be made. These will provide an important reference point for the data analysis; any errors in determining the nonspecific background may affect quantification.

(d) Reference spots (fluorescently labeled protein) should be added to each array, we have at least one reference spot per each grid of spots. These are needed for signal normalization during scanning and for pin calibration.

(e) Colored spots should be added to ease array handling. These can be for example Coomassie Brilliant Blue or Coomassie-stained protein or a just add a low concentration Ponceau S to printing buffer to all proteins or peptides spotted. Colored spots will help to determine the correct membrane surface, distinguish front from the back of the membrane and identify array borders.

(f) If using contact spotting, pins should be either matched or calibrated.

26. Pin washing and reconditioning is very important for the avoidance of carry-over contaminations and for achieving high reproducibility of spotting. Pin washing procedures and buffers differ significantly from DNA gridding protocols.

27. If a large number of pins is available to the user, the simplest way would be to select those which result in the identical efficiency of protein transfer from the microwell plates to the membrane (array). If this is not possible, pins should be calibrated (by measuring the fluorescence in each spot), from multiple replicates and the values should be taken into account when interpreting the main assay results. Alternatively, calibration controls (fluorescence reference spots) should be included for each individual pin when spotting the antibodies

28. Multiple transfers should be made for each spot (i.e., the material spotted repeatedly onto the same spot on the membrane). This will dramatically increase the reproducibility of antibody transfer and increase the amount of the spotted antibodies (leading to the stronger and more reproducible signals and lesser variability between spots). We routinely use between six and ten transfers per spot. Lengthy transfer procedures should be avoided to prevent sample evaporation issues.

29. High humidity should be maintained inside the robot whilst spotting, especially for longer runs.

30. When using contact spotting, the volume transferred by the pins will depend on many parameters, such as sample viscosity, surface tension, cleanliness of the pins, contact time, and the material and porosity of the membrane. These are difficult to predict but easy to measure. We typically have values of ~20 nL per single transfer per pin.

31. Making small batches of arrays (up to ten arrays per batch) works best in our hands. Increasing the number of arrays further increases variations in the efficiency of transfer. This is probably due to the buildup of dry residue on the pins, which causes the changes. As a rule keep the total number of transfers between pin washes below ~50.

32. Because protein cross-linking with formaldehyde vapor occurs slowly, long incubation time is necessary. This will also ensure better reproducibility of the cross-linking. Alternatively, transfer spotted arrays (or membranes) into 0.003 % solution of freshly prepared glutaraldehyde and incubate overnight. Blocking the unreacted groups with glycine or Tris buffer is optional; we found no clear evidence for including this step, perhaps because blocking might be accomplished during the subsequent steps during incubation of the membranes in the blocking and assay buffers containing amino groups.

33. Because of the competitive nature of the assay, higher concentration of unlabeled peptide (test sample) will yield weaker fluorescent staining (higher degree of displacement of the labeled reference). At least two samples should be assayed, so relative concentrations of the assayed peptides can be compared between the two samples, or between one unknown sample and one known or polled reference sample. Labeled peptides' concentrations may be high, ideally should be above their K_D values. A typical individual assay should have approximately 1:1 (50 %) displacement. One control assay mixture should contain only labeled peptides with no unlabeled peptide added (no displacement array). Another control assay mixture (complete displacement array) should contain a 100-fold excess of the unlabeled peptides. The unlabeled and labeled peptides should be mixed prior to the incubation with the arrays.

34. The protocol described here is most suitable for running a number of different affinity assays and for relative quantification of the peptide levels. The pooled serum sample will serve as a reference sample. Alternatively any one of the samples can be used, e.g., any normal serum sample. The concentration (or dilution) of the unlabeled proteolytic peptides should be approximately equivalent to the concentration of pooled labeled peptides. This will result in the most accurate measurements (50 % displacement). Before running large series, it is worth running a pilot experiment to check that addition of the unlabeled test sample does not reduce the fluorescent signal more than twice on average. Use two identical slides, make the assay mixture for two arrays, but only add unlabeled serum to one of the arrays (use equivalent volume of 9 % BSA in PBS for the other array).

35. Alternatively, transfer 100 μL of the assay mix to a small petri dish, place the microarray face down on top of the drop of incubation mix. Close the Petri dish; Incubate at room temperature in the dark

36. We use 50 mL Falcon tubes for washes. For convenience and to prevent handling mistakes, we use sets of three tubes for each array, filled with 50 mL of the washing buffer. The arrays are transferred from one flask to another at pre-set intervals. Optionally membranes can be further rinsed in deionised water prior to the next step.

37. In competitive assays a higher readout would indicate lower competition for the immobilized binding site from the unlabeled sample and therefore lower concentration of the competing unlabeled peptide. Lower fluorescence would indicate indicates increased competition for binding sites (higher concentration of the matching peptide in the test sample).

References

1. Soloviev M, Finch P (2006) Peptidomics: bridging the gap between proteome and metabolome. Proteomics 6:744–747

2. Soloviev M, Finch P (2005) Peptidomics, current status. J Chromatogr B 815:11–24

3. Scrivener E, Barry R, Platt A et al (2003) Peptidomics: a new approach to affinity protein microarrays. Proteomics 3:122–128

4. Soloviev M, Terrett J (2005) Practical guide to protein microarrays: assay systems, methods and algorithms. In: Schena M (ed) Protein microarrays. Jones and Bartlett, Sudbury, MA, pp 43–56

5. Barry R, Diggle T, Terrett J, Soloviev M (2003) Competitive assay formats for high-throughput affinity arrays. J Biomol Screen 8:257–263

6. Zhang F, Dulneva A, Bailes J et al (2010) Affinity peptidomics: peptide selection and affinity capture on hydrogels and microarrays. Methods Mol Biol 615:313–344

7. Artimo P, Jonnalagedda M, Arnold K et al (2012) ExPASy: SIB bioinformatics resource portal. Nucleic Acids Res 40:W597–W603

8. Dietrich JB (1985) Antibodies against synthetic peptides—their interest in biology. Annee Biol 24:129–152

9. Hopp TP, Woods KR (1981) Prediction of protein antigenic determinants from amino-acid sequences. Proc Natl Acad Sci U S A 78:3824–3828

10. Westhof E, Altschuh D, Moras D et al (1984) Correlation between segmental mobility and the location of antigenic determinants in proteins. Nature 311:123–126

11. Hofmann HJ, Hädge D (1987) On the theoretical prediction of protein antigenic determinants from amino acid sequences. Biomed Biochim Acta 46:855–866

12. Pellequer JL, Westhof E (1993) PREDITOP: a program for antigenicity prediction. J Mol Graph 11:204–210

13. Ponomarenko JV, Bourne PE (2007) Antibody-protein interactions: benchmark datasets and prediction tools evaluation. BMC Struct Biol 7:64

14. Flower DR (2007) Immunoinformatics and the in silico prediction of immunogenicity. An introduction. Methods Mol Biol 409:1–15

15. Ponomarenko J, Bui HH, Li W et al (2008) ElliPro: a new structure-based tool for the prediction of antibody epitopes. BMC Bioinformatics 9:514

16. Hancock DC, O'Reilly NJ (2005) Synthetic peptides as antigens for antibody production. Methods Mol Biol 295:13–26

Chapter 6

Low-Cost Peptide Microarrays for Mapping Continuous Antibody Epitopes

Ryan McBride, Steven R. Head, Phillip Ordoukhanian, and Mansun Law

Abstract

With the increasing need for understanding antibody specificity in antibody and vaccine research, pepscan assays provide a rapid method for mapping and profiling antibody responses to continuous epitopes. We have developed a relatively low-cost method to generate peptide microarray slides for studying antibody binding. Using a setup of an IntavisAG MultiPep RS peptide synthesizer, a Digilab MicroGrid II 600 microarray printer robot, and an InnoScan 1100 AL scanner, the method allows the interrogation of up to 1536 overlapping, alanine-scanning, and mutant peptides derived from the target antigens. Each peptide is tagged with a polyethylene glycol aminooxy terminus to improve peptide solubility, orientation, and conjugation efficiency to the slide surface.

Key words Antibody epitope, Epitope mapping, Linear epitope, Continuous epitope, Pepscan, Peptide array, Peptide microarray, Aminooxy-tagged peptide

1 Introduction

Antibodies are an important class of therapeutics and essential tools for biomedical research. Most human vaccines to date elicit neutralizing antibodies that correlate directly with protection [1]. The molecular interactions between antibody and antigen, the epitope on the antigen where the antibody binds, and the subclass of the antibody have significant impact on the biological activities of the antibodies [2]. Antibody epitopes are usually defined as continuous (linear) or discontinuous epitopes. By definition, discontinuous epitopes are present only on folded antigens whereas continuous epitopes are preserved even if the antigens are unfolded and denatured. Some discontinuous epitopes are formed partly by a linear component; therefore antigen denaturation reduces but does not abolish antibody binding. Some epitopes involve moieties acquired by posttranslational modifications of the antigens, e.g., N-linked glycans [3]. A number of methods are available for mapping antibody specificity. The commonly used methods are mostly

Marina Cretich and Marcella Chiari (eds.), *Peptide Microarrays: Methods and Protocols*, Methods in Molecular Biology, vol. 1352, DOI 10.1007/978-1-4939-3037-1_6, © Springer Science+Business Media New York 2016

based on binding competition between the antibody samples and monoclonal antibodies (mAbs) to known epitopes of the antigens, and binding to antigens with the putative epitopes deleted by mutagenesis [4–6]. Further analysis includes determination of the atomic structures of antigen-antibody complexes to define the chemical bonds involved [3, 7]. Pepscan analysis, the use of a library of overlapping peptides spanning the entire length or a region of the antigen polypeptide, is also a popular method to map continuous epitopes and the linear component of partially discontinuous epitopes. This assay has gained popularity in vaccine research as it provides detailed information on antibody responses that are usually not obtained by standard immunoassays [8–13]. For example, pepscan assays can help monitor changes in antibody specificities when the antigens are modified, or when different adjuvants and/or immunization conditions are used. Another potential usage is to study cross-reactive antibody responses to antigenically diverse viruses, e.g., HIV, HCV, and influenza viruses, by using overlapping peptide libraries consisting of multiple viral serotypes/genotypes. However, pepscan assay in the ELISA format is not practical for studying large number of samples from infected or vaccinated subjects. With this application in mind, we have adopted the pepscan assay to the microarray format to improve throughput, minimize sample use, and lower the cost of the assays [14]. For an overview of the method, *see* Fig. 1.

2   Materials

2.1   Array Peptide Synthesis Components

1. Cellulose paper, Whatman filter paper 50 cut to approximately 10 cm × 12 cm.

2. Chemical-resistant polypropylene membrane wash box (IntavisAG, Cologne, Germany, pt. no. 30.120).

3. Fmoc-β-alanine.

4. Anhydrous N,N-dimethylformamide (DMF).

5. Diisopropyl carbodiimide (DIC).

6. 1-Methyl imidazole (NMI).

7. Anhydrous methanol.

8. Drierite desiccant.

9. A custom handheld hole puncher with 5/32″ diameter round hole size.

10. An 18-gauge needle for picking up membrane disc.

11. 0.6 M standard side-chain protected Fmoc-amino acids, Fmoc-2-(aminooxy) acetic acid, Fmoc- and the Fmoc-PEG2-Suc-OH, and Fmoc β-alanine: Quantities are weighed out into 50 mL conical tubes and the appropriate amount of anhydrous

N-methylpyrrolidone (NMP) is added. The tubes are closed and vortexed at maximum speed until the dissolution of the Fmoc-amino acid derivatives. Some may require an hour or more of vortexing to completely dissolve.

12. Activated 3 Å molecular sieves.

13. 5 % capping solution: 10 mL acetic anhydride is added to 190 mL anhydrous DMF.

14. 1.5 M 1-hydroxybenzotriazole hydrate (HOBt): 3.03 g of HOBt is added to a graduated 15 mL conical tube and filled to the 15 mL mark with anhydrous NMP.

15. 1.1 M diisopropyl-carbodiimide (DIC): 2.55 mL of DIC is added to a graduated 15 mL conical and filled to the 15 mL mark with anhydrous NMP.

16. 20 % piperidine: 10 mL of piperidine is added to a 50 mL conical and filled to the 50 mL mark with anhydrous NMP.

17. 13 mL reaction tubes (IntavisAG, Cologne, Germany).

18. 384-Well plate (well volume 252 μL).

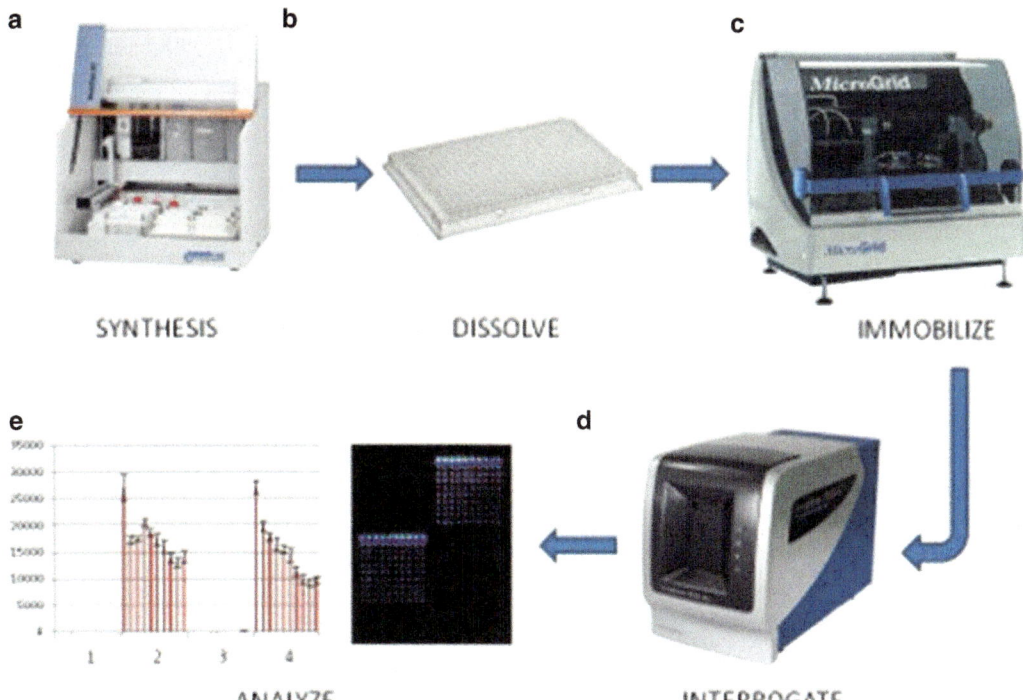

Fig. 1 Method overview: (**a**) Linear peptides are synthesized onto solid support using Multipep RS. (**b**) Released and deprotected peptides are dissolved and stored in microtiter plates. (**c**) Diluted peptides are immobilized onto NHS-ester-functionalized glass slides using a contact microarray printer. (**d**) Printed arrays are interrogated by the antibody samples and scanned for binding interactions. (**e**) Photometric signals are analyzed and plotted

19. Dichloromethane (DCM).

20. Desiccator containing Drierite desiccant.

21. Post-synthesis side-chain deprotection cleavage cocktail: 80 % trifluoroacetic acid (TFA), 3 % triisopropylsilane (TIPS), 2 % methyl sulfide, 5 % deionized water, 10 % dichloromethane (DCM). 64 mL of the cleavage cocktail is required for one 384-well plate.

22. Peptide cleavage from membrane: Anhydrous ammonia gas lecture bottle and Drierite desiccant.

2.2 Peptide Microarray Printing and Usage Components

1. Dimethyl sulfoxide (DMSO).

2. TFA.

3. Milli-Q water.

2.2.1 Peptide Dilution

2.2.2 Plate Setup

1. Printing buffer: 3× SSC buffer (Life Technologies, NY), pH 8.0 containing 0.005 % Tween-20 and 0.1 % poly-vinyl alcohol (PVA) (*see* **Note 1**). Store at 4 °C for a long term but must be at room temperature before use to avoid crystallization.

2. Matrix standard clear 384-well small-volume microplates.

3. Matrix universal polystyrene lid use with 96-, 384-, and 1536-well plates.

4. Sealing film.

5. Dissolved peptides in synthesis plate.

2.2.3 Printing

1. Nexterion Slide H (SCHOTT, Jena, Germany) with polymer layer activated with *N*-hydroxysuccinimide (NHS) esters which covalently binds amine groups: Store sealed in bags at –20 °C for a long term. The slides must be removed from freezer for at least 4 h (usually overnight) before opening bags and printing.

2. Stealth Micro Spotting Pins (SMP4B) (ArrayIt Corporation, Sunnyvale, CA) or acceptable equivalent: Stored in arrayer head or in the manufacturer's box with tip protection intact.

3. Nitrogen gas (*see* **Note 2**).

2.2.4 Humidification/ Immobilizing

1. Large Pyrex dish (18″ × 8″).

2. Test tube rack.

3. Plastic wrap.

2.2.5 Numbering, Blocking, and Storage

1. Black lab marker (*see* **Note 3**).

2. Metal rack/50 slide holder.

3. Blocking buffer: 50 mM ethanolamine in 50 mM borate buffer, pH 8.0. Stored at 4 °C but warm up to room temperature before use. To prepare a 4 L batch, 4 L of 50 mM boric acid solution is stirred constantly with a magnetic stir bar and the pH monitored in a large flask. Ethanolamine from a 16.54 M stock solution is added to borate buffer to reach the desired 50 mM concentration. Finally, add concentrated sodium hydroxide to bring the pH up to 8.0.

4. Glass Pyrex loaf dish with fitted glass lid ($12'' \times 6''$).

5. Glass spinning slide dishes.

6. Drierite Anhydrous Calcium Sulfate, Indicating: Stored at room temperature in tray on the bottom of desiccator boxes. Replaced when color changes.

2.2.6 Sample Incubation

1. Corning PBS 10× stock solution, pH 7.3–7.5 at room temperature.

2. Tween 20 detergent: Stored at room temperature.

3. Skim milk powder: Stored at room temperature.

4. Super PAP PEN hydrophobic slide marker or acceptable alternative: Stored at room temperature lying horizontally on its side.

5. Kimwipes.

6. Primary and secondary antibodies (depending on the assay performed).

7. Plastic wrap.

8. Aluminum foil.

3 Methods

Using an Intavis Multipep RS peptide synthesizer for automation of coupling and deprotection cycles, linear peptides bearing an N-terminal aminooxy moiety and a PEG2 linker are synthesized onto a modified solid support. Following side-chain deprotection and peptide cleavage, released peptides are transferred into microtiter plates, desiccated, and stored at –20 °C until use. Dried peptides are rehydrated/dissolved in DMSO and water and diluted in printing buffer for microarray printing. Using a contact microarray printing robot, the peptide library is immobilized covalently onto NHS-ester-functionalized glass slides. Printed arrays can then be probed using a variety of reagents including, but not limited to, sera, monoclonal antibodies, proteins, and cofactors. Reagents used to interrogate the peptide arrays must be detectable using a fluorescent reporter or be directly tagged with a fluorophore in the visible spectrum. Readout of detection is accomplished using standard confocal microarray scanners and subsequently analyzed for photometric signals which can be plotted as bar graphs or sorted into tables for comparative epitope binding. This method improves upon previous peptide microarray fabrication by reducing the amount of peptide input necessary for creating the arrays as well as integration of an orthogonal linker which reduces the formation of undesired bonds from reactive side-chain groups to the array surface.

3.1 Synthesis of Array Peptides

Peptides are synthesized from C- to N-terminus onto the surface of Fmoc-β-alanine-functionalized cellulose membranes using HOBt/DIC coupling using a MultiPep RS peptide array synthesizer (IntavisAG, Cologne, Germany) [15]. The instrument can simultaneously synthesize up to four 384-well plates at a time or 1536 different peptides. At the N-terminus of the peptide, a β-alanine and a PEG2-Suc-OH linker are added followed by an aminooxy terminus. The aminooxy moiety improves peptide solubility and allows for the oriented N-terminus coupling of the peptides to the NHS-ester microscope array slides (SlideH, Schott/Nexterion) [14].

3.1.1 Preparation of Fmoc-β-Alanine-Derivatized Cellulose Membrane [16]

1. Place the precut cellulose filter paper into the chemical-resistant polypropylene membrane wash box.

2. Dissolve 1.28 g of Fmoc-β-alanine in 20 mL anhydrous DMF in a 50 mL conical tube. To this add 748 μL DIC and 634 μL NMI. Mix well and transfer this solution to the chemical-resistant box with the cellulose filter paper.

3. Thoroughly soak the paper by tilting the box a few times, avoid air bubbles under the membrane, and ensure that membrane is slightly covered by solution.

4. Close the box, leave the membrane to sit for 1 h, then flip the membrane over, let it sit for an additional hour, then flip again, and incubate overnight.

5. Wash the membrane three times with anhydrous DMF; this is best accomplished by emptying the box of solution while holding the paper and adding fresh DMF to the box.

6. Wash the membrane in the same manner three times with anhydrous methanol, and then allow to air-dry.

7. Once dry, the derivatized membrane can be stored in a plastic bag with Drierite desiccant at −20 °C.

8. Prepare 5/32″ hole size membrane discs from the derivatized membrane using a handheld hole puncher.

3.1.2 Preparation of Reagents for Peptide Synthesis

1. The discs are placed individually into the MultiPep RS 384-well disc-plate holder, using an 18-gauge needle to pick up each disc and to push the disc into place in the well.

2. Once the entire 384-well plate is full, the rack is moved into position on the instrument and screwed into place in one of the two vacuum 384-well disc-plate rack holders on the instrument. Each rack can hold two 384-well plates. Make sure that the plates are oriented correctly with the A1 well placed in the appropriate position as shown in the image displayed in the software (*see* **Note 4**).

3. Using the MultiPep RS computer software, all peptides that are to be synthesized are entered into the program. The software then calculates the volume of each solvent and reagent required for the synthesis. In the software on the Show Report tab, under the Calculations drop-down menu, select Excess volume. A relative excess of 3 % for the amino acid derivatives and the reagents, and an absolute excess of 1 mL for the amino acid derivatives and 20 mL for the reagents, should be set to ensure that the needle delivery system does not run out of reagents during the synthesis. Under the Edit Method tab, select the PreActivate step in the synthesis cycles; an "excess volume of amino acid" should be set at 50 μL. This will make sure that there is enough activated amino acid to account for the dead volume in the reaction tubes during the in situ activation step.

4. Peptide sequences are typed into the MultiPep RS software in the following format: apbXXXXXXXXXXXXXXX (e.g., for a 15-amino acid peptide), where X = one of the 20 standard Fmoc-amino acid derivatives, a = the aminooxy derivative, p = PEG2-Suc derivative, and b = β-alanine derivative. These modified derivatives must be defined in the Derivative List, under the Edit Sequence tab, in the software.

5. Control peptides (shown below) can be used with the corresponding control antibody to measure synthesis quality; blank sample(s) can be used to monitor background binding. These three sample types are typically added at several locations throughout the 384-well plate. β-Alanine derivatives (b) are added to the N-terminal side of the control peptide sequences to adjust the peptide length to be the same as the rest of the peptides on the plate (*see* **Note 5**).

(a) apbbbbbbbbbbYPYDVPDYAA: Human influenza hemagglutinin (HA) peptide control.

(b) apbbbbbbbbbbDYKDDDDKGAA: FLAG peptide control.

(c) HAc: Blank control.

After deprotection in the first cycle, acetic anhydride caps the β-alanine on the surface of the membrane.

6. Place the 50 mL tubes of dissolved side-chain-protected Fmoc-amino acids, Fmoc-2-(aminooxy) acetic acid, Fmoc- and the Fmoc-PEG2-Suc-OH, and Fmoc β-alanine directly in

the rack on the instrument. The rack is propped up by, e.g., a piece of polypropylene underneath, so that the dispensing needle can approach the bottom of the tubes to minimize unused volume.

7. The 5 % capping solution, 1.5 M HOBt, 1.1 M DIC, and 20 % piperidine are placed on the instrument in the appropriate positions. Throughout the synthesis, the 1.5 M HOBt and 1.1 M DIC solutions are replaced daily with freshly made batches.

8. Anhydrous NMP is added to the MultiPep RS bottle position on the instrument (*see* **Note 6**).

9. Based on the software list of reagents, the appropriate amount of anhydrous methanol is added to the methanol (6 L) bottle on the instrument.

10. Based on the software list of reagents, the appropriate amount of anhydrous DMF is added to the DMF (6 L) bottle on the instrument.

11. Fresh 13 mL reaction tubes are placed in the appropriate position on the instrument for the in situ activation of the Fmoc-amino acid derivatives with DIC/HOBt. This occurs just prior to the coupling step.

3.1.3 Synthesis of Array Peptides

1. Once all plates, tubes, solvents, and reagents are loaded onto the instrument, the delivery dispensing needle is homed and checked for proper positioning in the *x*, *y*, and *z* coordinate system on all racks on the instrument used in the synthesis. The needle lockdown screw should always be left loose during this step, so that if something is incorrect, the needle will not be bent by the actuator arm. After this, the lockdown screw must be tightened during the synthesis.

2. The needle and the solvent delivery manifold are checked for proper flow of both anhydrous methanol and DMF. Under the Manual tab, each can be selected and primed with either solvent.

3. During peptide synthesis, piperidine deprotection and coupling steps vary somewhat throughout the process. In each deprotection step, 5 μL of Piperidine is delivered to the membrane disc. In each coupling step, 0.5 μL of HOBt, 0.5 μL DIC, 0.001 μL NMP, and 1 μL of Fmoc-amino acid derivative are combined in the reaction tube, where in situ activation of the amino acid derivative occurs forming the activated HOBt-ester of the corresponding amino acid derivative. This mixture is then delivered to the membrane disc for coupling to the N-terminus of the growing peptide chain. As an example method, an 18-amino acid-long peptide with the last three

modified derivatives would use a 21-cycle method, as follows:

Cycles 1–6: Three deprotection steps, 15 min each, and two coupling steps, 40 and 60 min are used. Acetic anhydride capping step is included.

Cycles 7–13: Three deprotection steps, 15 min each, and three coupling steps, 40, 40, and 60 min are used. Acetic anhydride capping step is included.

Cycles 14–20: Four deprotection steps, 15 min each, and three coupling steps, 40, 50, and 60 min are used. Acetic anhydride capping step is included.

Final cycle: Four deprotection steps, 15 min each, and three coupling steps, 40, 50, and 60 min are used. The acetic anhydride capping step is deleted from this cycle.

4. During the initial wash steps of DMF and methanol, an extraction time of 30 s is used to allow solvent to puddle in the plate. Later wash steps use an extraction of 90 s, which reduces puddling. These settings work well for puddling, but not overfilling the plate during the wash steps. A final extraction of 900 s removes all solvents and readies the plate for the next step.

5. The Start button may be selected in the software to begin the synthesis. The synthesis will generally take 1–2 weeks, depending on the number of peptides and plates being synthesized.

6. All reagents and solvents should be watched daily for their use throughout the synthesis. If the level starts to get close to the minimal volume requirement for the tube, fresh reagent or solvent should be added.

7. After the synthesis is complete, the membrane discs are transferred to a Nunc DeepWell 384-Well Plate for side-chain deprotection and peptide recovery. The MultiPep RS 384-well disc plate is first unscrewed from the vacuum plate rack holder. The DeepWell plate is placed directly over the disc-plate holder, such that the wells align directly with each other, and the plates are taped together. The MultiPep RS plate is then flipped with the DeepWell plate on bottom and the discs are punched from the backside into the DeepWell plate with any plastic tool (e.g. a liquid dispensing manifold supplied with the MultiPep RS instrument). Once transfer appears to be complete, remove the tape and separate the plates very slowly (*see* **Note 7**). Using a pipette tip, 18-gauge needle, or plastic tool to push any discs sticking on the side down into the well so that solution will cover them in subsequent steps.

8. Add 150 μL of the deprotection cleavage cocktail to each well containing a membrane disc to deprotect side chains of membrane bound peptides (*see* **Note 8**). Cover the 384 DeepWell

plate with regular aluminum foil (*see* **Note 9**) and incubate that plate for 4 h in a chemical fume hood. Aspirate the TFA cleavage cocktail from the 384 DeepWell plate into a Nalgene polypropylene flask, then, dispose of properly (*see* **Note 10**).

9. Rinse the membrane discs in the 384 DeepWell plate two times with 150 μL DCM and two times with 150 μL anhydrous methanol. Use multi-channel aspirator to remove solvent washes and dispose of properly.

10. Allow membrane discs in 384 DeepWell plate to air dry in a chemical fume hood for 30 min.

11. Ammonia gas is used to cleave the peptides from membrane support [17] (*see* **Note 11**). Place the 384 DeepWell plate into the Wheaton Dry Seal Desiccator. Apply vacuum to evacuate the air inside and close the valve. Attach tubing to the desiccator from a lecture bottle of anhydrous ammonia gas and open the valve to the lecture bottle and then the valve to the desiccator. Fill the desiccator with the ammonia gas, then close both valves and remove the tubing. Fill a standard balloon with the ammonia gas and attach to the desiccator then open the desiccator valve. Allow discs to incubate in the ammonia gas overnight. Then, vent the desiccator, remove the lid, and allow the gas to dissipate for 20 min in the fume hood before handling.

12. Remove the plate from the desiccator and add 100 μL of 0.05 % TFA solution to each well to recover the peptides. Allow to incubate at room temperature for 30 min. Pipette mix the 100 μL up and down a few times then transfer to a fresh 384 DeepWell Nunc plate. Repeat this procedure one more time to give a total 200 μL of eluted peptide solution. The 384 DeepWell plate is then transferred to a Savant SpeedVac Plus system and evaporated to dryness.

3.2 Peptide Microarray Printing and Usage

3.2.1 Dissolving Peptides for Stock Solution

1. Using a liquid-handling robot or multichannel pipettor, add 12.5 μL of DMSO to every well of the peptide synthesis plate that contains synthesized peptide. Once all wells have DMSO, cover the plate with sealing film and leave at room temperature for 24 h to allow the peptides to dissolve.

2. Shake the plate on a vortex instrument equipped with a plate holder attachment for 10 min at medium speed. Following the shaking, spin the plate down to precipitate the droplets to the bottom of the wells using a centrifuge equipped with microplate holders for 1 min at a speed of 1000 RPM.

3. Add 12.5 μL of MilliQ water to every well of the plate. Cover the plate with sealing film, vortex the plate for 10 min at medium speed, and centrifuge the plate as above.

4. At this point, peptides in solution should be used immediately to prepare printing stocks/plates or sealed with aluminum sealing film and stored at –20 °C until ready for printing.

3.2.2 Printing
Plate Setup

The design and layout of the final peptide microarray should be completed before this process is started.

1. Using a liquid-handling robot or multichannel pipettor, add 7.5 μL of printing buffer to every well of a printing plate. Spin the plate down as above to precipitate the buffer to the bottom of the wells.

2. Transfer 2.5 μL of dissolved peptide to the 7.5 μL printing buffer in each of the printing plate, and mix three times at 5 μL pipetting volume to fully homogenize the solution (*see* **Note 12**).

3. Spin the plate briefly as above. As Tween and PVA tend to create bubbles during the mixing process, break any bubbles in each well using a pipette tip. The printing plate is now ready and should be used as soon as possible to avoid potential peptide breakdown.

3.2.3 Printing

The authors make the assumption that the reader has access to a microarrayer and the knowledge of its use. Our lab uses a MicroGrid II 600 by Digilab, but any comparable machine would be acceptable. The slides are sensitive to humidity and their surface can be hydrolyzed by moisture over the course of time. This should be taken into account, especially long prints at high humidity.

1. Place arrayer pins in the printing head according to the correct configuration of the desired layout.

2. Remove slides from bags, open boxes, and remove any dust particles or small pieces of plastic that might be on their surfaces by blowing ultra-high purity Nitrogen gas on each slide. Anything that is on the printing surface of the slide could potentially get picked up by the pins and inhibit the ability to print. Place the desired number of slides, coated side up, on the arrayer stage.

3. Program the arrayer with the appropriate conditions, load the 384 well plates into the printer, and begin. Periodically check that all the pins are printing by shining a flashlight onto the stage so the spots are visible with the glare. Printing should be kept at between 55 and 65 % humidity.

3.2.4 Humidification/
Immobilizing

1. Once the printing is finished, the slides must undergo a temporary immobilization step, also called humidification. This includes the slides being placed in a chamber with 100 % humidity immediately after printing for a period of 30 min. The chamber can be constructed by simply placing a few very wet paper towels flat in the bottom of a large Pyrex glass dish, the slides on a rack with the print side up, and sealing the dish with plastic wrap to trap in the moisture.

2. After the humidification step, the slides are numbered and blocked immediately.

3.2.5 Numbering,
Blocking, and Storage

Marking, blocking, and storing the slides are important steps to prepare the slides for incubation with antibody samples [18]. At this stage, the spots are still visible as small crystals left by the printing buffer. Blocking buffer will be used to quench any unreacted NHS residues present on the slides. Once the blocking step is completed, the crystals will be washed away and the grids will no longer be visible on the slide surface. Therefore, the print area on the slides must be checked and marked prior to blocking. This is also when slide numbering should occur. Barcodes or a series of numbers can be assigned to each slide for identification. The orientation of the print must be marked clearly when the slides are bordered and numbered.

1. Using a black non-indelible lab marker on the back side (not the print surface), mark brackets around each of the four corners of the printed array and number, if desired.

2. After each slide is marked, carefully place each slide in a slide staining rack appropriate for the number of slides to be blocked. Place the rack inside a slide staining dish and add blocking buffer just to cover the tops of the slides. Place the lid on and shake gently for 1 h.

3. Remove the slides from the liquid and, in a separate dish filled with ddH$_2$O, dip the entire rack of slides ten times up and down in the water to rinse off any residual blocking buffer. Dispose of the blocking buffer in the appropriate waste container.

4. Transfer the slides to a centrifuge equipped with hanging basket that can accommodate the staining racks and spin dry the slides for 5 min at 200 rcf.

5. Once the slides are dry, they can be used immediately or stored at −20 °C in a sealed bag for 3 months.

3.2.6 Sample Incubation

For every batch of printed slides, quality control will be performed on the first, middle and last slides, or two to three randomly picked slides. Our peptide arrays contain positive and negative peptide controls for this purpose. Following an initial QC, antibody samples can be studied using the peptide microarrays. Incubation is performed in a humidification chamber. Biohazardous materials (e.g., serum samples potentially contaminated with infectious agents) can be pretreated to inactivate the infectious agents (*see* **Note 13**) and should be handled in a biosafety cabinet with appropriate decontamination procedures. Sample volumes vary based on the array surface area and can be further reduced by performing incubation beneath a cover slip.

1. Delineate the print area with hydrophobic marker on the outside of print area, indicated by the brackets drawn during the

blocking steps, and let it dry for several minutes. Alternatively, a commercial mask may be used. This step helps limit the incubation area and preserve sample volume.

2. Block non-specific binding surface of the arrays with an appropriate volume of Blotto solution (PBS, 1 % skim milk powder, and 0.05 % Tween-20). Incubate for at least 5 min.

3. Dilute the antibody samples in Blotto to the desired concentration. For monoclonal antibodies, we typically start at initial probe concentrations of 10 and 1 µg/mL. For plasma or sera, a dilution between 1:50 to 1:500 generally gives informative results (*see* **Note 14**).

4. Pipette the appropriate volume of sample onto the print surface. Cover the humidification chamber and incubate for 1 h.

5. For non-hazardous samples, discard samples by simply pouring it off the slides over a waste container. Wash the slides one at a time by dipping them four times in PBS with 0.05 % Tween, then four times in PBS only, and finally four times in ddH$_2$O. Wipe the back of the slide with a Kimwipe and return the slides to the humidification chamber.

6. For infectious samples, pipet the liquid into a container with decontaminating agents (e.g. 10 % Chlorox). Wash the slides by pipetting, instead of dipping, sufficient volume of the wash buffers onto the slide surface and incubate 1 min between each wash as above.

7. Dilute the secondary detecting reagent to an appropriate concentration in Blotto. We typically start with a concentration of 10 µg/mL. The final working concentration will be determined based on the results.

8. Add the appropriate volume of secondary reagent to the array surface in the humidification chamber and incubate for 1 h. Place the chamber in the dark to avoid photobleaching of the fluorescent label on the secondary reagents.

9. Wash the slides as above. Dry the slides by centrifugation or with a gentle stream of ultra high purity nitrogen gas (*see* **Note 15**).

3.2.7 Scanning and Data Analysis

The peptide arrays can be scanned at different settings dependent on the scanning instrument used. Our lab uses an InnoScan 1100 AL confocal slide scanner with a 24-slide auto loader. The slides are scanned at increasing values of Photo Multiplier Tube (PMT) enhancement and laser power at a selected setting to produce an image containing as many signals as possible within the dynamic range. It is assumed that the reader is familiar with creating a mapping file to create the data analysis grids (GeneID, GenePix Array List GAL file, etc.), made by the printing robot or manually in

Microsoft Excel [18]. We assume that users are also familiar with the use of microarray image analysis software to extract signal data using the software.

1. Load the dried slides into the auto loader. Make sure the print surface is facing the laser source.

2. Using the scanner software, set up the instrument to scan with the desired parameters. The settings can be tested and optimized although each scan degrade the fluorescent signals of the samples to some extent. The images are saved in a multi-image or individual TIFF file.

3. Once scanning is completed, transfer the images to a computer with the image processing program installed. Our lab uses Mapix image analysis software, included with the InnoScan 1100 AL scanner. Using a previously constructed GAL file template, measure the array and save the data results (*see* **Note 16**).

4. The output file, in the Text file format, can be opened in Microsoft Excel or any tabular data analysis software. The data can be quickly sorted, organized, and graphed using a multi-step macro for all slides in the same print format.

5. The data for monoclonal antibodies should immediately identify the continuous epitopes. Binding to consecutive overlapping peptides will also help define the boundary of the epitopes. The results should be validated in a second assay, e.g., peptide inhibition of antibody binding to the antigen in an immunoassay (e.g., ELISA).

6. The data analysis of polyclonal antibodies will require one or multiple controls. It will also be necessary to estimate the sensitivity of the microarrays by dose titration of a monoclonal antibody with known affinity to one of the peptides (*see* **Note 17**). For vaccinated laboratory animals that have not been exposed to many different antigens, the pre-immunization blood sample (pre-bleed) should be a good control to identify non-specific peptides and to define cutoff signals for the assays (usually set at 2× or 3× binding signals of the negative control). For other polyclonal samples (e.g., vaccinated or infected humans), the signals tend to be noisy with multiple potentially non-specific binding peptides. In some cases, "prebleed" samples may also not be available for background determination. To understand such binding patterns, non-specific binding peptides and background signals for each peptide can be estimated using a panel of 20 or more normal blood donor samples. Highly nonspecific or cross-reactive peptides (e.g., those with extremely high hydrophobicity, or derived from antigens sharing high similarity between organisms) should be noted in the data analysis. Positive hits should be validated in a second assay, e.g., peptide inhibition of antibody binding to the antigen in an immunoassay (e.g., ELISA).

4   Notes

1. To prepare a stock of printing buffer, first make about 500 mL of a 3× SSC buffer. By weight, add 0.1 % PVA to the 3× SSC (for 500 mL, 0.5 g). To completely dissolve the PVA, it will be necessary to heat the SCC solution to near boiling on a heat plate. Use a magnetic stir bar to keep the solution in motion while dissolving the PVA. Once the PVA is completely dissolved, add 1 %, by volume of TFA (for 500 mL, 5 mL). Using a 10 % stock solution of Tween-20 in ddH$_2$O, add the appropriate amount to the phosphate buffer to reach an end result of 0.005 % Tween-20 content. With a pH meter and sodium hydroxide, adjust the final pH of the printing buffer to pH 8.

2. Only ultrahigh-purity Nitrogen gas must be used on slides, pins, and any other pieces of equipment involved in printing. If a lower quality gas is used, the oils and debris will cause endless problems with pins sticking and printing to be defective.

3. Do not use Sharpie or other regular permanent marker when marking and numbering slides. While the product description may claim permanent, the ink will gradually soak off in the blocking buffer resulting in a mess and wasted slides. VWR lab markers have shown to withstand the buffer for such an extended amount of time.

4. If only one 384-well plate is being synthesized, a second empty plate should be loaded into the second position on the rack and a 384-well siliconized rubber matting (IntavisAG, Cologne, Germany) taped to the top to decrease loss of vacuum from the side not used in the synthesis. Sealing up of unused wells should also be done if only a portion of a plate will have peptides synthesized on it. Loss of vacuum pressure on the plate(s) will cause the solutions delivered to it not to be evacuated during the solvent removal vacuum steps and ultimately synthesis failure.

5. Typically, all peptides synthesized should be the same length. They all must end in the aminooxy derivative, which must remain uncapped after the synthesis, so it is ready for coupling to the SlideH (Schott/Nexterion). For this reason, the post-coupling acetic anhydride capping step is turned off in the last cycle. This capping step can only be turned off for a particular cycle in the synthesis, which will apply to all plates on the synthesizer. Peptides of different length can be synthesized, but then the capping step must be turned off completely for the remainder of the synthesis past the aminooxy coupling cycle for the shortest peptide on any of the plates.

6. Anhydrous solvents were either purchased as anhydrous or purchased in the non-anhydrous form but allowed to stand

overnight over standard activated 3 Å molecular sieves (Sigma-Aldrich, St. Louis, MO). Just before use the solvents were either filtered or decanted to ensure dust was not present. The presence of dust will clog the dispensing needle on the MultiPep RS.

7. It is imperative that the discs all go into their proper well or peptide identification for whole or partial plate can be ruined at this point. Since the 384 DeepWell plate goes face down on top of the Multipep RS plate, which is then flipped. The peptide order will be inverted, the A1 peptide from the MultiPep RS plate will go into the P1 well of the DeepWell plate.

8. This step and all subsequent manipulations with the cleavage cocktail should be performed in a chemical fume hood. Methyl sulfide has a characteristic disagreeable odor and trifluoro acetic acid is a strong acid, so caution should be taken in handling.

9. The cleavage cocktail will destroy other standard types of 384-well seals.

10. A multi-channel aspirator should be constructed using a Nalgene Aspirator Vacuum Pump, a Nalgene Polypropylene Vacuum Flask and Nalgene 689 Polypropylene Tubing.

11. The following work with ammonia gas should take place in a chemical fume hood.

12. The peptide concentrations can vary between different batches and should be titrated after each synthesis. For printing, it is not recommended to have >12.5 % DMSO in the diluted peptides.

13. For example: heat inactivation, addition of nonionic detergents. Check biosafety guidelines for specific pathogens.

14. To perform a direct comparison between samples, the samples should be first titrated against the antigens in an immunoassay, preferably to both native and denatured antigens. The samples are diluted based on the equivalent antibody titers of the samples.

15. Using a stream of N_2 gas may create streaks, due to drying effects, on the slide surface. Any visible streaks could be interpreted as background by the image analysis software.

16. Peptide arrays can be printed in triplicate or more dependent on the total number of peptides and the space available on the microscopic slide. For arrays that have more than quadruplicate prints, spots with the lowest (and highest for hexaplicate prints) signals of the entire array may be excluded to reduce variability in the assays.

17. This will help determine the concentration of antibodies recognizing the same peptide required to produce a signal in this assay.

Acknowledgement

This work was supported by NIH grants AI079031 and AI106005 (M.L.). This is TSRI manuscript number 28094.

References

1. Plotkin SA (2010) Correlates of protection induced by vaccination. Clin Vaccine Immunol 17:1055–1065

2. Strohl WR (2009) Therapeutic monoclonal antibodies: past, present and future. In: An Z (ed) Therapeutic monoclonal antibodies: from bench to clinic. Wiley, Hoboken, NJ, pp 3–50

3. Garces F, Sok D, Kong L, McBride R, Kim HJ, Saye-Francisco KF, Julien JP, Hua Y, Cupo A, Moore JP, Paulson JC, Ward AB, Burton DR, Wilson IA (2014) Structural evolution of glycan recognition by a family of potent HIV antibodies. Cell 159:69–79

4. Moore JP, Sodroski J (1996) Antibody cross-competition analysis of the human immunodeficiency virus type 1 gp120 exterior envelope glycoprotein. J Virol 70:1863–1872

5. Law M, Maruyama T, Lewis J, Giang E, Tarr AW, Stamataki Z, Gastaminza P, Chisari FV, Jones IM, Fox RI, Ball JK, McKeating JA, Kneteman NM, Burton DR (2008) Broadly neutralizing antibodies protect against hepatitis C virus quasispecies challenge. Nat Med 14:25–27

6. Giang E, Dorner M, Prentoe JC, Dreux M, Evans MJ, Bukh J, Rice CM, Ploss A, Burton DR, Law M (2012) Human broadly neutralizing antibodies to the envelope glycoprotein complex of hepatitis C virus. Proc Natl Acad Sci U S A 109:6205–6210

7. Kong L, Giang E, Nieusma T, Kadam RU, Cogburn KE, Hua Y, Dai X, Stanfield RL, Burton DR, Ward AB, Wilson IA, Law M (2013) Hepatitis C virus E2 envelope glycoprotein core structure. Science 342:1090–1094

8. Janvier B, Archinard P, Mandrand B, Goudeau A, Barin F (1990) Linear B-cell epitopes of the major core protein of human immunodeficiency virus types 1 and 2. J Virol 64:4258–4263

9. Tomaras GD, Binley JM, Gray ES, Crooks ET, Osawa K, Moore PL, Tumba N, Tong T, Shen X, Yates NL, Decker J, Wibmer CK, Gao F, Alam SM, Easterbrook P, Abdool Karim S, Kamanga G, Crump JA, Cohen M, Shaw GM, Mascola JR, Haynes BF, Montefiori DC, Morris L (2011) Polyclonal B cell responses to conserved neutralization epitopes in a subset of HIV-1-infected individuals. J Virol 85:11502–11519

10. Li XQ, Qiu LW, Chen Y, Wen K, Cai JP, Chen J, Pan YX, Li J, Hu DM, Huang YF, Liu LD, Ding XX, Guo YH, Che XY (2013) Dengue virus envelope domain III immunization elicits predominantly cross-reactive, poorly neutralizing antibodies localized to the AB loop: implications for dengue vaccine design. J Gen Virol 94:2191–2201

11. Zhao S, Qi T, Guo W, Lu G, Xiang W (2013) Identification of a conserved B-cell epitope in the equine arteritis virus (EAV) N protein using the pepscan technique. Virus Genes 47:292–297

12. Ruwona TB, Giang E, Nieusma T, Law M (2014) Fine mapping of murine antibody responses to immunization with a novel soluble form of hepatitis C virus envelope glycoprotein complex. J Virol 88:10459–10471

13. Uchtenhagen H, Schiffner T, Bowles E, Heyndrickx L, LaBranche C, Applequist SE, Jansson M, De Silva T, Back JW, Achour A, Scarlatti G, Fomsgaard A, Montefiori D, Stewart-Jones G, Spetz AL (2014) Boosting of HIV-1 neutralizing antibody responses by a distally related retroviral envelope protein. J Immunol 192:5802–5812

14. Ruwona TB, McBride R, Chappel R, Head SR, Ordoukhanian P, Burton DR, Law M (2014) Optimization of peptide arrays for studying antibodies to hepatitis C virus continuous epitopes. J Immunol Methods 402:35–42

15. IntavisAG—MultiPep RS: Parallel Peptide Synthesis, User Manual

16. Winkler DF, Hilpert K, Brandt O, Hancock RE (2009) Synthesis of peptide arrays using SPOT-technology and the CelluSpots-method. Methods Mol Biol 570:157–174

17. Hilpert K, Winkler DF, Hancock RE (2007) Peptide arrays on cellulose support: SPOT synthesis, a time and cost efficient method for synthesis of large numbers of peptides in a parallel and addressable fashion. Nat Protoc 2:1333–1349

18. Busch J, McBride R, Head SR (2010) Production and application of glycan microarrays. Methods Mol Biol 632:269–282

Chapter 7

The Peptide Microarray-Based Resonance Light Scattering Assay for Sensitively Detecting Intracellular Kinase Activity

Tao Li, Xia Liu, Dianjun Liu, and Zhenxin Wang

Abstract

The peptide microarray technology is a robust, reliable, and efficient technique for large-scale determination of enzyme activities, and high-throughput profiling of substrate/inhibitor specificities of enzymes. Here, the activities of cyclic adenosine monophosphate (cAMP)-dependent protein kinase A (PKA) in different cell lysates have been detected by a peptide microarray-based resonance light scattering (RLS) assay with gold nanoparticle (GNP) probes. Highly sensitive detection of PKA activity in 0.1 μg total cell proteins of SHG-44 (human glioma cell) cell lysate (corresponding to 200 cells) is achieved by a selected peptide substrate. The experimental results also demonstrate that the RLS assay can be employed to evaluate the chemical regulation of intracellular kinase activity.

Key words Peptide microarray, Intracellular kinase activity, Gold nanoparticle, Resonance light scattering, Chemical regulation

1 Introduction

Protein kinases (an enzyme family with more than 500 members) modulate virtually all cellular processes in living systems through catalytic transfer of a phosphate group from adenosine triphosphate (ATP) to serine, threonine, or tyrosine residues of substrate proteins [1–10]. Dysregulated protein phosphorylation has been linked to numerous serious diseases including cancer, cardiovascular disease, and other pathologies, making kinases among the most important targets for therapeutic molecules [1–12]. Numerous assays/techniques including radioactive [γ-^{32}P]ATP labeling assay (the gold standard of testing kinase activity), optical assay, electrochemical assay, and mass spectrometric technique have been employed for detecting activities of pure kinase and/or kinase in cell lysate [13–27]. Among these assays, peptide microarray is an ideal platform for studying kinase functionality and inhibition since

Marina Cretich and Marcella Chiari (eds.), *Peptide Microarrays: Methods and Protocols*, Methods in Molecular Biology, vol. 1352, DOI 10.1007/978-1-4939-3037-1_7, © Springer Science+Business Media New York 2016

microarray allows undertaking thousands of phosphorylation reactions simultaneously [22, 23, 25–27]. For instance, we have developed a peptide microarray-based spectroscopic assay with triple readout principles (fluorescence, surface-enhanced Raman scattering (SERS), and RLS) for screening both serine kinase (i.e., PKA) and tyrosine kinase (i.e., leukocyte-specific protein tyrosine kinase, LCK) inhibitors from a commercial kinase inhibitor library [27]. Inhibition efficiencies (IC_{50} values) and the inhibiting type (type I) of these inhibitors have been successfully evaluated by the peptide microarray-based spectroscopic assay.

Herein, a peptide microarray-based RLS assay with GNP probes has been employed to detect intracellular PKA activities of different cell lines and monitor the chemical-mediated PKA activity fluctuation of living cell. This approach will strongly supplement existing kinomics techniques and provide a valuable tool with high-throughput format for screening kinase inhibitors at living cell level.

2   Materials

All solutions and buffers were prepared with Milli-Q water (18.2 MΩ. cm at 25 °C) and analytical grade reagents at room temperature unless otherwise specified. The reagents were stored according to the manufacturers' requirements.

1. Aldehyde 3-D slides were purchased from CapitalBio Ltd. (Beijing, China).

2. α-Catalytic subunit of cyclic adenosine 5′-monophosphate (cAMP) dependent protein kinase (PKA) (New England Biolabs Ltd., Ipswich, MA, USA) (*see* **Note 1**).

3. Biotinylated anti-phosphoserinen antibody (anti-phosphoserine-biotin). (Sigma-Aldrich Co., St. Louis, MO, USA) (*see* **Note 2**).

4. Inhibitor H89 (EMD Chemicals Inc., Gibbstown, NJ, USA) (*see* **Note 3**).

5. Peptides (*see* Table 1 for details) (ChinaPeptides Ltd., Shanghai, China).

6. SHG-44 (human glioma cell), HeLa (human cervical cancer cell), MCF-7 (human breast adenocarcinoma cell), SW-620 (human caucasian colon adenocarcinoma cell), and PC-12 (rat adrenal medulla pheochromocytoma cell) cells (Chinese Academy of Sciences Cell Bank, Shanghai, China).

7. Fetal bovine serum (FBS) (Gibco, Grand Island, NY, USA) (*see* **Note 4**).

8. Phosphate buffered saline 1 (PBS-1): 50 mM phosphate buffer plus 0.15 M NaCl, pH 7.5. Add about 100 mL water to a 1 L graduated cylinder or a glass beaker (*see* **Note 5**). Weigh 8.7 g NaCl, 8.95 g $Na_2HPO_4 \cdot 12H_2O$, and 3.9 g $NaH_2PO_4 \cdot 2H_2O$.

Table 1
Peptides and their PKA phosphorylation sites and phosphorylation efficiencies

Name	Sequence	Potential phosphorylation motifs	Phosphorylation efficiency[a]
CP	CGGALRRAGLG	–	–
SP1	CGGALRRASLG	$1 \times RRAS$[b]	0.2
SP2	CGGALRRASLGPP[c]	$1 \times RRAS$[b]	0.26
SP3	CGGALRRASLGAQ	$1 \times RRAS$[b]	0.15
SP4	CGGALRRASLGAQPP[c]	$1 \times RRAS$[b]	0.22
SP5	CGGALRRASLGRRAS	$2 \times RRAS$[b]	0.37
SP6	CGGALRRASLGRRASPP[c]	$2 \times RRAS$[b]	0.42
SP7	CGGALRRASLGAQLT	$1 \times RRAS$[b] and $1 \times T$[d]	0.24
SP8	CGGALRRASLGAQLTPP[c]	$1 \times RRAS$[b] and $1 \times T$[d]	0.28
SPT1	CGGALRRATLG	$1 \times T$[d]	0.06
SPT2	CGGALRRATLGPP[c]	$1 \times T$[d]	0.07

[a]The slope of corresponding peptide calibration line
[b]Based on the well-known PKA phosphorylation motif RRXS (Here, X = A; see http://www.cbs.dtu.dk/services/NetPhosK)
[c]The peptides with PP amino acid residues at C-terminals can form circle conformation
[d]Probable PKA phosphorylated site. The concentration of ATP in kinase solution is 50 µM and the amount of PKA is 250 U, respectively

Transfer the weighed NaCl, NaH_2PO_4, and Na_2HPO_4 to the cylinder/glass beaker. Add water to a volume of 900 mL. Mix and adjust pH with NaOH (*see* **Note 6**). Make up to 1 L with water.

9. Phosphate buffered saline 2 (PBS-2): 10 mM phosphate buffer, 0.137 M NaCl, pH 7.4. Weigh 8 g NaCl, 2.9 g $Na_2HPO_4 \cdot 12H_2O$, and 0.2 g KH_2PO_4 and prepare a 1 L solution as in previous PBS-1 step.

10. Tris buffer: 50 mM Tris–HCl, pH 7.5. Add about 100 mL water to a 1 L graduated cylinder or a glass beaker. Weigh 6.06 g Tris and transfer to the cylinder/glass beaker. Add water to a volume of 900 mL. Mix and adjust pH with HCl (*see* **Note 7**). Make up to 1 L with water. Store at 4 °C.

11. Spotting buffer: 0.3 M phosphate buffer, 0.2 M NaCl containing 20 µg/mL BSA, pH 8.5. Add about 10 mL water to a 100 mL graduated cylinder or a glass beaker. Weigh 2.9 g NaCl, 13.43 g $Na_2HPO_4 \cdot 12H_2O$, and 5.85 g $NaH_2PO_4 \cdot 2H_2O$ and prepare a 80 mL solution as in PBS-1 step. Weigh 2 mg BSA and transfer to the cylinder/glass beaker. Make up to 100 mL with water. Store at 4 °C.

12. Washing buffer-1: 50 mM phosphate buffer containing 1 % (w/v) BSA, pH 7.5. Add about 100 mL water to a 1 L graduated cylinder or a glass beaker. Weigh 8.95 g $Na_2HPO_4 \cdot 12H_2O$ and 3.9 g $NaH_2PO_4 \cdot 2H_2O$ and prepare a 900 mL solution as in PBS-1 step. Weigh 10 g BSA and transfer to the cylinder/glass beaker. Make up to 1 L with water.

13. Blocking buffer: 50 mM phosphate buffer, 0.15 M NaCl containing 1 % w/v BSA and 0.1 M ethanolamine, pH 7.5. Add 1 mL ethanolamine and about 80 mL PBS-1 to a 100 mL graduated cylinder or a glass beaker. Weigh 1 g BSA and transfer to the cylinder/glass beaker. Make up to 100 mL with PBS-1.

14. Probe buffer: 50 mM phosphate buffer, 0.15 M NaCl, supplemented with 0.1 % Tween-20 (v/v) and 1 % BSA (w/v), pH 7.5. Add 10 μL Tween-20 and about 8 mL PBS-1 to a 10 mL graduated cylinder or a glass beaker. Weigh 0.1 g BSA and transfer to the cylinder/glass beaker. Make up to 10 mL with PBS-1. Store at 4 °C.

15. Washing buffer-2: 20 mM Tris–HCl, 0.15 M NaCl, 10 mM EDTA, 1 mM EGTA with 0.1 % Triton X-100, pH 7.5. Add about 100 mL Tris buffer to a 1 L graduated cylinder or a glass beaker. Weigh 8.7 g NaCl, 2.92 g EDTA, and 0.38 g EGTA. Transfer the weighed reagents to the cylinder/glass beaker. Add 400 mL Tris buffer and 1 mL Triton X-100. Make up to 1 L with water.

16. PKA assay buffer: 50 mM Tris–HCl, 100 mM $MgCl_2$, 0.2 mM ATP, pH 7.5. Add about 100 mL Tris buffer to a 1 L graduated cylinder or a glass beaker. Weigh 2.03 g $MgCl_2 \cdot 6H_2O$ and 0.15 g DTT. Transfer the weighed $MgCl_2$ and DTT to the cylinder/glass beaker. Make up to 1 L with Tris buffer. Store at 4 °C.

17. Culture medium: Dulbecco's modified Eagle's medium (DMEM) supplemented with 10 % FBS. Add about 100 mL water to a 1 L graduated cylinder or a glass beaker. Weigh 13.4 g DMEM and 3.7 g $NaHCO_3$. Transfer the weighed DMEM and $NaHCO_3$ to the cylinder/glass beaker. Make up to 900 mL with water. Add 100 mL FBS to DMEM solution. Filter the culture medium by a 0.2 μm filter. Store at 4 °C (*see* **Note 8**).

18. Lysis buffer: PBS-2, 1 mM EDTA, 5 mM EGTA, 10 mM $MgCl_2$, 2 mM DTT, 1 mM sodium orthovanadate, 80 mM b-glycerophosphate, 3 μg/mL pepstatin A, 5 μg/mL aprotinin, and 1 mM phenylmethylsulfonyl fluoride (PMSF), pH 7.4. Add 10 mL PBS-2 to a 100 mL graduated cylinder or a glass beaker. Weigh 29.22 mg EDTA, 0.19 g EGTA, 0.2 g $MgCl_2 \cdot 6H_2O$, 30.85 mg DTT, 18.39 mg sodium orthovanadate, 0.69 g b-glycerophosphate, 300 μg pepstatin A, 500 μg aprotinin, and 17.42 mg phenylmethylsulfonyl fluoride (PMSF). Transfer the weighed reagents to the cylinder/glass beaker. Make up to 100 mL with PBS-2. Store at 4 °C.

3 Methods

3.1 Preparation of GNP Probes

1. 250 mL Tetrachloroaurate ($HAuCl_4$, 1 mM) was brought to a reflux while stirring (*see* **Note 9**).

2. 50 mL trisodium citrate solution (38.8 mM) was then added quickly, which resulted in a change in solution color from pale yellow to deep red.

3. The solution was refluxed for an additional 30 min, allowed to cool to room temperature, filtered through a 0.45 μm nylon filter, and stored at room temperature (*see* **Note 10**).

4. Avidin stabilized 13 nm GNPs (named as GNP probes) were prepared by stirring a mixture of 200 μL succinylated avidin (1 mg/mL in PBS-1) and 2 mL citrate stabilized 13 nm GNPs (3.8 nM) for 30 min at room temperature.

5. Excess protein was removed by repeated centrifugation at 13,000 rpm ($16,100 \times g$, three times) using an Eppendorf centrifuge. The as-prepared GNP probes were redispersed in PBS-1 and stored at 4 °C.

3.2 Preparation of Cell Lysates

1. SHG-44, HeLa, MCF-7, SW-620, and PC-12 cells were grown with fresh culture medium in a humidified 5 % CO_2 incubator for 24 h at 37 °C.

2. 10^6 cells in 0.3 mL lysis buffer were sonicated for 15 s at 0 °C (three times) by a JY88-IIN homogenizer.

3. The homogenates were centrifuged at 13,000 rpm ($16,100 \times g$) for 50 min at 4 °C using an Eppendorf centrifuge.

4. Supernatants were extracted and stored at 0 °C (named as extract-1).

5. The pellets were resuspended in 0.2 mL lysis buffer supplemented with 0.1 % Triton X-100, and sonicated for 15 s at 0 °C (three times), centrifuged at 13,000 rpm ($16,100 \times g$) for 30 min at 4 °C.

6. The supernatants were collected and mixed with extract-1 (named as extract-2) (*see* **Note 11**).

7. The cell lysates (extract-2) were diluted by desired volumes of PKA assay buffer for using in phosphorylation experiments, respectively. Total proteins of cell lysates were determined using the Bradford reagent (*see* **Note 12**). The concentration of cell lysate was defined by the amount of total proteins in the assay solution.

8. For PKA activation study, 10^6 cells were co-cultured with 10 μM Forskolin (Fsk) in 5 mL culture medium for 30 min, washed by 5 mL PBS-2 (two times), and lysated as previously described, respectively.

9. For H89 inhibition assay, 10^6 cells were co-cultured with H89 (3 μM) in 5 mL culture medium for 1 h, washed by 5 mL PBS-2 (two times), and lysated as previously described, respectively.

3.3 Peptide Microarray Fabrication and Phosphorylation

3.3.1 Screening Peptide Substrate of PKA

1. Eleven peptides including one control peptide (CP) and ten PKA substrates derived from kemptide (SP1 to SP8, SPT1 and SPT2, see Table 1 for details) were dissolved in 100 μL spotting buffer with desired concentrations (0.1 nM–1 mM), respectively.

2. 15 μL peptide solutions were transferred into 384-well microtiter plate wells, respectively.

3. Multiple copies of 11 peptides were spotted on commercial aldehyde 3-D slides (see **Note 13**) by a standard contacting procedure using a SmartArrayer 48 system with a Stealth Pin.

4. The slides were incubated under vacuum at 30 °C for 12 h.

5. The slides were washed with 30 mL washing buffer-1 for 5 min (two times), and then immersed in blocking buffer for 1 h to inactivate any free aldehyde groups on slide surfaces. Subsequently, the slides were washed with 30 mL washing buffer-2 for 10 min (three times) and 30 mL PKA assay buffer for 10 min.

6. The slides were then incubated with PKA which were diluted to the desired concentrations (0, 5, 15, 25, 35, 50, 75, 100, 150, 200, 250, and 300 U in 20 μL PKA assay buffer) for 1 h at 37 °C (see **Note 14**). Subsequently, the slides were rinsed with 30 mL washing buffer-2 for 5 min (three times), 30 mL washing buffer-2 without Triton X-100 for 5 min (three times), 30 mL water for 3 min (three times), and dried by centrifugation ($200 \times g$ for 40 s), respectively.

7. The slides were incubated with 200 μL anti-phosphoserine-biotin antibody solution (20 μg/mL in probe buffer) for 1 h at 37 °C, respectively. Then, a series of washing steps were applied to the slides: (1) 30 mL PBS-1 with 0.1 % Tween-20 for 5 min (three times), (2) 30 mL PBS-1 for 5 min (three times), and (3) 30 mL water for 3 min (three times), respectively.

8. The slides were incubated with 200 μL of GNP probes (5 nM) in PBS-1 for 2 h at 25 °C, respectively. The slides were subjected to a series of washing and drying steps as previously described in **step 7**, respectively.

9. 1 mL silver enhancer reagent (1:1 mixed solutions A ($AgNO_3$) and B (hydroquinone)) was applied to each slide for 8 min after being labeled with GNP probes (see **Note 15**). Then, slide was washed with 30 mL water (three times), and dried with centrifugation.

10. The slides were read by an ArrayIt SpotWare Colorimetric Microarray Scanner. The detection result is shown in Fig. 1 (see **Notes 16** and **17**).

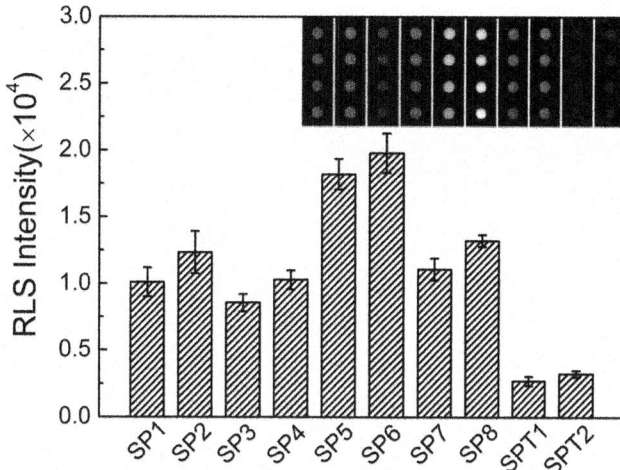

Fig. 1 RLS images (*inset*) and corresponding RLS intensities of ten peptide substrates. The concentrations of peptides in spotting solution are 10 μM, the concentration of ATP is 50 μM, and the amount of PKA is 250 U, respectively. The signals have been subtracted by the average RLS intensity of CP

3.3.2 Detecting Intracellular PKA Activity in Different Cell Lines

1. Multiple copies of SP6 (10 μM in spotting solution) were spotted and treated as described in Subheading 3.3.1, **steps 3–5**.

2. The slide was dried by centrifugation ($200 \times g$ for 40 s) and separated into 12 independent subarrays by PTFE grid.

3. SP6 on the subarray spot was phosphorylated by SHG-44 cell lysate, HeLa cell lysate, MCF-7 cell lysate, SW-620 cell lysate, and PC-12 cell lysate, respectively.

4. The subarrays were recognized with anti-phosphoserine-biotin antibody, labeled with GNP probes, treated with silver enhancer reagent, and detected as described in Subheading 3.3.1. Corresponding RLS images and data analysis are shown in Fig. 2 (*see* **Notes 16** and **18**).

3.3.3 Detection Sensitivity of RLS Assay

Here, the experimental procedure is the same as that of Subheading 3.3.2 except phosphorylation of subarray containing SP6 by desired concentrations (2, 5, 10, 20, 25, 30, 50, 80, 100, 120, and 150 μg/mL) of SHG-44 cell lysate. Corresponding RLS images and data analysis are shown in Fig. 3 (*see* **Notes 16** and **19**).

3.3.4 Studying Chemical Regulation of Intracellular Kinase Activity

In this case, the experimental procedure is the same as that of Subheading 3.3.2 except phosphorylation of subarray containing SP6 by H89 treated cell lysates and Fsk treated cell lysates, respectively. Corresponding RLS images and data analysis are shown in Fig. 4 (*see* **Notes 16, 20** and **21**).

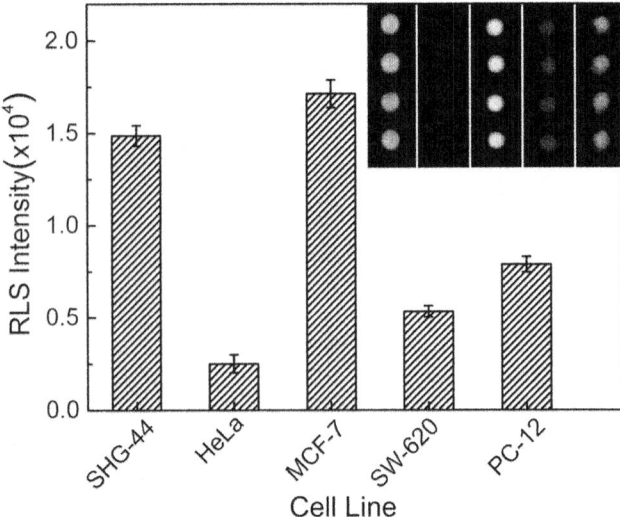

Fig. 2 RLS images (*inset*) and corresponding RLS intensities of five different cell lysates. The signal intensities of cell lysates were subtracted by that of blank lysis buffer. The concentration of SP6 in spotting solution is 10 μM, the concentration of ATP is 50 μM, and the concentrations of cell lysates are 100 μg/mL, respectively (Reproduced from [26] with permission from ACS (the American Chemical Society))

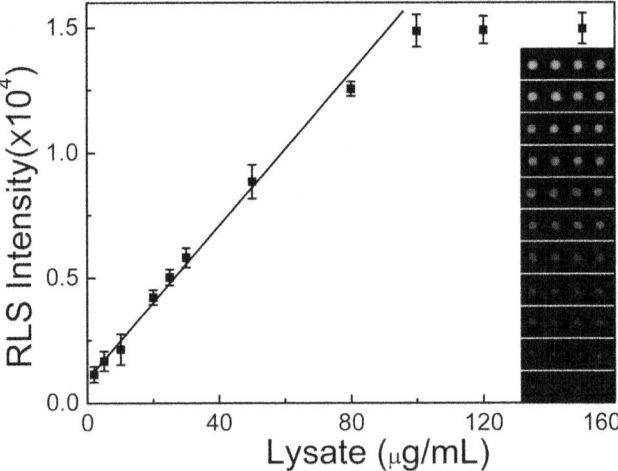

Fig. 3 RLS images (*inset*) and the corresponding plots of the integrated RLS intensity as a function of the concentration of SHG-44 cell lysate (2, 5, 10, 20, 25, 30, 50, 80, 100, 120, and 150 μg/mL) in the kinase assay buffer. The signal intensities of cell lysates were subtracted by that of blank lysis buffer. The concentration of SP6 in spotting solution is 10 μM and the concentration of ATP is 50 μM, respectively (Reproduced from [26] with permission from ACS)

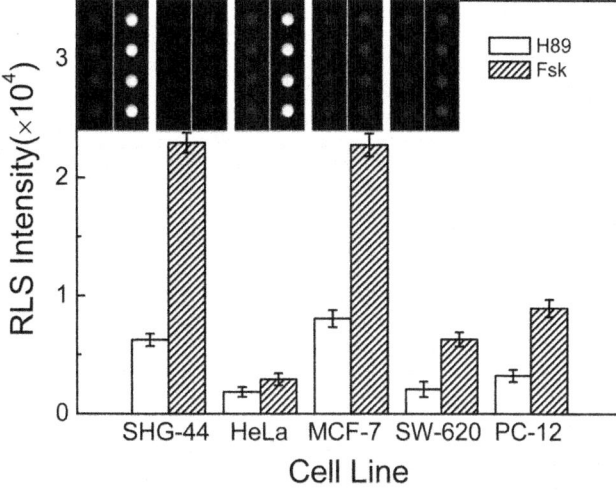

Fig. 4 RLS images (*inset*) and corresponding data analysis on H89 and Fsk regulation of intracellular PKA activity. The cells were co-cultured with 10 μM Fsk, or 3 μM H89, respectively. The concentration of SP6 in spotting solution is 10 μM, the concentration of ATP is 50 μM, and the concentrations of cell lysates are 100 μg/mL, respectively. The signals have been normalized to the average RLS intensity of corresponding untreated cell lysate

4 Notes

1. Store PKA at −20 °C and prepare PKA solution at 0 °C.

2. Store anti-phosphoserine-biotin at −20 °C and prepare anti-phosphoserine-biotin solution at 0 °C.

3. The H89 solution should be freshly prepared and used immediately.

4. The frozen FBS should be stored at 4 °C for 12 h, and then moved to room temperature until completely unfrozen.

5. Having magnetic stir bar at the bottom of the cylinder helps to dissolve phosphate salts and sodium chloride relatively easily, allowing the magnetic stir bar to go to work immediately.

6. Concentrated NaOH (10 M) solution can be used at first to narrow the gap from the starting pH to the required pH. From then on it would be better to use diluted NaOH (1 M) to avoid a sudden increase of pH above the required pH. Add NaOH solutions dropwise.

7. Concentrated HCl (10 M) solution can be used at first to narrow the gap from the starting pH to the required pH. Then 1 M HCl is used to avoid a sudden drop in pH below the required pH. Add HCl solutions dropwise.

8. The culture medium should be used as soon as possible.

9. The citrate stabilized 13 nm GNPs were synthesized by the traditional Turkevich-Frens method [28, 29]. All glasswares were cleaned in aqua regia (three parts HCl with one part HNO_3), rinsed with water, and then oven dried prior to use.

10. A typical solution of 13 nm GNPs in diameter exhibited a characteristic surface plasmon band centered at 518–520 nm.

11. The cell lysates were prepared by literature reported method [21].

12. The linear range of Bradford assay is about 0–2000 μg/mL, and the binding of dye (Brilliant Blue G) with protein is inhibited by detergents.

13. The aldehyde group on the glass surface reacts readily with the N-terminal of peptide to form a Schiff base linkage.

14. The PKA assay buffer without PKA was employed as blank sample in the phosphorylation reaction process.

15. GNPs exhibit the ability to resonantly scatter visible and near infrared light (generally named as resonance light scattering, RLS). This property is the result of the excitation of surface plasmon resonances (SPR) and is extremely sensitive to the size, shape, and aggregation state of the particles. The RLS of small GNPs (<40 nm in diameter) can be enhanced after silver deposition.

16. The background originating from the slide was recorded and subtracted from each image prior to evaluation. The mean values and standard deviations of signals were determined for the 20 spot replicates per sample from single assay, respectively. The detection limit was determined to be the concentration where signal/standard deviation = 3 was reached. To determine the linear ranges of the curves, the range of concentrations that best fitted the linear equation $y = mx + b$ was specified. The phosphorylation efficiency is defined by the slope of PKA calibration line of the assay.

17. Under same experimental conditions, SP6 shows the strongest RLS signal intensity and the highest phosphorylation efficiency.

18. The intracellular activities of PKA in five cell lines follow the order of MCF-7 > SHG-44 > PC-12 > SW-620 > HeLa. The result of RLS assay is comparable with the results of literature reports [13, 30, 31].

19. The RLS signal is increased linearly by increasing the concentration of SHG-44 cell lysate in the kinase assay buffer from 5 to 100 μg/mL and began to saturate above 100 μg/mL, indicating a dynamic range of two orders of magnitude. PKA activity could be detected as low as 5 μg/mL cell lysate in the kinase assay buffer. Given a reaction solution volume of 20 μL, this means that the PKA activity in 0.1 μg cell lysate (i.e., approximately 200 cells) can be detected.

20. The intracellular PKA activity is inhibited by H89 and IC_{50} values of H89 are depended on intracellular PKA activity levels and the drug sensitivity of cell line.

21. The intracellular PKA activities are upregulated by Fsk and following the order of SHG-44 > MCF-7 > SW620 > HeLa > PC-12.

Acknowledgement

This work was supported by NSFC Grants 21075118 and 21205113.

References

1. Manning G, Whyte DB, Martinez R et al (2002) The protein kinase complement of the human genome. Science 298:1912–1934

2. Hutti JE, Jarrell ET, Chang JD et al (2004) A rapid method for determining protein kinase phosphorylation specificity. Nat Methods 1:27–29

3. Houseman BT, Huh JH, Kron SJ et al (2002) Peptide chips for the quantitative evaluation of protein kinase activity. Nat Biotechnol 20:270–274

4. Noble MEM, Endicott JA, Johnson LN (2004) Protein kinase inhibitors: insights into drug design from structure. Science 303:1800–1805

5. Davies SP, Reddy H, Caivano M et al (2000) Specificity and mechanism of action of some commonly used protein kinase inhibitors. Biochem J 351:95–105

6. Hunter T (2000) Signaling—2000 and beyond. Cell 100:113–127

7. Espanel X, Walchli S, Ruckle T et al (2003) Mapping of synergistic components of weakly interacting protein-protein motifs using arrays of paired peptides. J Biol Chem 278:15162–15167

8. Knight ZA, Lin H, Shokat KM (2010) Targeting the cancer kinome through polypharmacology. Nat Rev Cancer 10:130–137

9. Zhang J, Yang PL, Gray NS (2009) Targeting cancer with small molecule kinase inhibitors. Nat Rev Cancer 9:28–39

10. Lapenna S, Giordano A (2009) Cell cycle kinases as therapeutic targets for cancer. Nat Rev Drug Discov 8:547–566

11. Force T, Kolaja KL (2011) Cardiotoxicity of kinase inhibitors: the prediction and translation of preclinical models to clinical outcomes. Nat Rev Drug Discov 10:111–126

12. Dancey J, Sausville EA (2003) Issues and progress with protein kinase inhibitors for cancer treatment. Nat Rev Drug Discov 2:296–313

13. Shults MD, Janes KA, Lauffenburger DA et al (2005) A multiplexed homogeneous fluorescence-based assay for protein kinase activity in cell lysates. Nat Methods 2:277–284

14. Rhee HW, Lee HS, Shin IS et al (2010) Detection of kinase activity using versatile fluorescence quencher probes. Angew Chem Int Ed 49:4919–4923

15. Lipchik AM, Parker LL (2013) Time-resolved luminescence detection of spleen tyrosine kinase activity through terbium sensitization. Anal Chem 85:2582–2588

16. Kubota K, Anjum R, Yu YH et al (2009) Sensitive multiplexed analysis of kinase activities and activity-based kinase identification. Nat Biotechnol 27:933–940

17. Holt LJ, Tuch BB, Villén J et al (2009) Global analysis of cdk1 substrate phosphorylation sites provides insights into evolution. Science 325:1682–1686

18. Breitkreutz A, Choi H, Sharom JR et al (2010) A global protein kinase and phosphatase interaction network in yeast. Science 328:1043–1046

19. Tarrant MK, Rho HS, Xie Z et al (2012) Regulation of CK2 by phosphorylation and O-GlcNAcylation revealed by semisynthesis. Nat Chem Biol 8:262–269

20. Xue L, Wang WH, Iliuk A et al (2012) Sensitive kinase assay linked with phosphoproteomics for identifying direct kinase substrates. Proc Natl Acad Sci U S A 109:5615–5620

21. Xu XH, Liu X, Nie Z et al (2011) Label-free fluorescent detection of protein kinase activity

based on the aggregation behavior of unmodified quantum dots. Anal Chem 83: 52–59

22. Martić S, Gabriel M, Turowec JP et al (2012) Versatile strategy for biochemical, electrochemical and immunoarray detection of protein phosphorylations. J Am Chem Soc 134: 17036–17045

23. MacBeath G, Schreiber SL (2000) Printing proteins as microarrays for high-throughput function determination. Science 289:1760–1763

24. Wang ZX, Lee J, Cossins AR et al (2005) Microarray-based detection of protein binding and functionality by gold nanoparticle probes. Anal Chem 77:5770–5774

25. Sun LL, Liu DJ, Wang ZX (2007) Microarray-based kinase inhibition assay by gold nanoparticle probes. Anal Chem 79:773–777

26. Li T, Liu X, Liu DJ et al (2013) Sensitive detection of protein kinase A activity in cell lysates by peptide microarray-based assay. Anal Chem 85:7033–7037

27. Li T, Liu DJ, Wang ZX (2010) Screening kinase inhibitors with a microarray-based fluorescent and resonance light scattering assay. Anal Chem 82:3067–3072

28. Turkevich J, Stevenson PC, Hillier J (1951) A study of the nucleation and growth processes in the synthesis of colloidal gold. Discuss Faraday Soc 11:55–75

29. Frens G (1973) Controlled nucleation for the regulation of the particle size in monodisperse gold suspensions. Nat Phys Sci 241:20–22

30. Sigoillot FD, Sigoillot SM, Guy HI (2004) Breakdown of the regulatory control of pyrimidine biosynthesis in human breast cancer cells. Int J Cancer 109:491–498

31. Nishizuka Y (1984) The role of protein kinase C in cell surface signal transduction and tumour promotion. Nature 308:693–698

Chapter 8

Anomalous Reflection of Gold: A Novel Platform for Biochips

Amir Syahir, Kin-ya Tomizaki, Kotaro Kajikawa, and Hisakazu Mihara

Abstract

The importance of protein detection system for protein functions analyses in recent post-genomic era is rising with the emergence of label-free protein detection methods. We are focusing on a simple and practical label-free optical-detection method called anomalous reflection (AR) of gold. When a molecular layer forms on the gold surface, significant reduction in reflectivity can be observed at wavelengths of 400–500 nm. This allows the detection of molecular interactions by monitoring changes in reflectivity. In this chapter, we describe the AR method with three different application platforms: (1) gold, (2) gold containing alloy/composite ($AuAg_2O$), and (3) metal-insulator-metal (MIM) thin layers. The $AuAg_2O$ composite and MIM are implemented as important concepts for signal enhancement process for the AR technique. Moreover, the observed molecular adsorption and activity is aided by a three-dimensional surface geometry, performed using poly(amidoamine) or PAMAM dendrimer modification. The described system is suitable to be used as a platform for high-throughput detection system in a chip format.

Key words Protein-peptide interactions, Label-free detection method, Anomalous reflection (AR) of gold, Gold platform, Surface chemistry, Glass substrate

1 Introduction

In the development of convenient platforms for biochip, three major components have to be taken into consideration. These are (1) Detection methods, (2) Surface chemistry, and (3) Capture agent. Many capture agents such as antigens, antibodies, and peptides are commercially available. Detection methods and surface chemistry, however, are still being rapidly developed.

Anomalous reflection (AR) of gold is an application of a phenomenon, in which the reflectively of gold surface for blue or purple light has been reduced upon the adsorption of an ultrathin dielectric layer (Fig. 1) [1]. The AR technique has been shown to be applicable for the detection of biomolecular interactions, and the thickness of molecular layer on the gold surfaces can be quantitatively estimated.

Marina Cretich and Marcella Chiari (eds.), *Peptide Microarrays: Methods and Protocols*, Methods in Molecular Biology, vol. 1352, DOI 10.1007/978-1-4939-3037-1_8, © Springer Science+Business Media New York 2016

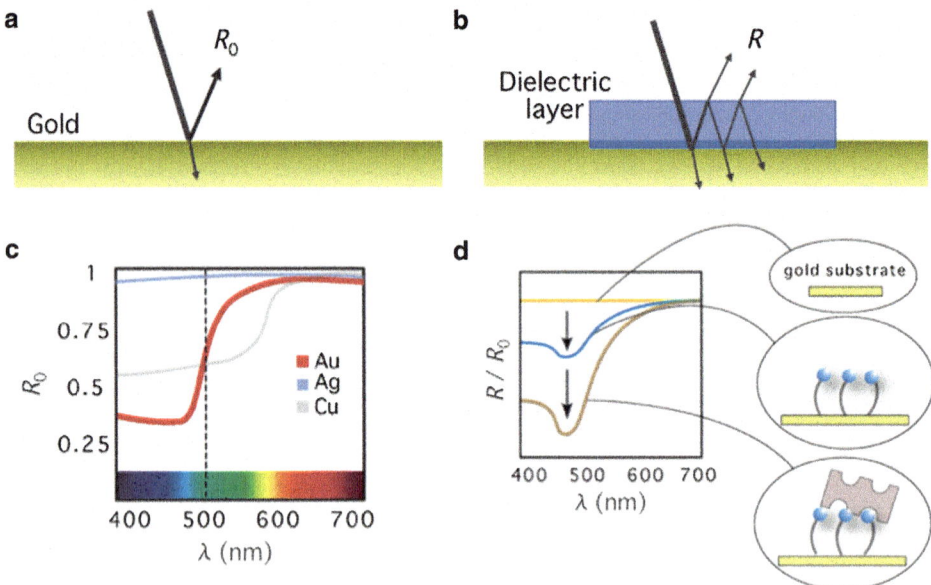

Fig. 1 Light reflectivity scheme of the AR technique. (**a**) Spectra of reflectivity for various metals (below) shows that gold absorbs light with shorter wavelength. (**b**) Decrease in normalized reflectivity can be observed as molecular layer formed on top of the gold thin layer. (**c**) Reflectivity, R_0, of a semi-infinite Au, Ag, and Cu layer within visible range. (**d**) Schematic drawing for AR detection signal, R/R_0, with recording molecular events

We have demonstrated the AR analyses to detect the interactions between peptides and proteins (or proteins displayed on T7 phage), in which alkanethiol-based self-assembled monolayers (SAM) were employed as biomolecular interface [2, 3]. The AR and another label-free technique, called quartz crystal microbalance (QCM), have been successfully combined to evaluate viscoelasticity of proteins of interests on the gold surface [4].

The merits of the AR technique are as follows: (1) Amounts of proteins bound onto gold surfaces can be quantitatively estimated by calculation using the transfer matrix technique [5]. (2) Unlike other plasmonic techniques that stringently require thin layer of gold, the AR detection method allows using semi-infinite gold thin film (practically over 100 nm). (3) The AR technique is tolerance in incidence angle of light source, allowing the use of incoherent light such as light emitting diode (LED). (4) The AR technique does not require any bulky optical setup, which made it portable and easy to miniaturize. However, the spatial resolution for the AR method is generally ten times smaller than that of surface plasmon resonance (SPR) technique [6]. Although the AR technique shows somewhat less in sensitivity, the merits from (2) to (5) are quite attractive over the ATR-based SPR. Therefore the AR technique could provide a promising platform for high-throughput biomolecular detection [7, 8]. The illumination of LED light at normal incidence and direct collection of the reflected light made this method practically the most suitable for the chip-format detection, as demonstrated before [6].

In this chapter, we describe the AR of gold method as three different platforms: (1) gold, (2) gold containing alloy/composite (AuAg$_2$O) [9], and (3) metal-insulator-metal (MIM) thin layers [10, 11]. In addition, observation on protein adsorption and activity is aided by a three-dimensional surface geometry, performed using PAMAM dendrimer modification [11, 12]. The described system is suitable to be used as a platform for high-throughput detection system in a chip format. Wide range of molecular binding assays with surface-bound peptides as capturing agent has been reported.

Utilization of peptides as capturing agent is preferred because of its stability and robustness. Watanabe et al. used the AR method in solution to capture a real-time observation of calmodulin (CaM) interacted with surface-bound α-helical L8K6 peptide [2]. In another research, we managed to obtain nucleic acid related proteins from cDNA library of *S. cerevisiae*, which were displayed on phages, using nucleobase amino acids (NBA) containing peptides as capturing agent [3].

On the other hand, Syahir et al. used the AR method in dry condition to confirm peptides and proteins immobilization on gold surfaces. Protein-peptide interactions, e.g., CaM interacted with surface-bound LKC and LEC peptides, and antibody-antigen model interaction of anti-FLAG monoclonal antibody interacted with surface-bound FLAG peptide were assayed, and kinetics parameters were afforded using Langmuir's adsorption isotherm [11]. In both measurements, the peptides were immobilized on a PAMAM platform to facilitate the interactions. The calculated surface kinetics parameters (K_{surf}) were found comparable with the one using SPR method.

2 Materials

2.1 Platform

Glass slide, Au (99.99 %) grains, Ag wire, Cr grains, and PMMA: poly(methyl methacrylate), tungsten boats (Nilaco, Japan); SF106W (0.1 × 6.0 × 80 mm) and SF307W (0.3 × 7.0 × 100 mm).

2.2 Self-Assembled Monolayer (SAM)

1-Mercaptoundec-11-yl-tri(ethyleneglycol) (PEG3-OH), 1-mercaptoundec-11-yl-hexa(ethylene glycol) acetic acid *N*-hydroxy-succinimide active ester (PEG6-NHS), 11′-dithiobis(succinimidyl undecanoate) (DSU), 1-amino-8-octanethiol (AOT).

2.3 Capture Agent

Amine-ended starburst PAMAM G4 dendrimers (Sigma-Aldrich Japan), 1,2-Bis(2-aminoethoxy)ethane (diamine), 2-aminoethanol, maleimide-PEG$_2$-*N*-hydroxysuccinimide ester

(Mal-PEG$_2$-OSu) (Quanta BioDesign, USA) [*only Mal-PEG2-OSu from Quanta BioDesign*]. 4-(4,6-dimethyl-1,3,5-triazin-2-yl)-4-methylmorpholinium (DMT-MM, Kokusan Chemical, Japan), biotin-dPEG$_4$-acid (Toyobo, Japan), sulfosuccinimidyl d-biotin (Biotin-OSu, Dojindo, Japan), biotin. FLAG peptide, LKC and LEC α-helical peptide, and monosaccharide-modified peptides were synthesized using FMOC solid phase synthesis method.

2.4   Protein

Avidin (egg white), calmodulin (CaM), garden pea (*Pisum sativum*) agglutinin (PSA), *Ricinus communis* agglutinin-120 (RCA).

2.5   Solvent and Buffer

Distilled water, ethanol, acetone, dimethylformamide (DMF), phosphate-buffered saline, PBS (10 mM phosphate, 150 mM NaCl, pH 7.4).

---

3   Methods

3.1   Preparation of the AR Platform

1. Cut glass slide in 8 mm × 15 mm dimension. Dip it in water (MilliQ) containing Extran® MA02 for 15 min under ultra-sonication, followed by sonication in 2-propanol, and then acetone for 15 min each to remove contaminants (*see* **Note 1**). Dry in an oven at 100 °C for 30 min, and allow a natural cooling process.

2. Mount the cleaned glass slides in a vapor-thermal deposition system VE-2030 (vacuum chamber). There are two tungsten boat (left and right) divided with a divider. Place Cr grain (weighted 0.1 g) on one tungsten plate and Au (weighted 0.5 g) on another tungsten plate.

3. Vacuum the chamber at 10^{-5} Pa [13].

4. Melt the Cr slowly to deposit 0.1 nm Cr thin layer. Monitor the evaporation using a 5.0 MHz QCM device placed parallel to the glass slide.

5. Consequently, evaporate Au to acquire a 150 nm gold thin film (*see* **Notes 2** and **3**).

6. Off the vacuum chamber and allow natural cooling for 30 min at vacuumed condition.

7. Store the glass slide under dry and clean condition.

3.2   Preparation of Au-Ag$_2$O Composite Thin-Layer Platform

1. Follow the AR platform procedures 1–2.

2. Instead of one tungsten boat for Au/Ag deposition, set two tungsten boats, SF106W and SF307W, in a parallel circuit.

3. Place 0.3 g Ag wire (cut into 2 cm pieces) on SF106W, and 0.3 g Au grains on SF307W.

4. Follow the AR platform procedures 3–4.

5. Consequently, melt Au and Ag (simultaneously) at 150 nm in thickness (*see* **Notes 4** and **5**).

6. Measure the reflectivity of Au, Ag, and Au-Ag alloy thin layer using MCPD-3000 normalized by a silver thin-layer (150 nm) reflectivity (Fig. 2).

7. Switch off the vacuum chamber following the AR platform procedures 6–7.

8. Expose the Au-Ag alloy to low-pressure mercury lamp (wavelength = 186 nm) from a distance of 5 cm to allow UV-ozone oxidation (*see* **Note 6**). The Au-Ag alloy is now oxidized to Au-Ag_2O composite thin layer [14–16].

9. Measure the dielectric constant, ε, of Au, Au-Ag alloy and Au-Ag2O samples with an ellipsometer using a Xe lamp source, at $\theta = 75$ of incident angle. Phase difference, Δ, and amplitude ratio, Ψ, results can be obtained from this measurement. The complex reflectance ratio of R_p/R_s is denoted as $\rho = (\tan\Psi)e^{i\Delta}$, where $\tan\Psi$ and Δ are amplitude and a phase, respectively. We assumed that substrate is homogeneous and isotropic and the ambient air is transparent. Thus, according to the two-phase calculation dielectric constant of the substrate, $\varepsilon = \sin^2\theta + \sin^2\theta \tan^2\theta [(1-\rho)/(1+\rho)]$ [17].

10. Simulate a contour plot of ΔR_1 (AR signal for 1 nm molecular layer) in the function of ε (*see* **Note 7**).

11. Map the measured ε value (real and imaginary) of Au, Au-Ag alloy, and Au-Ag_2O composite with the simulated contour plot of ΔR_1 (Fig. 3). From the ε-ΔR_1 mapping, we can see that ε of the substrate is moved to higher signal intensity area (from ε_{Au} to ε_{Ag2O}).

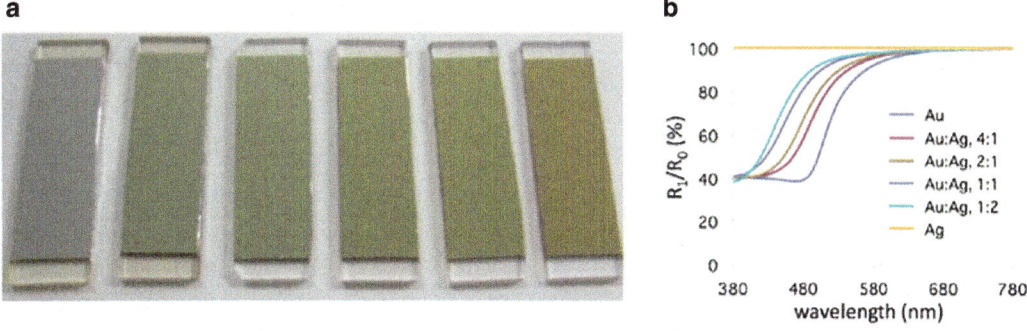

Fig. 2 (a) Au-Ag alloy thin layer on top of glass substrate. From *right*-to-*left* are Au:Ag = 1:0, 4:1, 2:1, 1:1, 1:2, and 0:1 substrate. **(b)** Reflectivity of each spectra was normalized with an Ag thin layer reflectivity. Reprinted with permission from [9]. Copyright 2014 Elsevier

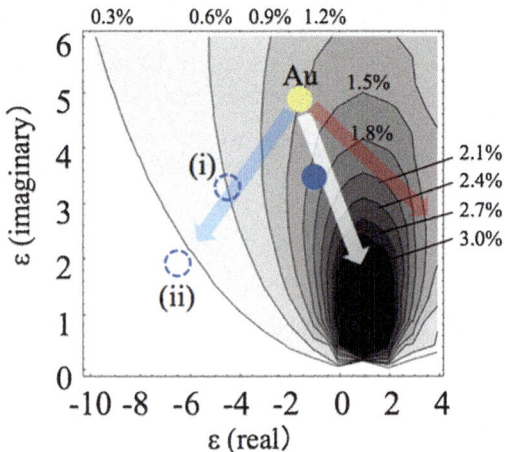

Fig. 3 Intensity of ΔR (% shown at *border*) in *white*-to-*black tone color* within a range of dielectric constant. Plotted are ε of Au (*filled yellow*), Au-Ag 2:1 (*i*), Au-Ag 1:1 (*ii*), and Au-Ag$_2$O (*filled blue*). *White arrow* directed to targeted area of high sensitivity. The ε of Au was successfully moved to that of Au-Ag$_2$O. *Red arrow* shows theoretical ε_{Au}-ε_{Ag2O} diagonal. Reprinted with permission from [9]. Copyright 2014 Elsevier

3.3 Preparation of Au or Ag-Based Metal-Insulator-Metal (MIM) Thin-Layers Platform

1. Follow the AR platform procedures 1–7. For Ag-based MIM, use Ag wire pieces weighted 0.5 g instead of Au grains (at **step 2**).

2. Dissolve PMMA grains in 2-ethoxyethyl acetate. The concentrations used for PMMA are 1.7, 2.3, 3.0, 3.3, 4.3, and 5.0 wt%. In order to dissolve the PMMA, stir with magnetic stirring bar in a sealed Erlenmeyer flask for 12–24 h.

3. Using a spin coating machine set the Au substrate (or Ag for Ag-based MIM) ready for a spin at 4032 g-force in 1 min. Take 50 μL of the PMMA solution. Deposit the PMMA solution and consequently start the spin coating [18]. Following these procedures, the produced PMMA thin layers are 30, 40, 55, 70, 90, and 110 nm in thickness, respectively, according to spectrum fitting [10, 11].

4. Mount the PMMA-coated substrate in the vapor-thermal deposition system and deposit a 30 nm Au thin layer at the same condition (i.e., in the AR platform procedures 5–6).

5. Undo the vacuumed chamber, and store the glass slide in a dry and clean storage. Optional: Measure and differentiate the reflectivity of all the MIM platforms using MCPD-3000 normalized by a silver thin layer (150 nm) reflectivity.

3.4 Three-Dimensional Surface, and Capture Agent Modification

Three-dimensional nano-roughness surface structure on a solid supported platform promotes protein adsorption by decreasing steric hindrance of the target protein, and increasing the degree of freedom for a surface-bound capture agent [12, 19–21].

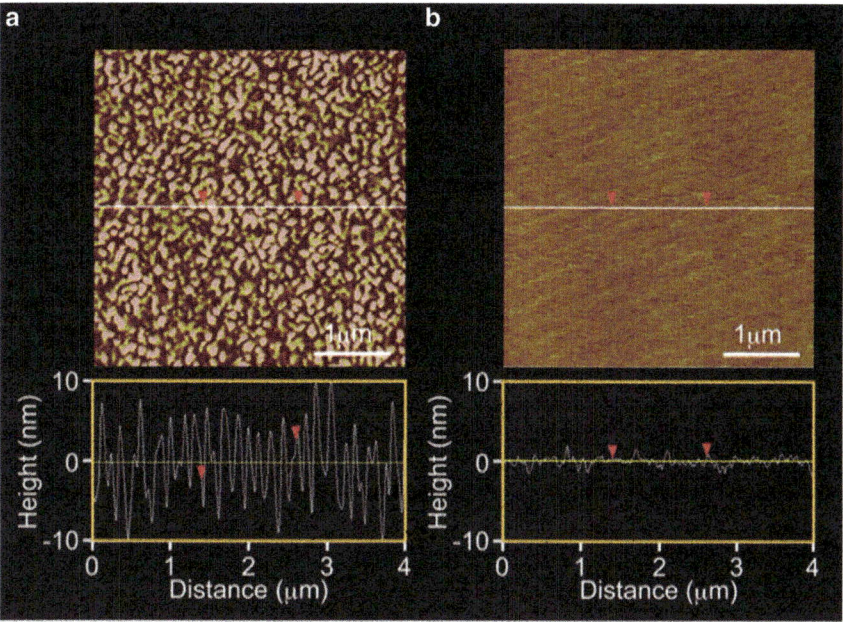

Fig. 4 AFM images for gold surfaces (4 × 4 mm) modified with (**a**) PAMAM G4 on the surface of $\chi_{NHS} = 1$ % and (**b**) diamine on the surface of $\chi_{NHS} = 50$ %. Representative line scans are shown in the bottom. In order to demonstrate the difference in the surface topology, the image was plotted in the same data scale. Reprinted with permission from [12]. Copyright 2009 American Chemical Society

A nano-roughness surface structure (Fig. 4) can be constructed by using a sphere-shaped molecule like PAMAM dendrimer (2–5 nm in diameter). Depending on its generation, PAMAM has many peripheral primary amines available that can be easily activated by a surface chemistry reaction. We modified the platform with PAMAM G4 molecule with standard steps that can be applied for all AR, Au-Ag₂O composite, and MIM platforms.

1. Immerse the platform in a mixture of PEG3-OH and PEG6-NHS at a total thiol concentration of 1.0 mM in ethanol over 1 h at room temperature. For AR and MIM platforms, the best ratio for our purpose is PEG3-OH:PEG6-NHS = 1:100 [12] (*see* **Notes 8** and **9**). In a simple protein immobilization assay (Fig. 5b), the substrate is immersed in DSU (0.1 mM in ethanol) for 1 h at room temperature, giving an active ester terminated SAM onto the surface [11].

2. Wash the SAM-modified gold substrate with a streaming ethanol onto the platform (1 mL × 5), and dry the platform with a stream of air.

3. Immobilize PAMAM G4 dendrimer molecule onto the SAM-modified surface at 50 °C for 1 h [22] (*see* **Note 10**).

4. Wash the obtained substrate with methanol (1 mL × 5), and dry with a stream of air.

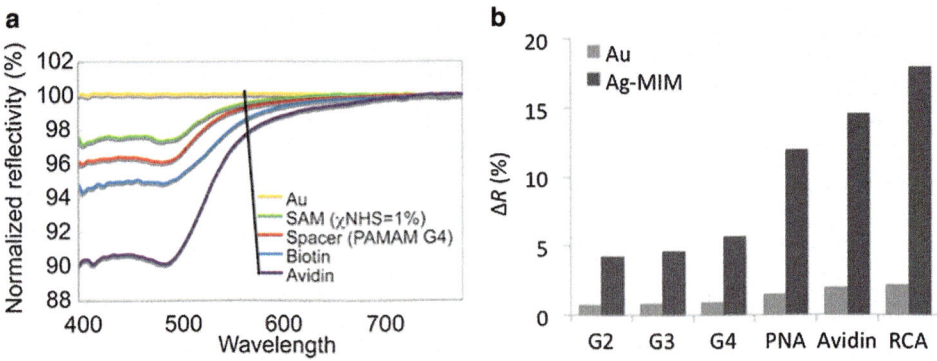

Fig. 5 (**a**) An example of a stepwise detection using AR technique. Curves depict every layer from Au to target protein (avidin). (**b**) Subtracted value of target biomolecule. The bars compare AR measurement technique with the new Ag-MIM based technique. The parameters for Ag-MIM used are 30 and 90 nm for the *top* Au layer (d_2) and the *middle* PMMA layer (d_3), respectively. (**a**) and (**b**) were adapted with permission from [11] and [12], respectively. Copyright 2009 American Chemical Society, and 2012 John Wiley and Sons

5. For capping procedure, incubate the substrate with 2-aminoethanol (in excess concentration, e.g., 10 mM in ethanol), for 2 h at room temperature to terminate non-reacted succinimidyl active esters on the surface.

6. Repeat the washing and drying step (**step 4**).

7. The three-dimensional amine surface of PAMAM G4 (confirmed by AFM measurement (Fig. 4)) is now ready to be functionalized.

8. PAMAM surface can be activated using one of the below procedures:

   (a) Coupling surface-amine with carboxylic acid-functionalized capturing agent: Incubate biotin-PEG4-acid (10 mM in water) onto the PAMAM G4-terminated gold surface with DMT-MM (10 mM in DMF) at 50 °C for 4 h [23, 24].

   (b) Coupling surface-amine with NHS-functionalized capturing agent, e.g., NHS-biotin. Incubate NHS-biotin solution (1.0 mM in water) onto the PAMAM G4-terminated gold substrate for 1 h [11].

   (c) Coupling via a maleimide linker/cysteine-functionalized peptide: Incubate the PAMAM dendrimer functionalized substrate into Mal-PEG₂-OSu linker solution (1.0 mM in ethanol) for 1 h. After a washing and drying process, incubate 100 μL cysteine-functionalized peptide (100 μM in PBS) onto the PAMAM G4-terminated gold substrate for 30 min [11].

Washing and drying processes are followed after each activation step.

3.5 On-Chip assay

1. Drop a protein solution (100 µL in PBS, pH 7.4) onto the capture agent (peptide)-modified gold substrate and incubate for 1 h. In the case of free peptide interaction with surface-bound proteins, drop peptide solutions (100 µL in PBS, pH 7.4) [11].

2. Wash the protein (or peptide)-associated gold surface with PBS, pH 7.4 (200 µL×1), then MilliQ (200 µL×2), and dry with a stream of air.

3. Mount the obtained glass slide in a hand-made AR apparatus equipped with a MCPD-3000 spectromultichannel photodetector (Otsuka Electronics), a PHL-150 halogen lamp (Mejiro Precision), and a multi-mode optical fiber (1.2 mm in diameter, Otsuka Electronics) [12].

4. Bring the end face of the optical fiber to the glass surface on a stage as close as possible to the glass surface (approximately 2 mm).

5. Set up the MCPD-3000. The typical exposure time is 200 ms, and the number of runs for accumulation is 30 times.

6. Adjust the light intensity at maximum intensity of 80 % for bare gold platform.

7. Do reference measurement (bare gold platform).

8. Do a "dark measurement" as you will be directed.

9. Take a measurement of bare gold platform (standard reference); it should result 0 % (±0.02 % of standard measurement error) reflectivity for all visible range ($\lambda = 380$–780 nm).

10. Take the protein amount measurement. Collect reflectivity data at 20 different points within the same slide (middle area) and compared with that of the standard reference.

11. Measure the immobilized molecular layers (SAM, PAMAM dendrimer, capture agent, protein) stepwise as depicted in Fig. 5a. The amount of a particular layer (e.g., protein) can be determined by subtracting the ΔR of the lower layer (Fig. 5b).

12. For in situ measurement, time-based mode of measurement is selected. After setting up the reaction chamber equipped with the AR substrate (Fig. 6a), follow **steps 4–8** (*see* **Notes 11** and **12**).

13. Take the current measurement as reference, and then click "run". Total measurement time is set at 30 min. After 5 min of stable signal (at 100 %), 10 µL protein sample is injected to the reaction chamber. The result (Fig. 6b) is then acquired.

The measurement technique can be applied in both dry and in-solution (real-time) situation (Figs. 5 and 6, respectively). The sensitivity of the AR (Au-only substrate) is improved by several factors with Au-Ag$_2$O composite thin layer and the MIM substrates. It is also shown that the improved substrate can potentially be used to detect the presence of small molecules (Fig. 7).

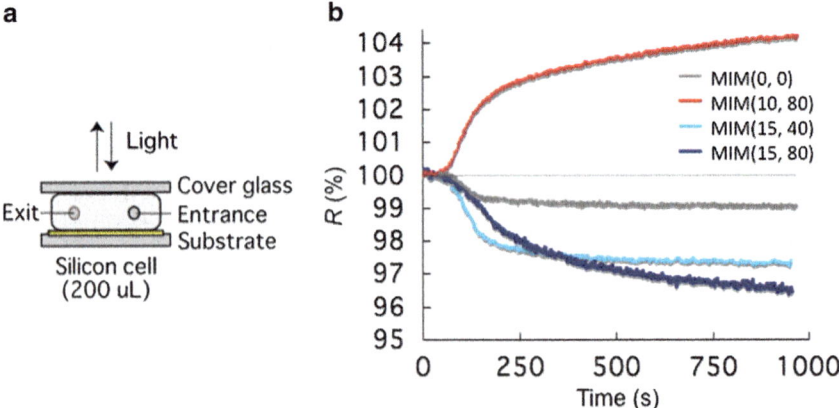

Fig. 6 (a) Schematic drawing of the in situ measurement. A self-made silicon chamber (200 μL) with an entrance and exit point is used. The chamber is covered by a glass slide on the top and the substrate is placed at the bottom. The overall system is kept airtight with two clips. (**b**) Biotin-avidin interactions. Protein was injected at approximately 100 s. Adapted with permission from [10]. Copyright 2010 American Chemical Society

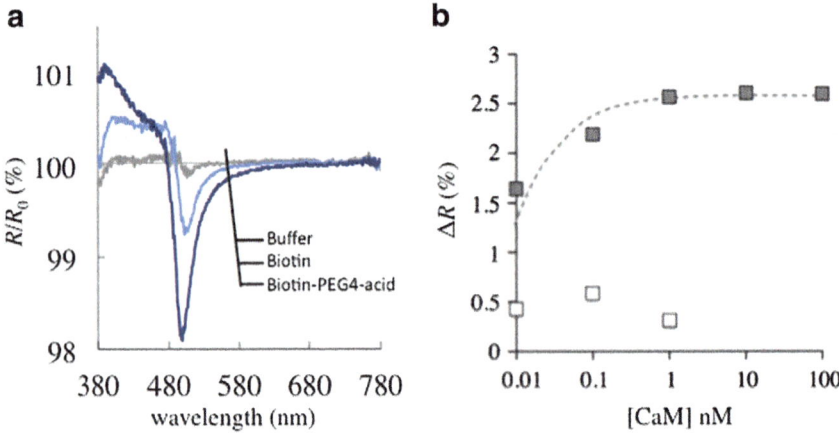

Fig. 7 (a) Detection of biotin and its derivative using Ag-MIM (30, 90). Buffer solution, biotin molecule (Mw. 244), biotin-PEG4-acid molecule (Mw. 458) were tested in sequence. (**b**) Monitoring of protein interactions (Calmodulin) with surface-bound LKC (specific interactions, *filled square*) and LEC (less-specific interactions, *open square*) peptides. Reprinted with permission from [11]. Copyright 2012 John Wiley and Sons

3.6 Calculation of Surface Coverage

1. Estimate the mass per unit area, ρ_A (ng/mm^2) of the captured protein. ρ_A is defined as ΔR/sensitivity (%/ng/mm^2), where the sensitivity is 1.2, and 5.9, for AR (Au-only substrate) and Ag-MIM (30, 90) respectively [11, 12].

2. The amount of immobilized adsorbates (nmol/mm^2) was calculated to be $\Delta R \cdot d/(1.2 \cdot Mw)$ and $\Delta R \cdot d/(5.9 \cdot Mw)$ for Au and Ag-MIM (30, 90) substrates, respectively, where d is density of the adsorbates. For calculation convenience, sometimes d is regarded as 1 [10].

Table 1
Examples of calculated density, *d*, radius, *r*, and maximum attainable surface coverage, Θ_{max}

Name	kDa	*d* (g/cm³)	*r* (Å)	Calculated Θ_{max}	
				pmol/mm²	ng/mm²
PAMAM G2	3.3	1.52	9.5	0.526	1.74
PAMAM G3	6.9	1.5	12.2	0.318	2.19
PAMAM G4	14	1.46	15.6	0.195	2.73
PNA	50	1.41	24.1	0.082	4.09
Avidin	64	1.41	26.2	0.069	4.44
RCA	120	1.41	32.3	0.046	5.47

3. Calculate the monolayer coverage of protein by comparing the measured ρ_A with a theoretically maximum attainable number of immobilized molecules per millimeter square, M for a two-dimensional plane surface. $M = 0.9 \times 10^{14}/(\pi \cdot r^2)$, considering hexagonal close-packed arrangement (90 % is considered full coverage) [25]. The protein radius (Å), $r = 10^8(3Mw/4\pi d N_A)$ [25], and density (g/cm³), $d = 1.41 + 0.145 \cdot \exp(-Mw_{(kDa)}/13)$ [26]. Example of full coverage for 1–150 kDa protein is tabulated in Table 1 with the assumption of a sphere-shaped molecules.

4   Notes

1. Glass slides were arranged in a slide rack and placed into a 250 mL beaker. Aluminum foil cover is used when washed with acetone.

2. Pressure of the vacuumed chamber will slightly increase while evaporating the metal (Amonton's law), so wait until the pressure reaches 10^{-5} Pa before the next thermal evaporation.

3. The thickness of deposited metal layer was monitored using a QCM device that is set next to the glass slide. The QCM deposition rate for Cr and Au is 0.05 Hz/nm and 0.167 Hz/nm, respectively. Deposit Au at a constant evaporation speed of 0.2 Hz/s. The Au deposition rate relation slope (Δf-thickness) is formulated from a self-made SPR spectrum fitting, using mathematica (Wolfram) software.

4. The ratio of electrical resistance produced in SF106W to SF307W tungsten boat is 1:29, according to resistance calculation of

$R = \rho \times (l/A)$, where ρ is specific electrical resistance, l is length, and A is cross-section area of the tungsten boat. Au and Ag were observed to melt at almost the same time.

5. The thickness was calculated from Δf value (QCM) based on Au and Ag density, with an assumption of 1:1 composition.

6. The process resulted a self-limiting oxidized layer, in which oxidation did not occur throughout the interior of the alloys as suggested in the literatures [17]. The thin layer spectrum may be taken stepwisely to monitor changes before and after oxidation.

7. A three-layer model is constructed for the reflectivity calculations, consisting of ambient (refractive index, $n_a = 1.0$), biomolecular layer (refractive index, $n_b = 1.5$; thickness d_b), and the Au substrate having complex refractive index ($N_c = n + ik$). The coefficient reflection (normal incidence angle), rij, at the i–j interface is described as $rij = (Ni - Nj)/(Ni + Nj)$. When the biolayer is present, the reflection coefficient, r, at a wavelength λ is described as $r = [r_{ab} + \varphi_b^2 r_{bc}]/(1 - \varphi_b^2 r_{ab} r_{bc})$, where $\varphi_b = \exp(2\pi n_b d_b i/\lambda)$. The reflectivity in the absence of the biolayer, denoted by R_0, is given by $R_0 = |r_{ac}|^2$, and the reflectivity when 1 nm thick biolayer present is denoted by $R_1 = |r|^2$, when $d_b = 1$. The rate of reflectivity change for 1 nm biolayer, i.e., detection signal/nm, with respect to R_0 is denoted by ΔR_1, where ΔR_1 is defined as $\Delta R_1 = (R_0 - R_1)/R_0$. While $\varepsilon = n^2$, $\varepsilon \Delta R_1$ is contour-plotted (Fig. 3) via a self-coded Mathematica (Wolfram) software.

8. The surface chemical reactions were done in a 2.0 mL Eppendorf® tube. The SAM components were first mixed in a different chamber before appropriate amount (e.g., 1.2 mL) was taken into the reaction tube. PEG3-OH is used as spacer to provide enough room for bulky-headed PEG6-NHS.

9. As an alternative to PEG3-OH/PEG6-NHS system, one might want to consider the usage of DSU-SAM (1.0 mM in ethanol), and react it using the same condition. This applies especially for Au-Ag$_2$O composite thin layer because of a lower SAM coverage.

10. The surface chemical reactions were done in a 2.0 mL Eppendorf® tube. The volume used for dendrimer solution is 800 μL of 1.0 %wt in methanol. The temperature 50 °C is to avoid shape distortion and promote surface collision [22]. As for a flat-surface model, use a diamine instead of PAMAM G4.

11. The Reaction chamber is made by soft silicon with inner dimension, e.g., length × width × deep = 8 × 5 × 5 mm (total volume of 200 μL).

12. Cover with a glass cover and incubate for 1 h at room temperature (4 °C to room temperature).

References

1. Watanabe M, Kajikawa K (2003) An optical fiber biosensor based on anomalous reflection of gold. Sens Actuat B 89:126–130

2. Watanabe S, Usui K, Tomizaki K-Y, Kajikawa K, Mihara H (2005) Anomalous reflection of gold applicable for a practical protein-detecting chip platform. Mol Biosyst 1:363–365

3. Watanabe S, Tomizaki K-Y, Takahashi T, Usui K, Kajikawa K, Mihara H (2007) Interactions between peptides containing nucleobase amino acids and T7 phages displaying S. cerevisiae proteins. Biopolymers 88:131–140

4. Manaka Y, Kudo Y, Yoshimine H, Kawasaki T, Kajikawa K, Okahata Y (2007) Simultaneous anomalous reflection and quartz-crystal microbalance measurements of protein bindings on a gold surface. Chem Commun 3574–3576

5. Bethune DS (1989) Optical harmonic generation and mixing in multilayer media: analysis using optical transfer matrix techniques. J Opt Soc Am B 6:910–916

6. Fukuba S, Naraoka R, Tsuboi K, Kajikawa K (2009) A new imaging method for gold-surface adsorbates based on anomalous reflection. Opt Commun 282:3386–3391

7. Tomizaki K-Y, Usui K, Mihara H (2010) Protein-protein interactions and selection: array-based techniques for screening disease-associated biomarkers in predictive/early diagnosis. FEBS J 277:1996–2005

8. Tomizaki K-Y, Usui K, Mihara H (2008) Proteins: array based techniques. In: Begley TP (ed) Wiley encyclopedia of chemical biology, vol 4. Wiley, Hoboken, NJ, pp 144–157

9. Syahir A, Kajikawa K, Mihara H (2014) Enhanced refractive index sensitivity for anomalous reflection of gold to improve performance of bio-molecular detection. Sens Actuat B 190:357–362

10. Syahir A, Mihara H, Kajikawa K (2010) A new optical label-free biosensing platform based on a metal-insulator-metal structure. Langmuir 26:6053–6057

11. Syahir A, Kajikawa K, Mihara H (2012) Sensitive detection of small molecule-protein interactions on a metal-insulator-metal label-free biosensing platform. Chem Asian J 7:1867–1874

12. Syahir A, Tomizaki K-Y, Kajikawa K, Mihara H (2009) Poly(amidoamine)-dendrimer-modified gold surfaces for anomalous reflection of gold to detect biomolecular interactions. Langmuir 25:3667–3674

13. Kakati KK, Wilman H (1973) The development of oriented crystal growth during condensation of gold, silver and copper films in vacuum, and its systematic dependence on the residual gas pressure and adsorption, and the film thickness, atomic mobility and chemical reactivity. J Phys D Appl Phys 6:1307–1317

14. Ramanathan S, Chi D, McIntyre PC, Wetteland CJ, Tesmer JR (2003) Ultraviolet-ozone oxidation of metal films. J Electrochem Soc 150:F110–F115

15. Chen ZY, Liang D, Ma G, Frankel GS, Allen HC, Kelly RG (2010) Influence of UV irradiation and ozone on atmospheric corrosion of bare silver. Corros Eng Sci Technol 45:169–180

16. Chang C-L, Ramanathan S (2008) A theoretical approach to investigate low-temperature nanoscale oxidation of metals under UV radiation. J Electrochem Soc 155:H620–H624

17. Palik ED (1985) Handbook of optical constants of solids. Academic, San Diego

18. Akimov AV, Mukherjee A, Yu CL, Chang DE, Zibrov AS, Hemmer PR, Park H, Lukin MD (2007) Generation of single optical plasmons in metallic nanowires coupled to quantum dots. Nature 450:402–406

19. Ouyang H, DeLouise LA, Miller BL, Fauchet PM (2007) Label-free quantitative detection of protein using macroporous silicon photonic bandgap biosensors. Anal Chem 79:1502–1506

20. Bonanno LM, DeLouise LA (2007) Steric crowding effects on target detection in an affinity biosensor. Langmuir 23:5817–5823

21. Rechendorff K, Hovgaard MB, Foss M, Zhdanov VP, Besenbacher F (2006) Enhancement of protein adsorption induced by surface roughness. Langmuir 22:10885–10888

22. Tokuhisa H, Zhao M, Baker LA, Phan VT, Dermody DL, Garcia ME, Peez RF, Crooks RM, Mayer TM (1998) Preparation and characterization of dendrimer monolayers and dendrimer–alkanethiol mixed monolayers adsorbed to gold. J Am Chem Soc 120:4492–4501

23. Kunishima M, Kawachi C, Morita J, Terao K, Iwasaki F, Tani S (1999) 4-(4,6-dimethoxy-1,3,5-triazin-2-yl)-4-Methyl-morpholinium chloride: an efficient condensing agent leading to the formation of amides and esters. Tetrahedron 55:13159–13170

24. Gemma E, Hulme AN, Jahnke A, Jin L, Lyon M, Müller RM, Uhrín D (2007) DMT-MM

mediated functionalisation of the non-reducing end of glycosaminoglycans. Chem Commun 2686–2688

25. Lahiri J, Isaacs L, Tien J, Whitesides GM (1999) A strategy for the generation of surfaces presenting ligands for studies of binding based on an active ester as a common reactive intermediate: a surface plasmon resonance study. Anal Chem 71:777–790

26. Fischer H, Polikarpov I, Craievich AF (2004) Average protein density is a molecular-weight-dependent function. Protein Sci 13:2825–2828

Chapter 9

High-Throughput Peptide Screening on a Bimodal Imprinting Chip Through MS-SPRi Integration

Weizhi Wang, Qiaojun Fang, and Zhiyuan Hu

Abstract

Screening of high affinity and high specificity peptide probes towards various targets is important in the biomedical field while traditional peptide screening procedure is manual and tedious. Herein, a bimodal imprinting microarray system to embrace the whole peptide screening process is presented. Surface Plasmon Resonance imaging (SPRi) and matrix-assisted laser desorption ionization time of flight mass spectrometry (MALDI-TOF-MS) are combined for both quantitative and qualitative identification of the peptide. The method provides a solution for high efficiency peptide screening.

Key words OBOC, Microarray, Bimodal imprinting chip, MALDI-TOF-MS, SPRi

1 Introduction

The development of affinity ligands for biomolecular recognition has facilitated many applications in disease diagnostics and therapy [1–4]. As excellent small molecular ligands, peptides own many favorable properties, such as good cell penetrability, low immunogenicity, good biocompatibility, and easy to synthesis [5–7]. During the last 20 years, combinatorial library screening methods have driven rapid development of many novel affinity-peptide probes [8–10]. Among them, the "One bead one compound" (OBOC) combinatorial peptide libraries are composed of random peptide beads that are generated using chemical synthesis and have been utilized to discover specific peptide ligands [11–13]. However, traditional OBOC peptide screening is labor intensive and time-consuming. Generally, during the traditional OBOC screening process, there are four rate-limiting steps: the first one is the manually isolation of positive hits from the high-throughput library, the second one is the identification of the large amounts of beads, the third one consists in re-synthesizing the positive peptides, and the last one is the determination of the affinity of each peptide towards the target protein one by one. In modern combinational chemistry

Marina Cretich and Marcella Chiari (eds.), *Peptide Microarrays: Methods and Protocols*, Methods in Molecular Biology, vol. 1352, DOI 10.1007/978-1-4939-3037-1_9, © Springer Science+Business Media New York 2016

screening, several methods were developed to accelerate peptide screening process. Using specific labeling and conjugation assay, OBOC peptide beads can be trapped by magnetic field instead of being picked out manually, which have improved the screening efficiency [14, 15]. High-throughput peptide beads can also be coded by nanoparticles such as quantum dots with SERS signals so that each peptide can be detected and sequenced by a specific SERS spectrogram [16]. In addition, microarray chips patterned by printers were adopted as high-throughput tools for screening [17–20].

SPR imaging technique provides a rapid approach for high-throughput affinity measurements [21] Furthermore, the combination of SPRi (Surface Plasmon Resonance imaging) and mass spectrometry (MS) has been developed as an attractive technique in analytical chemistry. However, it is difficult to apply MS to on-bead peptide screening [22–25]. Nowadays, with the development of MEMS (micro-electromechanical system) techniques and microfluidic approaches, peptide synthesis and screening was realized in miniaturized chips [12, 26–28]. However, an integrated system consisting of peptide selection, sequencing, and affinity characterization in a high-throughput manner is yet to be developed.

Our previous works have been focusing on peptide synthesis and screening based on microfluidic chips [29–31]. Various peptides could be synthesized simultaneously on a microchip and be in situ detected. Herein, we present a lab-on-chip system which embraced the whole peptide screening process, including single-bead trapping, sequencing, in situ peptide transference, and high-throughput affinity analyses. MALDI-TOF-MS (matrix-assisted laser desorption ionization time of flight mass spectrometry) and SPRi were used on two different microarray chips. The peptide microarray was transferred from the microwell MS chip onto the SPRi chip as through the imprinting process.

As shown in Figs. 1 and 3, at the beginning of the screening process, a high-throughput peptide library with the capacity of 10^4–10^5 peptides was developed. Target protein was biotinylated. Then the peptide library was interacted with the biotinylated protein. Among them, the positive peptide would be conjugated with the protein which would show the biotin signal. Streptavidin-coated magnetic beads will be used to interact with the positive beads. Finally, the positive peptide beads were isolated by magnetic field. Then, a silicon chip with microwell array was fabricated using MEMS technique. Each well was designed with the suitable size to fit one solid phase peptide bead in order to trap beads in a one-well-one-bead manner. At a later stage, the peptides were partially photocleaved from the beads in situ and peptide in each well was transferred onto the SPR microchip through a face to face contact. Finally, the microwell array was utilized for single-bead MALDI-TOF-MS detection and the imprinted spot array was

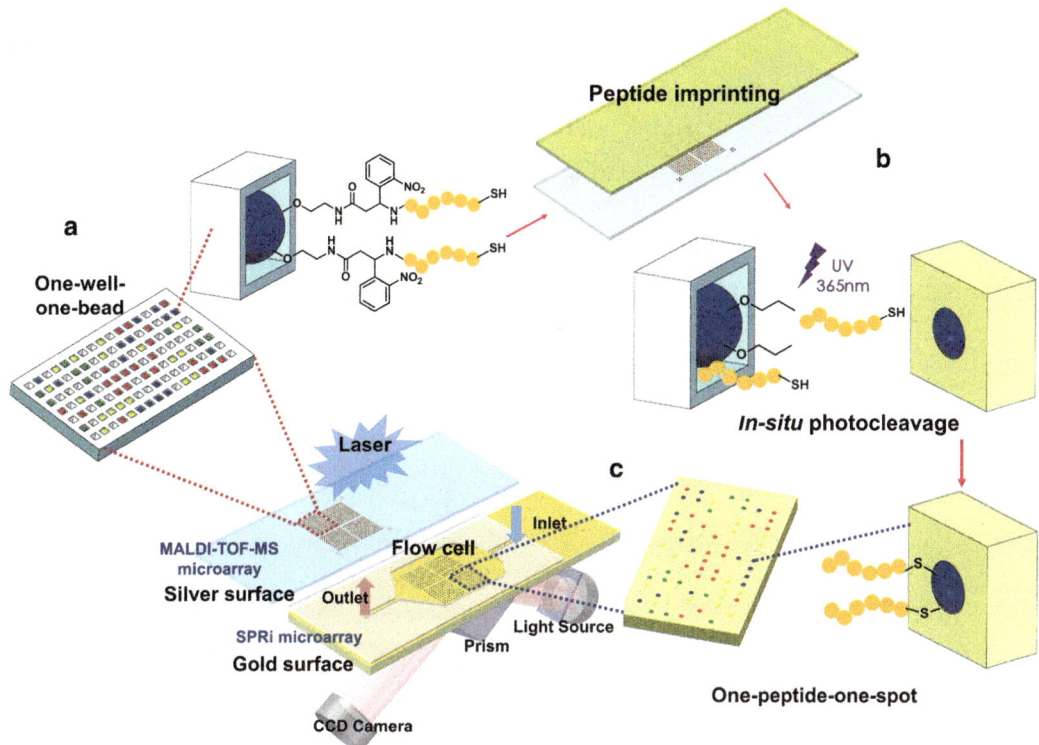

Fig. 1 Overview of integrated OBOC peptide library screening. (**a**) Peptide beads were trapped into the microwell array for MALDI-TOF-MS detection; (**b**) Peptides were in situ photocleaved and imprinted onto the SPRi chips; (**c**) SPRi detection. Reprinted with permission from [36]

employed for SPRi detection. In our experimental design, each spot of the imprinted microarray for SPRi detection has an area of about 0.04 mm². So the saturation mass of the peptide in such area is about 40 ng. In the experiment, the overnight incubation was carried out in order to realize a sufficient imprinting onto the SPRi chip. The peptide remains in the well were also sufficient for MALDI-TOF-MS detection (limit of detection: femtomole). In this way, peptides were patterned for both qualitative and quantitative detection on two different chips.

2 Materials

Prepare all solutions using deionized water (with a sensitivity of 18 MΩ cm at 25 °C) and analytical grade reagents (unless indicated otherwise). Prepare and store all reagents at room temperature (unless indicated otherwise). Diligently follow all waste disposal regulations when disposing waste materials.

2.1 Solid Phase Peptide Synthesis

1. Deprotection solution (DEP): 20 % piperidine in anhydrous DMF (N',N-dimethylformamide) (*see* **Note 1**).

2. Coupling solution for the activation to carboxyl group (ACT): 0.4 M N-methylmorpholine in anhydrous DMF.

3. Amino acid reagent: 0.4 mol Fmoc-Tyr(tBu)-OH and 0.4 mol 2-(1H-benzotriazole-1-yl)-1,1,3,3-tetramethyluronium hexafluorophosphate (HBTU) were mixed together as a white powder mixture, named Try; 0.4 mol Fmoc-Pro-OH and 0.4 mol HBTU were mixed together as a white powder mixture, named Pro; 0.4 mol Fmoc-Asp(OtBu)-OH and 0.4 mol HBTU were mixed together as a white powder mixture, named Asp; 0.4 mol Fmoc-Val-OH and 0.4 mol HBTU were mixed together as a white powder mixture, named Val; 0.4 mol Fmoc-Ala-OH and 0.4 mol HBTU were mixed together as a white powder mixture, named Ala; 0.4 mol Fmoc-Cys(Trt)-OH and 0.4 mol HBTU were mixed together as a white powder mixture, named Cys.

4. Kaiser test solution preparation: (a) solution 1: 5 g ninhydrin in 100 mL ethanol; (b) solution 2: 80 g phenol in 20 mL ethanol; (c) solution 3: 2 mL 0.001 M aqueous KCN (*see* **Note 2**) in 98 mL pyridine.

5. Solid phase support: Tentagel resin (Rapp Polymere, Germany) with the loading 0.53 mmol/g and the particle sizes of 200 μm was used for peptide synthesis (*see* **Note 3**).

6. Fmoc-3-Amino-3-(2-nitrophenyl) propionic acid (Advanced ChemTech, USA) was used as a photolabile linker and 0.4 mol Fmoc-3-Amino-3-(2-nitrophenyl) propionic acid and 0.4 mol HBTU were mixed together as a white powder mixture, named ANP.

7. Cleavage reagent: 92.5 % (v%) trifluoroacetic acid, 2.5 % (v%) Ethanedithiol, 2.5 % (v%) triisopropylsilane, and 2.5 % (v%) water.

2.2 Peptide Screening and Trapping

1. Protein reagents: protein AHA 10 μg/μL in 1×PBS (phosphate buffer).

2. ChromaLink™ Biotin Labeling Kits. The kits consists of ChromaLink biotin 0.5 mg, anhydrous DMF 1 mL, 10× modification buffer 2×1.0 mL, 10× PBS, 1.0 mL, 7K MWCO Zeba columns, collection tubes 1.5 mL, storage tubes 1.5 mL, bovine IgG 500 μg, biotinylated bovine IgG control 500 μg.

3. BCA™ determination Kits (*see* **Note 4**).

4. Magnetic beads: streptavidin-coated magnetic beads (1 μm) were suspended in 1× PBS.

2.3 Chip Imprinting and Identification

1. Matrix for MALDI-TOF-MS detection: CHCA (α-Cyano-4-hydroxycinnamic acid) was dissolved in solvent TA30. (TA30 is 30:70 [v/v] Acetonitrile (HPLC grade): water with 0.1 % TFA (HPLC grade)).

2. The peptide beads were suspended in the above matrix.

2.4 SPRi Detection

1. Protein AHA was diluted into 10, 5, 2.5, 1.25, and 0.625 $\mu g/mL$ with PBST (0.2 % tween 20 in PBS).

2. Blocking reagent: 5 % (m/v) non-fat milk in PBS.

3. Running buffer: PBST.

4. Regeneration buffer: 0.5 % (vol/vol) H_3PO_4 in water.

3 Methods

Carry out all procedures at room temperature unless otherwise specified.

3.1 Microarray Chip Fabrication

1. A silicon chip with microwell array was fabricated. To fabricate the microwell array chip, a silicon wafer (N/1-0-0, 75 mm (L) × 25 mm (L) ×0.5 mm (H)) was employed to fit the size of the target on the Bruker ULTRAFLEXTREME mass spectrometer.

2. The microwells with vertical sides were prepared by conventional soft lithography procedure. A sputtered aluminum layer (200 nm thick) was used as mask; microwells were formed by ICP (induced couple plasma) etching. Bosch techniques are utilized for dry etching [32]. The geometric parameters of microwells were 250 μm $(L)\times 250$ μm $(W)\times 200$ μm (D), which fit the sizes of microbeads.

3. After removing the aluminum layer by wet etching, a titanium layer (30 nm) was sequentially coated on the silicon wafer. Titanium serves as adhesive materials.

4. A silver layer (200 nm) was sequentially coated on the silicon wafer by sputtering (*see* **Note 5**) as the electrodes for mass detection.

5. Microwell array was divided into four regularly arranged sub-arrays with 10 × 10 wells of each, as shown in Fig. 2.

6. According to the principle of MALDI-TOF-MS, the distance between the target and the detector is the key parameter. For accurate measurement, extra external calibration wells with the same 3D dimension were designed and fabricated in the adjacent positions.

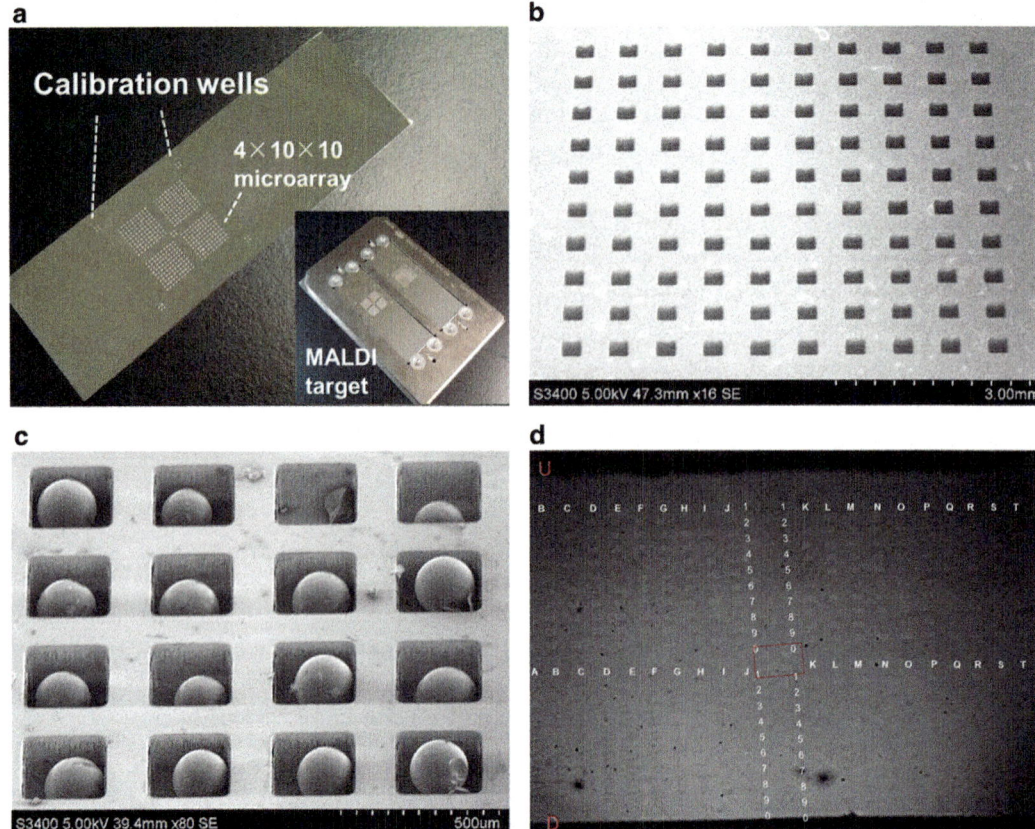

Fig. 2 (**a**) Fabricated silicon microwell chip with silver-sputtered surface (*upper left*); two chips were insetted in the MALDI-TOF-MS target (*lower right*). (**b**) SEM image of the microwell array. (**c**) SEM image of the one-well-one-bead trapping. (**d**) The SPR image of the imprinted microarray on gold surface chip. Reprinted with permission from [36]

3.2 Solid Phase Synthesis of Peptides PN, PL, and PH

1. To evaluate the effectiveness of this system, proof-of-principle experiments were performed. The human influenza hemagglutinin A (HA) peptide and its antibody AHA (Anti-HA) were chosen as the model ligand-receptor pair.

2. The sequence of the wild-type HA peptide (PH) is YPYDVPDYA which has a high binding affinity towards AHA. Meanwhile, we prepared peptide PYDVPDYA (PL) and YDVPDYA (PN) with the low binding affinity and no binding affinity towards AHA, respectively.

3. The three types of peptides were synthesized on monodisperse TentaGel [33] beads using Fmoc solid phase peptide synthesis strategy [34, 35]. Tentagel resins 100 mg were suspended in anhydrous DMF for 2 h until the resins were fully swollen in the solid phase peptide synthesis vessel.

4. Draw off the DMF in the vessel and DEP was added in the vessel and mixed on a shaker; the reaction time is 10 min.

5. Wash the resin with anhydrous DMF for six times.

6. Do Kaiser test (*see* **Note 4**); if the resin turns blue totally, continue for the next step and if the resin still remains colorless, repeat **step 2** until all the beads turn blue. Then draw off the DEP.

7. Dissolve ANP into 3 mL ACT and add into the vessel and mix on a shaker; the reaction time is 50 min.

8. Wash the resin with anhydrous DMF for six times.

9. Do Kaiser test (*see* **Note 4**); if the resin turns colorless totally, continue for the next step and if the resin still remains blue, repeat **step 5** until all the beads turn colorless. Then draw off the ACT.

10. Dissolve other amino acid reagent in succession according to the sequences of each peptide and add into the vessel in each peptide synthesis cycle until the peptide elongation was finished.

11. Add the cleavage reagents into the vessel and mix on a shaker; the reaction time is 40 min.

12. Wash the resins with dichloromethane and methanol in succession and store in PBS.

3.3 Solid Phase Synthesis of the OBOC Library

1. A peptide library was constructed with the sequence of CYPXXXXXXX with X standing for Y, P, D, V, and A randomly so that the capacity of the peptide library was 7×10^4 (5^7) and the redundancy of library was six. Therefore, high-throughput peptide screening was carried out from 10^5 candidate beads. Tentagel resins 300 mg were suspended in anhydrous DMF for 2 h until the resins were fully swollen in the solid phase peptide synthesis vessel.

2. Draw off the DMF in the vessel and DEP was added in the vessel and mixed on a shaker; the reaction time is 10 min.

3. Wash the resin with anhydrous DMF for six times.

4. Do Kaiser test; if the resin turns blue totally, continue for the next step and if the resin still remains colorless, repeat **step 2** until all the beads turn blue. Then draw off the DEP.

5. In order to realize the subsequent in situ cleavage and transference, photocleavable linker ANP (3-Amino-3-(2-nitrophenyl) propionic acid) at C-terminal and Cysteine at N-terminal were introduced. Dissolve ANP into 3 mL ACT and add into the vessel and mix on a shaker; the reaction time is 50 min.

6. Wash the resin with anhydrous DMF for six times.

7. Split the resins into five parts and each part was added into a new vessel, named vessel 1, vessel 2, vessel 3, vessel 4, and vessel 5.

8. Add Tyr/3 mL ACT, Pro/3 mL ACT, Asp/3 mL ACT, Val/3 mL ACT, and Ala/3 mL ACT into the five vessels, respectively, and mixed on a shaker; the reaction time is 50 min.

9. Do Kaiser test of each vessel independently in each vessel.

10. Wash the resin with anhydrous DMF for six times in each vessel.

11. Pool all the resins in all the vessels together into vessel 1.

12. Draw off the DMF in the vessel and DEP was added in the vessel and mixed on a shaker; the reaction time is 10 min.

13. Repeat **steps 6–12** until the OBOC library was finished.

14. Add the cleavage reagents into the vessel and mix on a shaker; the reaction time is 40 min.

15. Wash the resins with dichloromethane and methanol in succession and store in PBS.

3.4 Biotinylation of the AHA Protein

1. Place the spin column into a separate 1.5 mL microcentrifuge collection tube.

2. Place the spin columns in the microcentrifuge and spin at $1500 \times g$ for 1 min to remove the storage solution.

3. Carry out the first buffer exchange on the protein sample by loading the entire 100 μL of the protein sample to the top of the spin column.

4. Centrifuge the columns at $1500 \times g$ for 2 min to collect the eluate at the bottom.

5. Transfer the eluate to a clean tube.

6. Determine the protein concentration with BCA method.

7. Prepare a stock solution of ChromaLink biotin by resuspending the 0.5 mg vial of reagent into 100 μL of anhydrous DMF.

8. Mix the working solution thoroughly and ensure that the working solution is fully dissolved before use.

9. Immediately add the indicated volume of ChromaLink biotin working solution from the calculator output to the protein solution (90–100 μL) and pipette up and down to mix. Allow the reaction to incubate for 120 min.

10. Change the buffer in the tube and the protein was biotinylated.

11. Determination of protein concentration with BCA method. Purified AHA was biotinylated at an average 2.5:1 (biotin:AHA) molar ratio.

3.5 Peptide Beads Trapping and Screening

1. Peptide beads of PH, PL, and PN were incubated with biotinylated AHA for 2 h in 1× PBS (pH 7.4, 37 °C), washed (1× PBS, three times).

2. Peptide beads were incubated with streptavidin-coated immunomagnetic beads for 0.5 h in 1× PBS (pH 7.4, 37 °C). Due to the volume difference between the peptide bead and

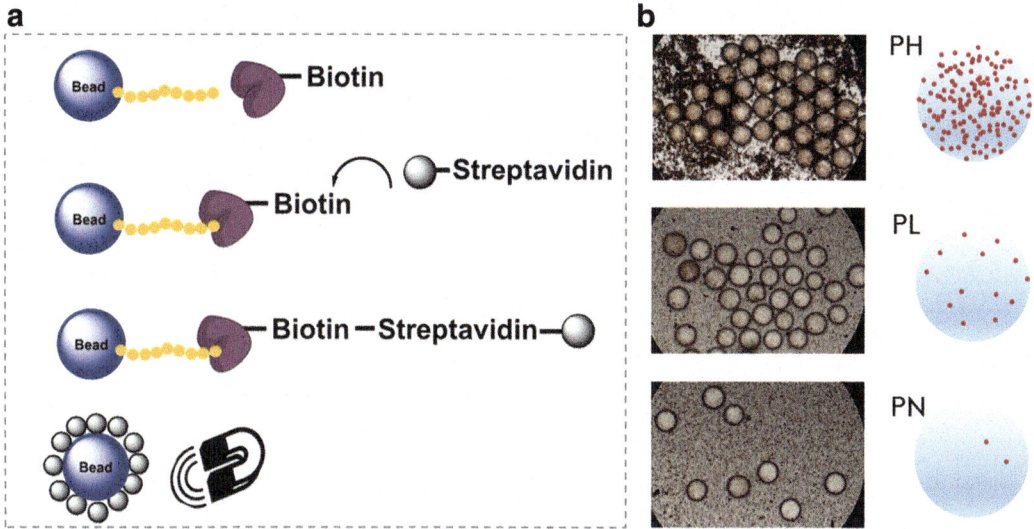

Fig. 3 (**a**) Principle of beads isolation by magnetic field. (**b**) Microscopic image of the interaction between the magnetic beads and the three types of peptides, PH, PL, and PN

the magnetic bead, positive peptide beads would be wrapped up by magnetic beads resulting in magnetism.

3. As shown in Fig. 3, PH beads interacted strongly with AHA as evidenced by large numbers of magnetic beads bound to PH, whereas fewer magnetic beads bound to PL and no magnetic beads bound to PN.

4. Three types of peptide beads were mixed and suspended in matrix solution, then dropped and scraped on the microarray area so that peptide beads fell into the microwells.

5. More than 80 % wells were filled with beads successfully and excess beads were scraped away and collected into a petri dish for the next round.

6. Because of the size of the microwell fits the peptide bead (220 μm in diameter after swelling in water solution), one-well-one-bead array pattern was achieved (Fig. 2c). This method performed an automated approach of isolation and patterning.

3.6 OBOC Peptide Sorting and Positive Beads Screening

1. Peptide beads of PH, PL, and PN were incubated with biotinylated AHA for 2 h in 1× PBS (pH 7.4, 37 °C), washed (1× PBS, three times).

2. Peptide beads were incubated with streptavidin-coated immunomagnetic beads for 0.5 h in 1× PBS (pH 7.4, 37 °C). Due to the volume difference between the peptide bead and the magnetic bead, positive peptide beads would be wrapped up by magnetic beads resulting in magnetism.

3. As shown in Fig. 3b, PH beads interacted strongly with AHA as evidenced by large numbers of magnetic beads bound to PH, whereas fewer magnetic beads bound to PL and no magnetic beads bound to PN.

4. Peptide beads were introduced into a Teflon tube (diameter, 1 mm, flow rate, 600 μL/min) with a magnet closely next to the outer wall of the tube.

5. The magnet was removed and trapped beads were flushed out and collected; collect the positive beads from the outlet.

6. The mixture of positive peptide beads and streptavidin-coated magnetic beads were suspended in a tube, and centrifuged for a very short time to collect the beads at the bottom.

7. Finally a real OBOC library was constructed and peptide screening from high-throughput library was carried out on the bimodal imprint chips. HA and AHA were still chosen as the model system.

3.7 Chip Imprinting and SPRi Detection of the Microarray

1. SPRi analysis was performed on a Plexera PlexArray® HT system (Plexera LLC, Bothell, WA) using bare gold SPRi chips (Nanocapture® gold chips, with a gold layer of 47.5 nm thickness).

2. Peptide beads were suspended in matrix solution and loaded into the wells of the microarray chip.

3. Bare gold SPRi chip was pressed on one unit with the gold surface down onto the microarray chip.

4. The chip was exposed to the UV light (36 W) for 20 min and then incubated at 4 °C overnight in a humid box. The SPRi chip was semi-transparent, ANP linker was photocleaved under the UV light (365 nm, 30 min), and free peptide was released into each well.

5. The two chips were flipped together; the released peptide covalently bound to the gold surface of the SPRi chip with the right orientation through the thiol group of cysteine at the N-terminal. Therefore, a portion of the peptide in each well was transferred and imprinted onto the SPRi chip (Fig. 2d). Every single spot was addressed on both chips.

6. The SPRi chip was washed and blocked using 5 % (m/v) nonfat milk in PBS overnight before use.

7. The SPRi analysis procedure follows the following cycle of injections: running buffer (PBST, baseline stabilization); sample (five concentrations of the protein, binding); running buffer (PBST, washing); and 0.5 % (vol/vol) H_3PO_4 in deionized water (regeneration).

8. Protein (APN) was diluted with PBST to concentrations of 10, 5, 2.5, 1.25, and 0.625 μg/mL. Real-time binding signals were recorded.

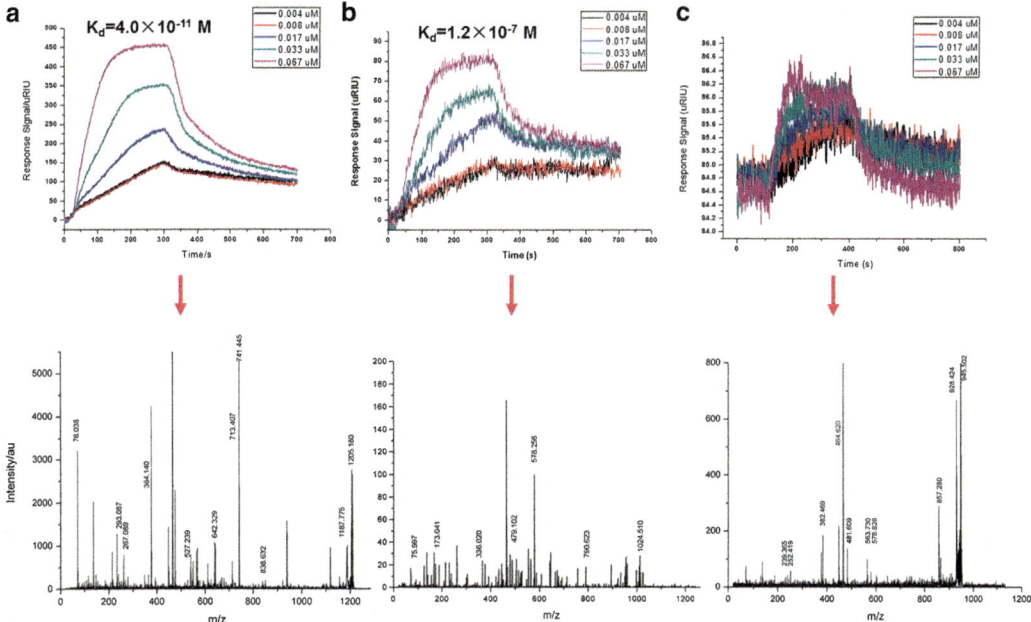

Fig. 4 (**a**) SPRi curve of PH and the corresponding ms/ms spectrum. (**b**) SPRi curve of PL and the corresponding ms/ms spectrum. (**c**) SPRi curve of PN and the corresponding ms/ms spectrum. Reprinted with permission from [36]

9. PH, PL, and PN could be characterized qualitatively and quantitatively through MS and SPRi simultaneously. The three types of TOF-TOF MS spectra were characterized by searching the MASCOT server after uploading the sequences list into the Swissport database.

10. Not only was the sequence of the three types of peptides obtained, but also the affinities of the peptides towards AHA protein could be calculated after the curve fitting. PH spots constantly showed a K_D of 7.8×10^{-10} mol/L, while PL with K_D of 1.2×10^{-7} mol/L and PN with very low affinity towards the AHA protein (Fig. 4).

3.8 MALDI-TOF Identification of the Peptide

1. MALDI-TOF-MS was used for identification of the peptides remained in the microwells. The silicon chip with peptide beads array was insetted in the target of MALDI-TOF-MS and used for in situ single-bead MS detection.

2. MALDI-TOF mass spectrometer was equipped with a nitrogen laser (wavelength 337 nm, laser pulse duration 3 ns) with reflectron and positive-ion modes. A 400 μm stepwise of the laser shooting was set and the mass spectra of each well could be obtained automatically.

3. The laser power energy was adjusted between 0 and 100 % to provide laser pulse energy between 0 and 100 μJ per pulse.

4. The mass spectra were typically recorded at an accelerating voltage of 19 kV, a reflection voltage of 20 kV, and with laser pulse energy of 60 μJ.

5. Each mass spectrum was acquired as an average of 500 laser shots.

6. TOF-TOF model was performed to get ms/ms spectra in order to realize the peptide sequencing. Mass spectra and the SPRi curves of the spots could be obtained respectively and shown in Fig. 4.

3.9 Detection and Identification of the Positive Beads in the OBOC Library

1. Using the magnetic trapping method, more than 400 positive beads were selected out from the OBOC library.

2. Positive beads were dropped into the silver-sputtered chip wells to form a $4 \times 10 \times 10$ microarray.

3. The peptide beads were photocleaved and consequently imprinted to achieve an SPRi array. Through the SPRi detection by Plexera® PlexArray HT system, data of each imprint spot were obtained and among them, seven peptides with decent binding affinity towards AHA were sorted out. SPRi curves are shown in Fig. 5.

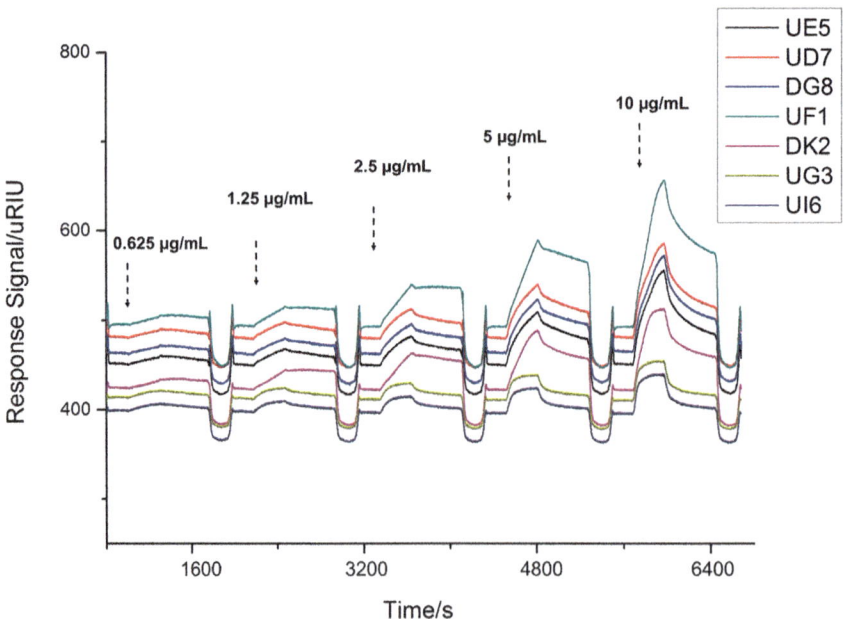

Fig. 5 SPRi curves of seven peptides sorted from the 7×10^4 OBOC library. The "y" axis indicates the change of reflective index (μRIU) and the "x" axis indicate the time (s). Each color of a line corresponds to a different peptide. Five concentrations of the AHA protein were injected into the system along with the time. The *dash line arrows* in the figure show the injection points and the injection concentrations of the AHA protein during the experiments. Reprinted with permission from [36]

Table 1
The affinities and the sequences of the seven peptides screened out of the high-throughput library

Spot code	Affinity	K_D (mol/L)	Sequence
UF1	Original ligand	5.2×10^{-9}	CYPYDVPDYA
DG8	High	1.5×10^{-8}	CYPAYAAAAD
UD7	High	1.4×10^{-8}	CYPAPAPPDV
UE5	High	2.0×10^{-8}	CYPAPPVPAD
DK2	Middle	7.1×10^{-7}	CYPAPVPAAA
UG3	Low	5.8×10^{-6}	CYPPPVAAVV
UI6	Low	4.8×10^{-6}	CYPPPPPAAD

4. The sequences of the seven peptides in the corresponding wells of the silver-sputtered chip were detected through the TOF-TOF-MS sequencing. The list of the peptide sequences and the corresponding K_D are reported in Table 1. Among the seven peptides, the original ligand HA (spot UF1) was identified with the highest affinity (5.2×10^{-9} mol/L). The other six peptides have affinity ranging from 10^{-6} to 10^{-8} mol/L.

4 Notes

1. Anhydrous DMF was prepared by drying overnight over 4A molecular sieve (10 mesh, ACROS) and then distilled under reduced pressure.

2. Note that potassium cyanide (KCN) is highly toxic!

3. Because the resin will be trapped in the uniform microwells, the tentagel resin should be as monodisperse as possible.

4. Kaiser test was carried out through the following steps: (a) A few resin beads are placed in a small test tube, and two to five drops of each solution are added; (b) The tube is placed in an oven and the reaction left to develop for 5 min in boiling water; (c) If the resin and solution turn blue, it means that there is free amino group; (d) If the resin and solution become colorless or light yellow, it means that the free amino group is protected.

5. Silver was used because it has high conductivity and good chemical stability.

Acknowledgements

We acknowledge funding from the National Natural Science Foundation of China (21305023, 31270875), Beijing Municipal Natural Science Foundation (2144058), Project of Chinese Academy of Science (YZ201217), State Key Development Program for Basic Research of China grant (2011CB915502), and International Cooperation Project (0102010DFB33880).

References

1. Hamieh S, Saggiomo V, Nowak P et al (2013) A "Dial-A-Receptor" dynamic combinatorial library. Angew Chem Int Ed 52:12368–12372

2. Tinberg CE, Khare SD, Dou J et al (2013) Computational design of ligand-binding proteins with high affinity and selectivity. Nature 501:212–216

3. Jung E, Kim S, Kim Y et al (2011) A colorimetric high-throughput screening method for palladium-catalyzed coupling reactions of aryl iodides using a gold nanoparticle-based iodide-selective probe. Angew Chem Int Ed 50:4386–4389

4. Oeljeklaus J, Kaschani F, Kaiser M (2013) Streamlining chemical probe discovery: libraries of "fully functionalized" small molecules for phenotypic screening. Angew Chem Int Ed 52:1368–1370

5. Canon F, Milosavljevic AR, Van Der Rest G et al (2013) Photodissociation and dissociative photoionization mass spectrometry of proteins and noncovalent protein-ligand complexes. Angew Chem Int Ed 52:8377–8381

6. Granier S, Manglik A, Kruse AC et al (2012) Structure of the delta-opioid receptor bound to naltrindole. Nature 485:400–404

7. Opitz CA, Litzenburger UM, Sahm F et al (2011) An endogenous tumour-promoting ligand of the human aryl hydrocarbon receptor. Nature 478:197–203

8. Israel MA, Yuan SH, Bardy C et al (2012) Probing sporadic and familial Alzheimer's disease using induced pluripotent stem cells. Nature 482:216–220

9. Jones SA, Shim S-H, He J et al (2011) Fast, three-dimensional super-resolution imaging of live cells. Nat Methods 8:499–508

10. Xie J, Liu G, Eden HS et al (2011) Surface-engineered magnetic nanoparticle platforms for cancer imaging and therapy. Acc Chem Res 44:883–892

11. Aina OH, Liu R, Sutcliffe JL et al (2007) From combinatorial chemistry to cancer-targeting peptides. Mol Pharm 4:631–651

12. Tan W-H, Takeuchi S (2007) A trap-and-release integrated microfluidic system for dynamic microarray applications. Proc Natl Acad Sci U S A 104:1146–1151

13. Kim Y-G, Shin D-S, Kim E-M et al (2007) High-throughput identification of substrate specificity for protein kinase by using an improved one-bead-one-compound library approach. Angew Chem Int Ed 46:5408–5411

14. Astle JM, Simpson LS, Huang Y et al (2010) Seamless bead to microarray screening: rapid identification of the highest affinity protein ligands from large combinatorial libraries. Chem Biol 17:38–45

15. Cho C-F, Amadei GA, Breadner D et al (2012) Discovery of novel integrin ligands from combinatorial libraries using a multiplex "beads on a bead" approach. Nano Lett 12:5957–5965

16. Kim J-H, Kang H, Kim S et al (2011) Encoding peptide sequences with surface-enhanced Raman spectroscopic nanoparticles. Chem Commun 47:2306–2308

17. Reddy MM, Wilson R, Wilson J et al (2011) Identification of candidate IgG biomarkers for Alzheimer's disease via combinatorial library screening. Cell 144:132–142

18. Zhang C-J, Tan CYJ, Ge J et al (2013) Preparation of small-molecule microarrays by trans-cyclooctene tetrazine ligation and their application in the high-throughput screening of protein-protein interaction inhibitors of bromodomains. Angew Chem Int Ed 52: 14060–14064

19. Lausted C, Hu Z, Hood L (2011) Label-free detection with surface plasmon resonance imaging. Methods Mol Biol 723:321–333

20. Bochet CG (2002) Photolabile protecting groups and linkers. J Chem Soc Perkin Trans 1:125–142

21. Lausted C, Hu Z, Hood L (2008) Quantitative serum proteomics from surface plasmon resonance imaging. Mol Cell Proteomics 7: 2464–2474

22. Remy-Martin F, El Osta M, Lucchi G et al (2012) Automated cancer marker characterization in human plasma using surface plasmon resonance in array combined with mass spectrometry (SUPRA-MS). In: Luxton R (ed) 2nd International conference on bio-sensing technology. pp 11–19

23. Remy-Martin F, El Osta M, Lucchi G et al (2012) Surface plasmon resonance imaging in arrays coupled with mass spectrometry (SUPRA-MS): proof of concept of on-chip characterization of a potential breast cancer marker in human plasma. Anal Bioanal Chem 404:423–432

24. Bellon S, Buchmann W, Gonnet F et al (2009) Hyphenation of surface plasmon resonance imaging to matrix-assisted laser desorption ionization mass spectrometry by on-chip mass spectrometry and tandem mass spectrometry analysis. Anal Chem 81:7695–7702

25. Amadei GA, Cho C, Lewis JD et al (2009) A fast, reproducible and low-cost method for sequence deconvolution of 'on beads' peptides via 'on target' Maldi-TOF/TOF mass spectrometry. J Mass Spectrom 45:241–251

26. Schirwitz C, Loeffler FF, Felgenhauer T et al (2013) Purification of high-complexity peptide microarrays by spatially resolved array transfer to gold-coated membranes. Adv Mater 25:1598–1602

27. Beyer M, Nesterov A, Block I et al (2007) Combinatorial synthesis of peptide arrays onto a microchip. Science 318(5858):1888

28. May M (2013) The clinical aspirations of microarrays. Science 15(6121):858–860

29. Zheng H, Wang W, Li X et al (2013) An automated Teflon microfluidic peptide synthesizer. Lab Chip 13:3347–3350

30. Wang W, Huang Y, Jin Y et al (2013) A tetralayer microfluidic system for peptide affinity screening through integrated sample injection. Analyst 138:2890–2896

31. Wang W, Huang Y, Liu J et al (2011) Integrated SPPS on continuous-flow radial microfluidic chip. Lab Chip 11:929–935

32. Cheung CL, Nikolic RJ, Reinhardt CE et al (2006) Fabrication of nanopillars by nanosphere lithography. Nanotechnology 17:1339–1343

33. Kofoed J, Reymond J-L (2007) Identification of protease substrates by combinatorial profiling on TentaGel beads. Chem Commun 4453–4455

34. Fields GB, Noble RL (1990) Solid-phase peptide-synthesis utilizing 9-fluorenylmethoxycarbonyl amino-acids. Int J Pept Protein Res 35:161–214

35. Merrifield RB (1963) Solid phase peptide synthesis.1. Synthesis of a tetrapeptide. J Am Chem Soc 85:2149–2154

36. Wang W, Li M, Wei Z et al (2014) Bimodal imprint chips for peptide screening: integration of high-throughput sequencing by MS and affinity analyses by surface plasmon resonance imaging. Anal Chem 86(8):3703–3707. doi:10.1021/ac500465e

Chapter 10

Analyzing Peptide Microarray Data with the R *pepStat* Package

Gregory Imholte, Renan Sauteraud, and Raphael Gottardo

Abstract

In this chapter we demonstrate the use of *R* Bioconductor packages *pepStat* and *Pviz* on a set of paired peptide microarrays generated from vaccine trial data. Data import, background correction, normalization, and summarization techniques are presented. We introduce a sliding mean method for amplifying signal and reducing noise in the data, and show the value of gathering paired samples from subjects. Useful visual summaries are presented, and we introduce a simple method for setting a decision rule for subject/peptide responses that can be used with a set of control peptides or placebo subjects.

Key words Normalization, False discovery rate, Data visualization, Baseline correction, Decision rule, Sliding mean, Smoothing, Background correction

1 Introduction

The peptide microarray assay has proven useful in immunology studies, where understanding antibody reactions has helped characterize vaccine responses, allergic reactions, and autoimmune disorders. A common slide design is the peptide tiling array, in which slide peptides have partially overlapping amino acid sequences drawn from a larger protein. In this chapter, we demonstrate how the *R* statistical computing environment [1] and the Bioconductor project [2] can be used to analyze peptide microarray data. In particular, we will demonstrate the *pepStat* package [3], which can be used to normalize, analyze, and visualize peptide microarray data. The *pepStat* package is an open-source software, freely available from the Bioconductor project (http://www.bioconductor.org), an open-source repository of *R* packages for bioinformatics.

1.1 Example Data   We use a specific dataset comprising 200 slides, prepared with pre- and post-treatment samples drawn from 100 subjects to illustrate the pepStat package. Among these 100 subjects, 80 subjects received an experimental vaccine regimen targeting human immunodeficiency

Marina Cretich and Marcella Chiari (eds.), *Peptide Microarrays: Methods and Protocols*, Methods in Molecular Biology, vol. 1352, DOI 10.1007/978-1-4939-3037-1_10, © Springer Science+Business Media New York 2016

128   Gregory Imholte et al.

virus (HIV) and the remaining 20 subjects received a placebo (see below for details). Thus we have 100 slides separately measuring each subject's baseline antibody responses, and 100 slides separately measuring each subject's antibody responses to treatment stimulus, for a total of 200 slides.

1.2 Slide Design

The proteins gp120 and gp41 are surface proteins on the HIV virion and are frequent targets of antibody response in naturally infected individuals. The trial vaccine consisted of a recombinant canarypox vector with a fabricated gp120 AIDSVAX B/E core boost [4], and the slides were designed to interrogate antibody responses against gp120 and gp41. To understand vaccine responses against a diverse panel of HIV strains, the slide design tiled consensus gp120/gp41 amino acid sequences from six HIV clades (A, B, C, D, CRF01, CRF02) and a consensus sequence, M [5]. The seven sequences were aligned against the HxB2 standard reference sequence to facilitate analysis and visualization. Following this alignment, each sequence was assigned a position number, corresponding to where the peptide's central amino acid aligned with the HxB2 reference sequence.

Each slide contains three identical subarrays printed in four-by-four grids of blocks. Each block was an 11 by 11 grid of probes. Thus, each subarray had 1936 total probes. Figure 1 illustrates the layout of blocks and probes for a single slide. Our HIV tiling array design contained 1423 unique peptide sequences representing gp120/gp41 proteins, each printed once per subarray. The remaining 513 subarray probes consisted of human-Ig controls, Cy3 landmarks, control peptides, and empty spots.

Sample Name: slide2

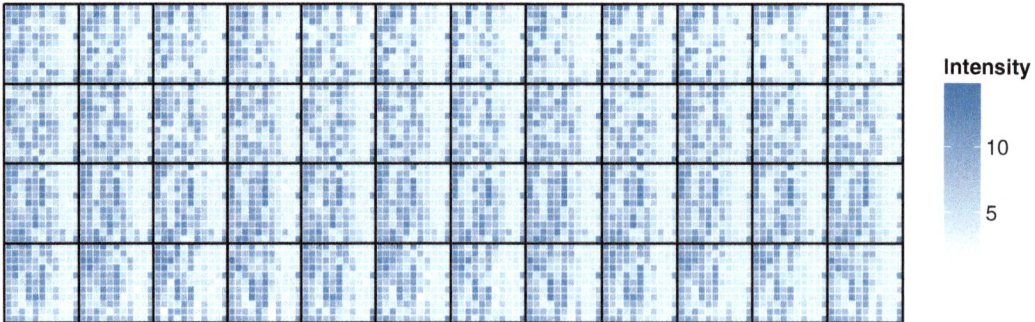

Fig. 1 Slide image for a single subject. Single boxes are colored according to log$_2$ intensity. *Black horizontal* and *vertical lines* separate slide blocks

2   Data Preprocessing

We outline the analysis of paired peptide microarray data with the *R* package *pepStat*. Our demonstration requires the installation of the free *R* statistical computing environment and the Bioconductor tool (http://bioconductor.org/install/). The software package *pepStat* and its dependencies are available from Bioconductor.

2.1   Reading GPR Files

The core of the *pepStat* package is the *peptideSet* class, which unites a variety of data sources such as expression information, sequence information, and slide metadata. Although *peptideSet* objects may be manually created, the *pepStat* package implements a function, *makePeptideSet*, to read in peptide microarray data stored in the GPR file format created by GenePix Pro software [6], connect the slide expression data with relevant slide metadata, and create a *peptideSet* class object. Given a directory containing GPR files, and a *mapping file* containing information about slide metadata, we can create a *peptideSet* object:

```
> library(pepStat)
> my_gpr_directory <- "~/directory/containing/gpr_files"
> my_mapping_file <- "~/filepath/to/mapping_file.csv"
> pset <- makePeptideSet(path = my_gpr_directory, mapping.
file = my_mapping_file)
```

Detailed information about how to structure a mapping file can be found in the help entry for *makePeptideSet*.

The function *makePeptideSet* extracts the F635 Median and B635 Median columns from GPR files, representing the median signal intensity of foreground pixels from probe pixels and local background pixels, respectively. The default background correction method is the *normexp* method, which has the benefit of always estimating a strictly positive corrected foreground intensity and has been shown to have favorable performance [7]. Background-corrected probe intensities are transformed to the $\log_2$ scale by default to reduce skewness, stabilize variance, and improve interpretation of results.

2.2   Accessing Data from peptideSet Objects

The *peptideSet* class combines the *GRanges* class from the package *GenomicRanges* [8] and the *ExpressionSet* class from the package *Biobase* [1]. The *peptideSet* class implements several *methods* (i.e., convenience functions) to access various data fields in a *peptideSet* object. Table 1 lists several convenient methods for the *peptideSet* class. The *exprs* method returns a matrix of probe intensities, where rows correspond to probes and columns correspond to slides. Information about probes can be accessed with the methods *ranges* and *values*. By default, *values* will access probes' names, sequences, and spatial information. Rows in the data frame returned by *values*

Table 1
peptideSet **class methods and their descriptions**

Method	Description
pData	Access slide metadata.
exprs	Access probe/peptide intensities.
ranges	Access *GRanges* probe/peptide sequence data and metadata.
values	Access *GRanges* probe/peptide metadata.
clade	Return logical matrix indicating peptide clade membership.
start, end, width	Return start, end, or width of peptides based on sequence data.
position	Return position of peptides, defined as *floor((start + end)/2)*.
peptide	Access peptide amino acid sequence for probes/peptides.
featureID	Access peptide/probe names.
"[i, j]"	Subset object by slides *i*, probes *j*.

correspond to rows in the matrix of probe intensities. The *pData* method returns a data frame with rows corresponding to slides, matching the columns in the matrix of probe intensities.

2.3 Plotting Slide Images

Visual inspection of microarray data can reveal spatial flaws that may not be obvious from data tables alone. The function *makePeptideSet* stores probes' spatial locations in the returned object, and the *pepStat* package provides two functions to plot probes' intensity values against slide spatial positions. The function *plotArrayImage* plots selected slides in a two-dimensional grid with color intensity proportional to background-corrected probe intensity. The function *plotArrayResiduals* exploits within-slide replication to give a clearer view of spatial trends among probe intensities. Points are colored according to a probe's deviation from the within-slide mean of replicates for that probe's type. Spatial patterns appear as continuous areas having similar color. The optional argument *smooth = TRUE* applies a two-dimensional smoother on the slide residuals to help identify broader spatial trends. Figure 1 demonstrates a slide image from a single pre-treatment subject.

```
> plotArrayImage(pset, array.index = 1:2)
> plotArrayResiduals(pset, array.index = 1:2)
> plotArrayResiduals(pset, array.index = 1:2, smooth = TRUE)
```

3   Data Summarization and Normalization

With microarray data, it is important to estimate and remove technical effects to improve the comparability of signal intensities across multiple slides. The extensive experimental protocol for peptide microarrays introduces several sources of variability. When we fail to account for these effects, we risk confounding true biological signals with technical effects arising from plate and sample preparation, such as batch effects, slide effects, and nonspecific binding effects. We can also exploit the overlapping nature of peptides in tiling array designs to reduce the noise and boost signal in our results. In this section we describe how to incorporate peptide sequence information and how to use the normalization procedure employed by *pepStat*.

3.1   Probe Summarization

With probe replication, we observe multiple fluorescence intensities against the same peptide sequence. Before further analysis, we summarize replicates within slide by their median with the function *summarizePeptides*. The median summary is robust against outliers and questionable observations. We refer to the median of probe intensities as the peptide intensity. During summarization we also add additional sequence information about our peptides. Our slide design incorporates peptides from seven aligned amino acid sequences representing different clades of HIV. Using the alignment information, we constructed a *GRanges* object containing information about each peptide's alignment position and clade membership. Our slide design is represented in the *pep_hxb2* data set contained in the package *pepDat*, available on Bioconductor. The following command executes the prescribed operations:

```
> pset_sum <- summarizePeptides(pset, summary = "median",
position = pep_hxb2)
```

Users can incorporate their own custom sequence information following instructions provided in the *pepStat* vignette.

3.2   Normalization

Batch and slide effects may arise in a variety of ways, such as subtle differences in sample concentrations, analyst procedures, scanner properties, or even plate manufacturing. Figure 2 shows box plots of the median peptide intensities on ten slides from pre-treatment subjects. In the absence of treatment effects, these box plots should approximately center around the same value. Instead we see differences as great as fourfold between the slides, indicating the presence of tremendous slide effects. Nonspecific binding refers to antibody binding events occurring against peptides unrelated to an antibody's epitope. Labeled secondary antibodies and subjects' primary antibodies can contribute nonspecific binding effects, and

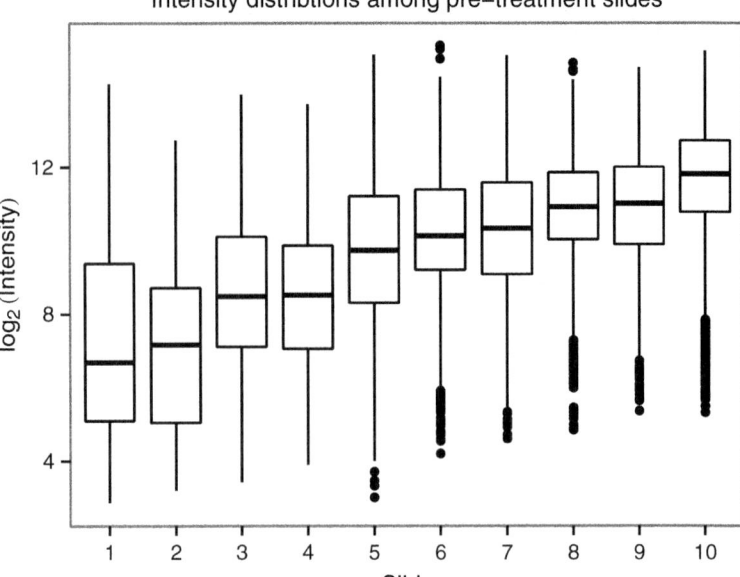

Fig. 2 Intensity distributions for summarized peptide intensities from pre-treatment subjects. Slides are ordered by median intensity. Large shifts in intensity distributions indicate a slide technical effect

pepStat normalization models nonspecific antibody binding as a function of peptide physiochemical properties. We use the amino acid z-scales [9], which are weighted combinations of 26 physiochemical properties. A peptide z-scale is obtained by summing the z-scale values of the amino acids it comprises. Figure 3 shows the nonlinear relationships between each peptide z-scale and intensity value.

The function *normalizeArray* implements a linear model that estimates peptide intensity as a function of a slide effect, quadratic z-scale effects, and a Student-*t*-distributed error term. Using *t-distributed* errors reduces the influence of outliers and signal-carrying peptides on the estimation of slide effects and nonspecific binding effects. This normalization method performs favorably compared to several other methods [10–12]. To accommodate differences in non-specific binding among subjects, this model is fit on a per-slide basis. We execute z-scale normalization with the command

> *pset_norm <– normalizeArray(pset_sum)*

and we refer to the resulting intensity values as normalized peptide intensities.

3.3 Sliding Mean Method

In a tiled slide design, peptides are drawn from a linear protein amino acid sequence in short, overlapping segments. The overlapping segments share amino acid sequences, and thus are similarly susceptible to binding from the same antibodies. If we order peptide

Pre–treatment peptide intensities for a
single subject against five Z–scales

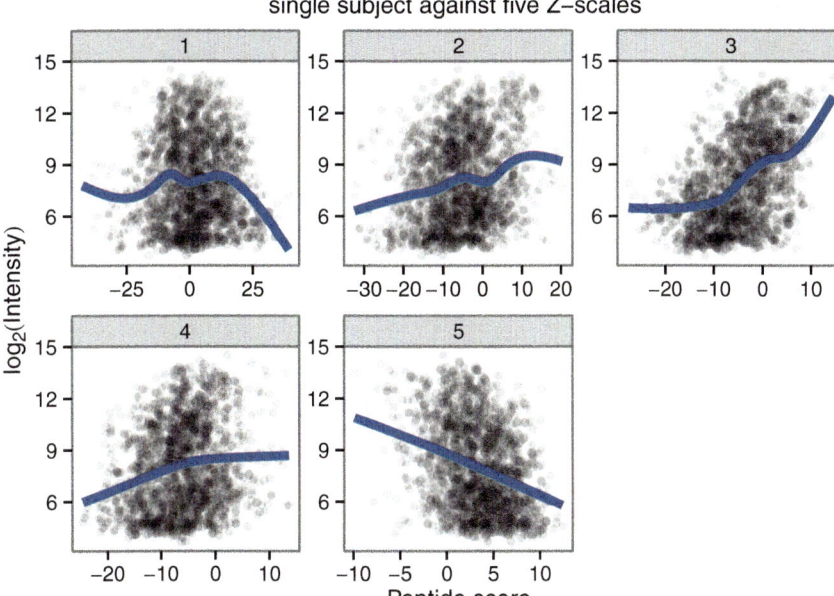

Fig. 3 Peptide intensity values from a pre-treatment slide are plotted against peptide z-scale values. Considerable trends indicate nonspecific binding effects related to peptide physiochemical properties

intensities and normalized peptide intensities by their position in the protein sequence, we detect strong patterns in estimated autocorrelation functions. Strong autocorrelation implies that neighboring peptides tend to share similar intensity values. Figure 4 demonstrates the reduction in autocorrelation induced by z-scale normalization. Although autocorrelation is reduced, the remaining spikes at lags one and two likely represent biological effects rather than technical effects. For our slide design, a position lag of one or two between peptides represents, on average, sharing a common sequence 12 or 9 amino acids in length. We exploit this remaining information by applying a sliding mean technique to boost signal and reduce noise.

The *pepStat* function *slidingMean* applies the mean function over a window generated by peptide positions. A peptide position can refer to a peptide's alignment within a baseline reference sequence, or to the relative ordering of a peptide within a sequence of overlapping peptides drawn from a single amino acid sequence. Peptide positions in our slide design are assigned according to alignment with the HIV HxB2 standard reference sequence. Within *slidingMean*, position is calculated as the rounded average of the *start* and *end* values from the *featureRange* slot of a *peptide-Set* object. The *position* method from Table 1 returns the values used in *slidingMean*. The *width* argument of *slidingMean* controls the width of the sliding mean window.

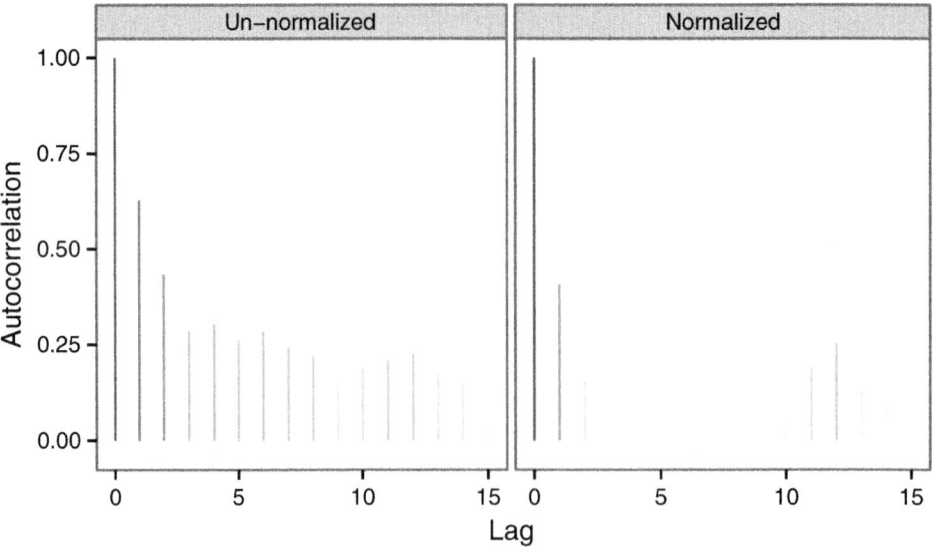

Fig. 4 Autocorrelation functions of peptide intensities from a single clade, before and after normalization. High autocorrelation in un-normalized data indicates similar binding effects for similar peptides. Normalization reduces autocorrelation, but does not eliminate the first few lags

The *slidingMean* function optionally smooths peptides by groups. Our slide design is based on seven amino acid sequences representing six HIV clades and a consensus sequence. Peptide clade membership is stored in the character vector *clade* of the *featureRange* slot, and membership in multiple clades is indicated with clade names separated by commas:

```
> ranges(pset_norm)$clade[74:76]
[1] "A" "C" "M,B,D"
```

Particular applications may choose to group peptides by factors other than clades, which may be accomplished by assigning the desired grouping factor to *clade* in the *featureRanges* slot of the *peptideSet* object. The *clade* method returns a logical matrix whose columns indicate group membership for each level of the grouping factor:

```
> clade(psset_rv)[74:76,]
A CRF01 CRF02 D B C M
WVTVYYGVPVWKDAE TRUE FALSE FALSE FALSE FALSE
FALSE FALSE
WVTVYYGVPVWKEAK FALSE FALSE FALSE FALSE FALSE
TRUE FALSE
WVTVYYGVPVWKEAT FALSE FALSE FALSE TRUE TRUE
FALSE TRUE
```

The argument *split.by.clade* controls whether smoothing is divided by clade or performed in aggregate.

$$pset\_sm\_all \quad <- \quad slidingMean(pset\_norm, \quad width=9, \quad split.$$
$$by.clade=FALSE)$$
$$pset\_sm\_split \quad <- \quad slidingMean(pset\_norm, \quad width=9, \quad split.$$
$$by.clade=TRUE)$$

4 Data Analysis

After normalizing slides and applying our sliding mean technique, we are prepared to analyze our data to discover binding trends induced by treatment. Additionally, we are interested in discovering which peptides experienced increased binding as a result of treatment. We discuss methods for setting detection thresholds, and give a few examples for useful visual summaries of results. The *Pviz* package [13], which builds upon *Gviz* [14], available on Bioconductor, provides a convenient framework for visualizing the results of peptide microarray experiments. *Pviz* and *Gviz* have many configuration options and we refer the reader to their respective user guides for details.

4.1 Baseline Correction for Nonspecific Binding

Nonspecific binding effects vary among subjects, but they are highly correlated for the same subject. We demonstrate this correlation using our placebo subjects. Placebo subject slide measurements contain no treatment-related biological signal, so we examine correlations between pre- and post-treatment smoothed intensities to check for consistent nonspecific binding effects. Figure 5 is a density plot of pre-/post-treatment pairs of smoothed peptide intensities among placebo subjects. The dashed line is the $y=x$ identity line. We see a strong linear association between the pre- and post-treatment measurements, indicating the presence of weak but consistent binding effects. Figure 6 shows the correlation of pre- and post-treatment smoothed intensities, calculated separately across 1423 peptides. All correlations are positive, well over half exceed a correlation of 0.6, suggesting that a difference of post-treatment minus pre-treatment measurements will often have lower variance than either, separately. For each subject-peptide combination, we subtract pre-treatment smoothed peptide intensity from the post-treatment value, and call the resulting value a response index.

4.2 Visualizing Antibody Responses

The *Pviz* defines plotting elements called tracks, which can carry information such as annotations, data summaries, or sequences. We show two visual summaries of our data using the *heatmap* and *line* types in the *DTrack* class, along with annotations formed with the *ATrack* class and an axis formed with the *ProteinAxisTrack*.

Figure 7 is a heatmap of response indices for all subjects. For this figure, we plot *aggregate* response indices created by smoothing without respect to clade membership. Rows represent observations

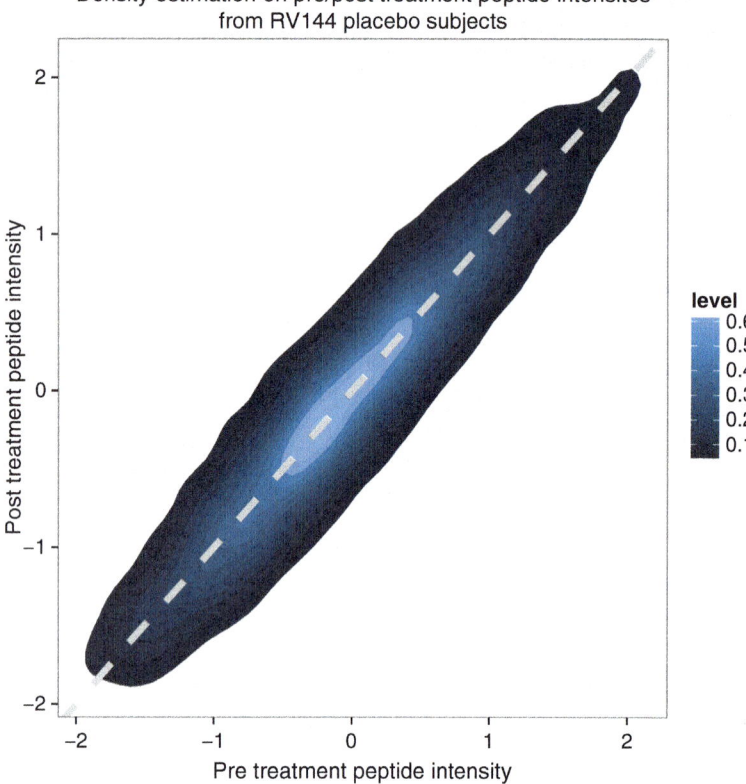

Fig. 5 A two-dimensional density estimation of the relationship between pre- and post-treatment smoothed peptide intensities

from a single subject, while columns are lined against position. The intensity of color corresponds to the positive magnitude of the response index for peptides at a given position. We detect a pattern of vertical bands prior to the V1 loop, inside the V2 and V3 loops, and near the end of gp120 among vaccinated subjects. Response indices among gp41 peptides and placebo subjects tend to be near zero, indicated by white or mild colors. Figure 8 uses the *clade* response indices created by smoothing with respect to clade membership. For the seven clades, we plot the average clade response index among vaccinated subjects against position. From such figures, we can detect differences in response indices among the clades. Here we observe relatively weak response indices in the V2 loop of the B clade, the gp120 terminus of the CRF02 clade, and the V3 loop of the CRF01/D/M clades.

4.3 Classifying Antibody Responses Within Subjects

Beyond aggregate summaries, subject level inference for peptide response may be of interest. Classifying subjects as responders or non-responders against a given peptide can enable further examination of relationships such as correlation between response status and other covariates. We opt to control the false discovery rate

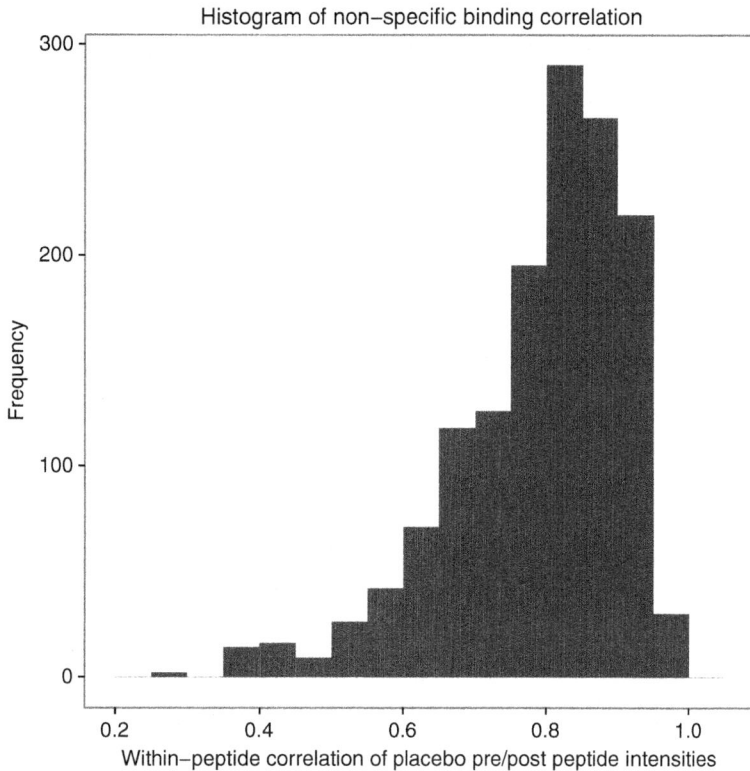

Fig. 6 Correlation between pre- and post-treatment smoothed peptide intensities estimated from 20 placebo subjects for each unique peptide

(FDR) in our classification [15]. We refer to a response classification as a call, and the FDR gives the expected proportion of calls that are false. We make calls by setting a threshold on response indices, and we have a few options available for setting thresholds that, on average, control the FDR to a desired level.

The *pepStat* function *makeCalls* implements a symmetry-based thresholding method, accessible with the argument *method = "FDR"*. This method is further described in (Imholte et al.). An alternative threshold may be set with the use of known placebo slides or control peptides. We illustrate choosing a threshold based on control peptides. For a given threshold t, we calculate the proportion of control peptides that we falsely call positive, and call this p_t. We estimate that the proportion of false calls in the control peptides is similar to the proportion of false calls made among non-control peptides. At a given threshold t we estimate the proportion of false discoveries FDRt as

$$\mathrm{FDR}_t = \frac{(\#\,\text{non-control peptide} \times n\text{subject}) \times p_t}{(\#\,\text{non-control response indices above threshold } t)}$$

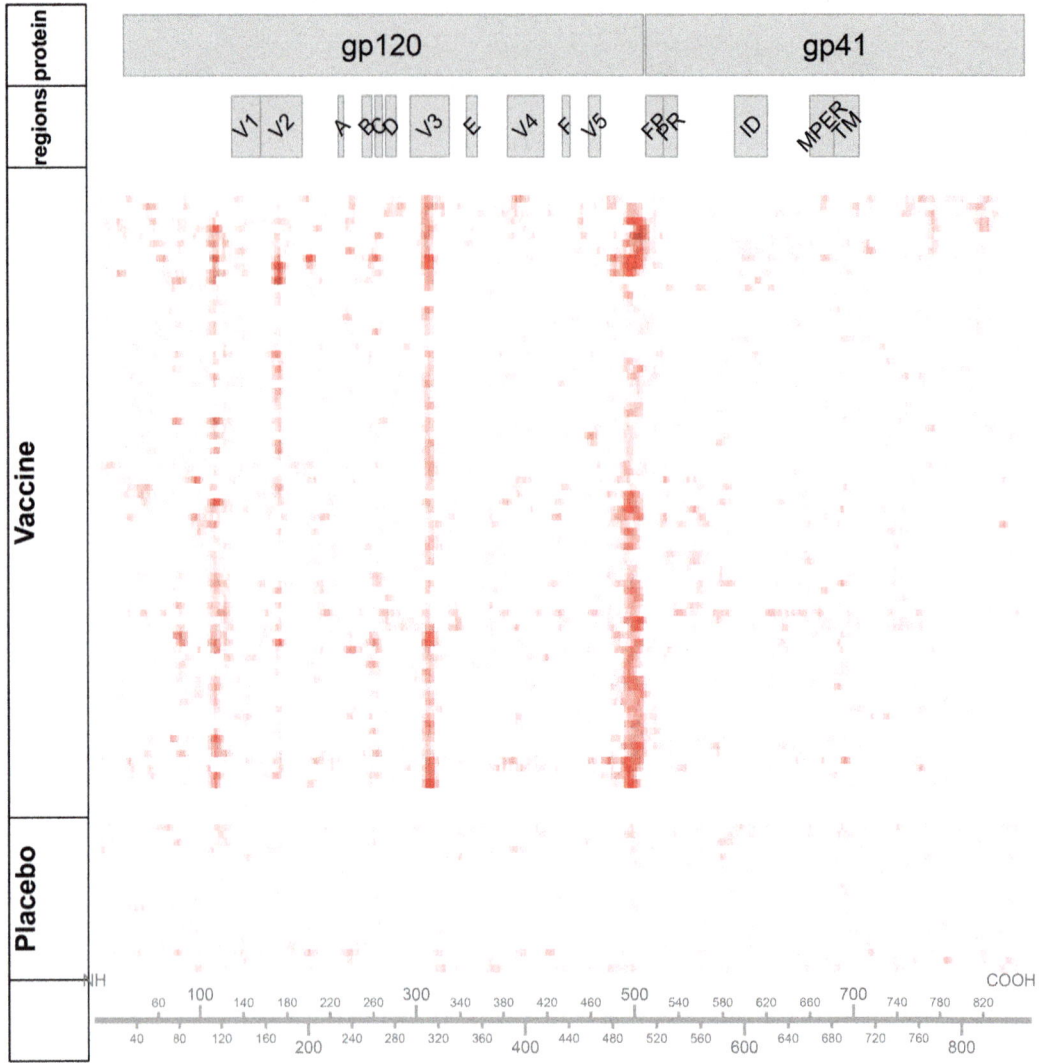

Fig. 7 A heatmap of response indices, plotted by position. *Rows* represent subjects. *Darker colors* represent higher response index values

In summary, the numerator estimates the number of false calls among non-control peptide calls, while the denominator is the total number of non-control calls. Searching over a range of t values, we choose a threshold t such that FDRt is closest to our desired FDR. We use gp41 peptides as control peptides, because the vaccine contained no gp41 insert. Using vaccine subjects only, we calculate that a response index threshold of $t = 1.8$ controls FDR at approximately 10 %. Figure 9 plots the proportion of calls among vaccinated subjects at each position, split by clade. Notably, call frequencies are relatively low in V2 positions for the B clade, V3 positions in clade CRF01, and C-terminus positions in CRF02.

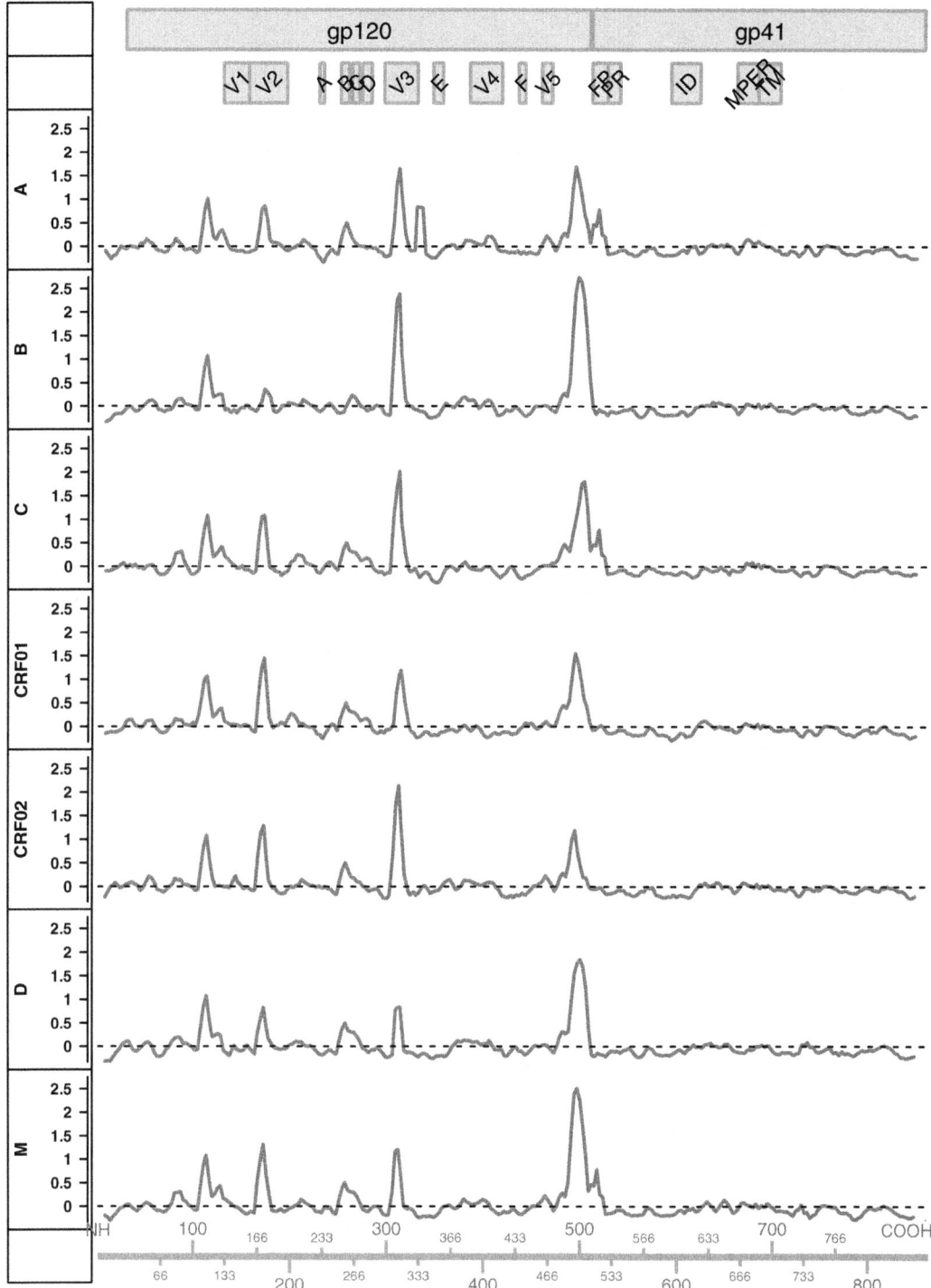

Fig. 8 Average response indices from vaccinated subjects, plotted by position

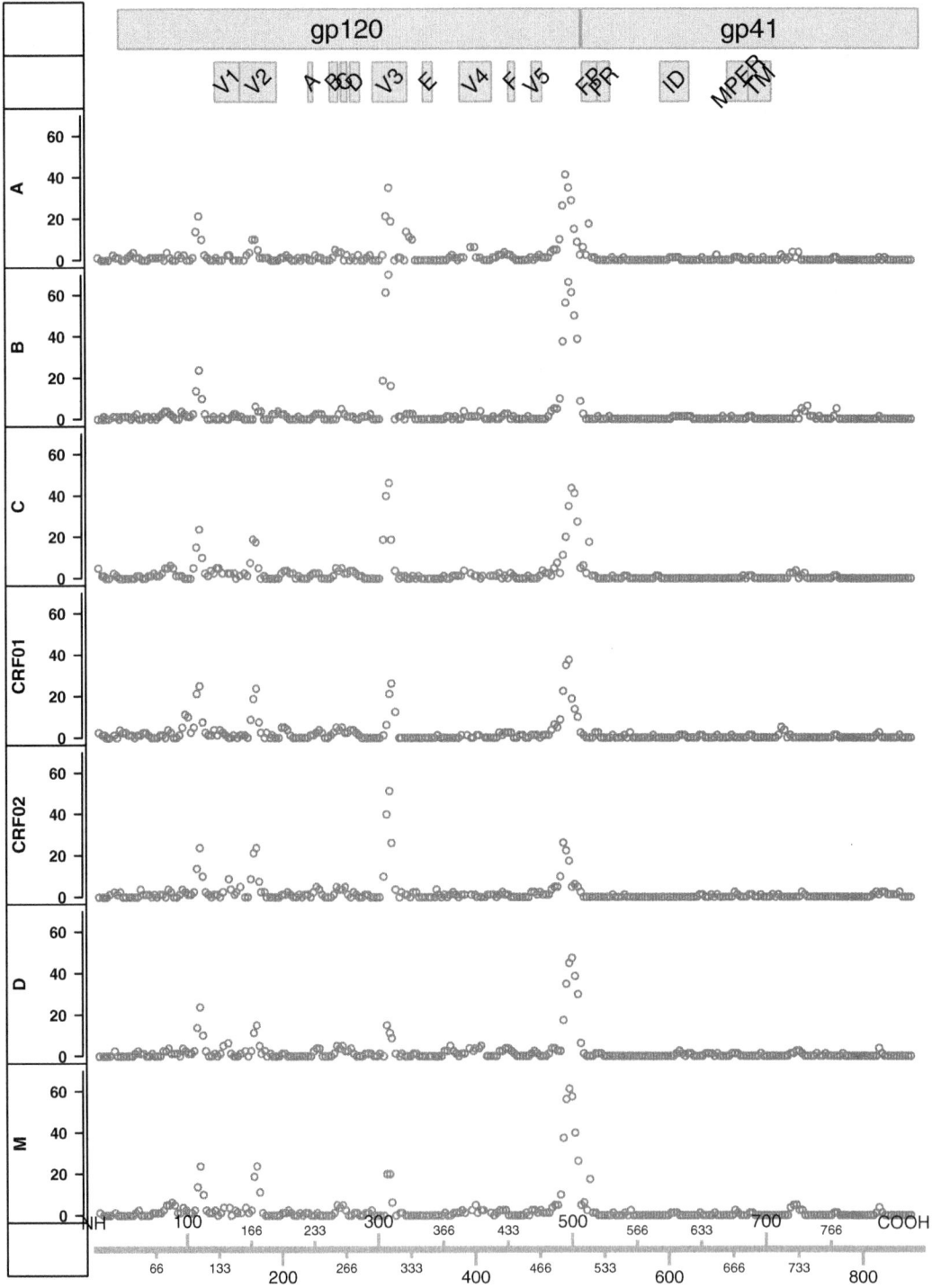

Fig. 9 Proportion of vaccinated subjects estimated to have responded against each position in different clades

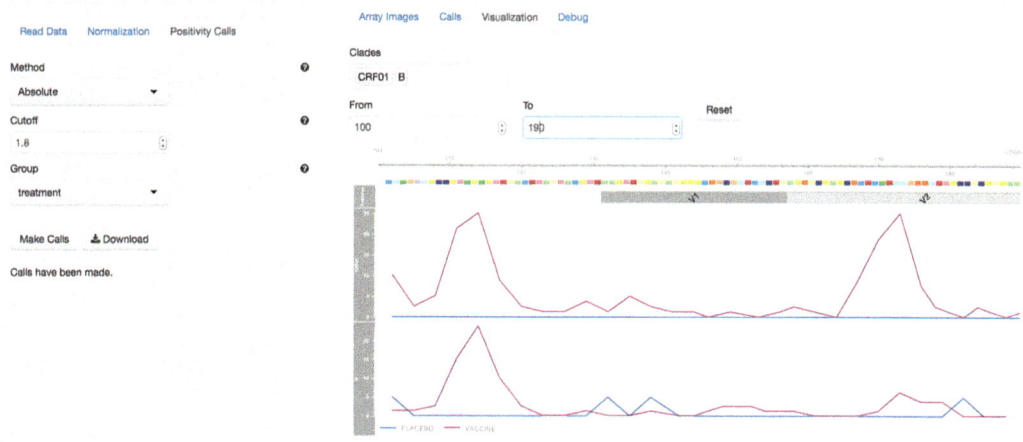

Fig. 10 A screenshot of *pepStat's* Shiny application running on our data

4.4 pepStat Shiny Application

The R *shiny* package [16] is a web application framework for creating graphical user interfaces to conduct interactive data analysis in R. Shiny applications are built on top of R packages, and provide users with a graphical alternative to traditional command line R scripting. The *pepStat* package includes a Shiny application to assist users with analysis of their own data sets, tracing the steps detailed above. The application allows users to quickly read and normalize data, generate calls, and plot response frequencies split by various metadata variables such as treatment. Figure 10 shows a screenshot of the *pepStat* Shiny application analyzing our data. We view a close up of B clade and CRF01 clade response rates against V1 and V2 loop positions. The application automatically draws *Pviz* tracks from our calls analysis, easily allowing us to compare response frequencies among different grouping factors, and to focus our attention on specific positions. We run the *pepStat* Shiny application with the following command:

> *shinyPepStat()*

References

1. R Core Team (2014) R: a language and environment for statistical computing. R Foundation for Statistical Computing, Vienna, Austria, http://www.R-project.org/

2. Gentleman R, Carey VJ, Bates DM et al (2004) Bioconductor: open software development for computational biology and bioinformatics. Genome Biol 5:R80

3. Gottardo R, Imholte G, Sauteraud R et al. (2014) pepStat: statistical analysis of peptide microarrays. R package version 1.1.0

4. Rerks-Ngarm S, Pitisuttithum P, Nitayaphan S et al (2009) Vaccination with ALVAC and AIDSVAX to prevent HIV-1 infection in Thailand. N Engl J Med 361(23):2209–2220

5. Gottardo R et al (2013) Plasma IgG to linear epitopes in the V2 and V3 regions of HIV-1 gp120 correlate with a reduced risk of infection in the RV144 vaccine efficacy trial. PLoS One 8:e75665

6. Molecular Devices (2014) GenePix Pro. http://mdc.custhelp.com/app/answers/detail/a_id/18792/~/genepix%C2%AE-pro-7-

microarray-acquisition-%26-analysis-software-download-page

7. Ritchie M, Silver J, Oshlack A et al (2007) A comparison of background correction methods for two-colour microarrays. Bioinformatics 23(20):2700–2707

8. Lawrence M, Huber W, Pagès H et al (2013) Software for computing and annotating genomic ranges. PLoS Comput Biol 9(8): e1003118. doi:10.1371/journal.pcbi.1003118

9. Hellberg S, Sjöström M, Skagerberg B et al (1987) Peptide quantitative structure-activity relationships, a multivariate approach. J Med Chem 30(7):1126–1135

10. Imholte G, Sauteraud R, Korber B et al (2013) A computational framework for the analysis of peptide microarray antibody binding data with application to HIV vaccine profiling. J Immunol Methods 395(1–2):1–13

11. Nahtman T, Jernberg A, Mahdavifar S et al (2007) Validation of peptide epitope microar-ray experiments and extraction of quality data. J Immunol Methods 328:1–13

12. Bolstad B, Irizarry R, Astrand M et al (2003) A comparison of normalization methods for high density oligonucleotide array data based on variance and bias. Bioinformatics 19:185

13. Sauteraud R, Jiang M, Gottardo R. Pviz (2014) peptide annotation and data visualization using Gviz. R package version 1.1.0

14. Hahne F, Durinck S, Ivanek R et al. Gviz (2012) plotting data and annotation information along genomic coordinates. R package version 1.8.0

15. Benjamini Y, Hochberg Y (1995) Controlling the false discovery rate: a practical and powerful approach to multiple testing. J R Stat Soc Series B 57(1):289–300

16. RStudio and Inc (2014) shiny: Web application framework for R. R package version 0.10.2.1. http://CRAN.R-project.org/package=shiny

Part II

Chemoselective Strategies to Peptide Immobilization

Chapter 11

Chemoselective Strategies to Peptide and Protein Bioprobes Immobilization on Microarray Surfaces

Alessandro Gori and Renato Longhi

Abstract

Ordered and reproducible bioprobe immobilization onto sensor surfaces is a critical step in the development of reliable analytical devices. A growing awareness of the impact of the immobilization scheme on the consistency of the generated data is driving the demand for chemoselective approaches to immobilize biofunctional ligands, such as peptides, in a predetermined and uniform fashion. Herein, the most intriguing strategies to selective and oriented peptide immobilization are described and discussed. The aim of the current work is to provide the reader a general picture on recent advances made in this field, highlighting the potential associated with each chemoselective strategy. Case studies are described to provide illustrative examples, and cross-references to more topic-focused and exhaustive reviews are proposed throughout the text.

Key words Peptide microarray, Oriented immobilization, Chemoselective reactions, Click chemistry

1 Introduction

Array-based technologies bring a unique potential for rapid and high-throughput analysis [1, 2]. Nevertheless, in order to guarantee the consistency and the reliability of the analytical data, the full exploitation of such powerful techniques relies on the effective and reproducible immobilization of bioprobe ligands onto the sensor surface. A key aspect in the manufacturing of microarrays is to preserve the biochemical properties of the immobilized biomolecule as well as to ensure its stable binding on the sensor surface throughout the experimental procedures.

Traditional schemes for biomolecule immobilization rely on non-covalent random absorption of the analyte on the sensor surface based on electrostatic and hydrophobic forces or, alternatively, on aspecific covalent binding [3–6]. In the latter case, covalent linkages are formed between bioprobe functional groups, mainly amine groups from lysine residues and sulfhydryl groups from cysteine side chains, and respectively cross-reactive groups present at the sensor surface. While covalent immobilization is less susceptible to

Marina Cretich and Marcella Chiari (eds.), *Peptide Microarrays: Methods and Protocols*, Methods in Molecular Biology, vol. 1352, DOI 10.1007/978-1-4939-3037-1_11, © Springer Science+Business Media New York 2016

those environment-related factors (pH, ionic strength) whose variation may compromise the stability of the probe attachment, as in the case of non-covalent interactions, the chemical conditions used during immobilization may be not always compatible with bioprobe structural and functional integrity [3].

In recent years, due to the growing awareness that peptide molecules play a key role in modulating a wide range of biological functions, along with their relative ease of synthesis and manipulation, peptides have progressively experienced an increasing application in the development of biosensors [7–9]. Nonspecific immobilization of peptide ligands on sensor surfaces potentially leads to heterogeneous presentation of the bioprobe (Fig. 1a) and, as a consequence, inherently mines the performance of the analytical assay, particularly in those cases where an excellent signal-to-noise ratio (SNR) is essential to lowering the influence of background signal levels. Spatially oriented immobilization strategies are therefore appealing to enable optimal exposure of the peptidic probe in order to guarantee the retainment of its full functionality with respect to the ligand-target interaction (Fig. 1b). To this end, a set of chemoselective strategies to specifically immobilize probe-surface linkage, suitable even in the context of highly functionalized molecule such as peptides, have been developed.

2   Bioorthogonal "Click" Reactions

Site-specific conjugation requires uniquely reactive functional groups. To meet the requirement of bioorthogonality, such reactive chemical handles should be non-native and non-perturbing and should give rise to selective reactions even in a complex biological context. Moreover, ideal bioprobe immobilization strategy should be high yielding and cost effective. The so-called "click" reactions well match these criteria. Indeed, the philosophy of "click chemistry" encompasses a wide range of chemical transformations (e.g., cycloadditions, nucleophilic substitutions, additions to carbon multiple bonds) mainly characterized by high conversion efficiency and selectivity, broad applicability, and biologically benign reaction conditions [10]. Given the often exquisite chemoselectivity they display toward common functional groups, click reactions have not surprisingly found extensive application in the realization of peptide bioconjugates and to build peptide-functionalized biomaterials [11]. Moreover, the fast reaction kinetics that commonly characterize this class of reactions well address the need for fast immobilization, particularly relevant to overcome intrinsically slow reactivity at the solution-solid support interface and thereby to reduce the bioprobe denaturation risks associated to prolonged reaction times. The most popular click reactions suitable for peptide-specific immobilization are herein presented and discussed.

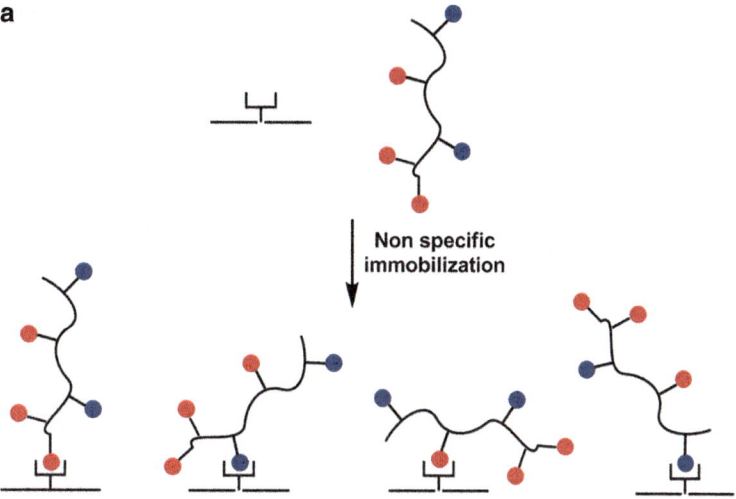

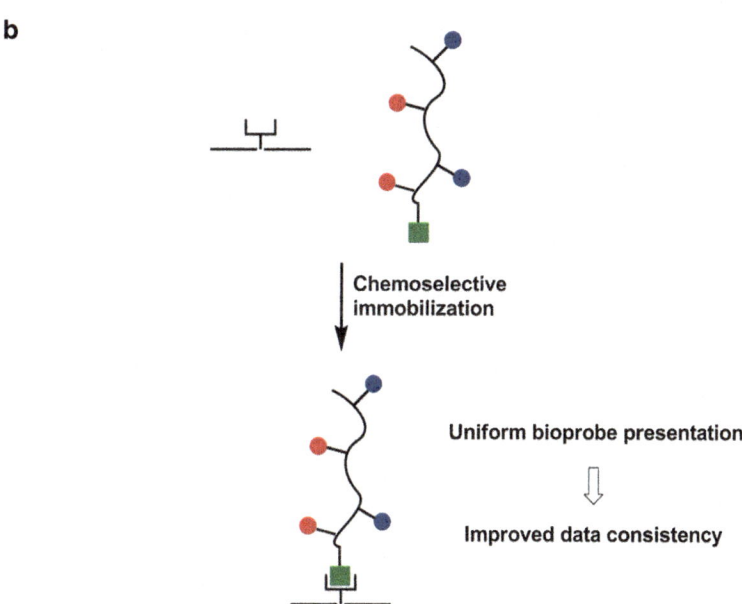

Fig. 1 Nonspecific versus chemoselective presentation. Nonspecific presentation may result in multiple forms of bioprobe display (panel **a**), potentially affecting optimal interaction with candidate target. In contrast, chemoselective immobilization results in univocal bioprobe presentation (panel **b**), which ensures ideal probe-target interaction and, thereby, reproducible data analysis

2.1 Azide-Alkyne Cycloaddition

Huisgen 1,3-dipolar cycloaddition of azides to alkynes is a well-known process, whose first reports date back to the early 1960s. However, this reaction gained an unprecedented popularity only after the introduction of the copper-catalyzed variant (CuAAC, Scheme 1a) [12, 13]. The efficiency and selectivity of this reaction combined with its wide scope, spanning from organic synthesis to peptidomimetics and bioconjugation techniques, have represented the basis for such a great success. CuAAC is indeed generally characterized by almost quantitative yields, fast to instantaneous reaction kinetics, wide tolerance to different reaction conditions, and, notably, nearly full orthogonality with common peptide and protein functional groups. Taken together, these features allow azido- or alkynyl-functionalized peptides to undergo selective and controlled conjugation, either to other functionalized biomolecules or solid supports. Additionally, the triazole moiety which is generated upon cycloaddition is remarkably inert under standard biological conditions [14], which guarantees the stability of the immobilization. It is then not surprising that CuAAC has been thoroughly exploited to obtain functionalized surfaces for analytical purposes.

Noteworthy, Lin and coworkers elegantly showed the impact of a site-specific bioconjugation strategy by selectively immobilizing the maltose-binding protein (MBP) to a glass surface through CuAAC and by evaluating its binding activity in comparison to randomly linked MBP. Results clearly evidenced that the specifically immobilized MBP preserved considerably higher binding activity than the randomly coupled MBP [15]. Further support to oriented immobilization advantages was recently reported by Zilio et al. in the functionalization of a Si-SiO$_2$ substrate using a clickable polymeric coating to enable the attachment of azido-modified peptides. Correct orientation of peptidic probes was found significantly favorable for optimal ligand-target antibodies interaction [16]. CuAAC can also serve the purpose to produce peptide-functionalized surfaces with tunable ligand concentration or patterned topology to quantitatively ascertain complex molecular interactions [17–20]. In a seminal report, the Becker group produced a functionalized surface where the peptide ligand density was tuned by means of a gradient concentration of alkyne functional groups. Following immobilization of azido-modified RGD peptides via CuAAC, the effect of peptide concentration on cell adhesion could be quantitatively evaluated on a single slide [20]. Intriguingly, the generation of ligands gradient can arise from the local generation of the Cu(I) catalyst, as demonstrated by Larsen an collaborators [19].

The only potential limitations associated to CuAAC arise from residual copper catalyst which, depending on the system, can lead to ligand or target denaturation and/or interference with analytical signal detection. To partially overcome this issue, while also improving reaction performance, Cu(I)-binding ligands have been developed [21]. Moreover, in recent years, catalyst-free strain-promoted azide-alkyne cycloaddition (SPAAC) has also emerged as a gold-standard technique for conjugation of peptide-based probes, particularly for in vivo applications (Scheme 1b) [22]. In this case, the driving force of the reaction is uniquely given by the

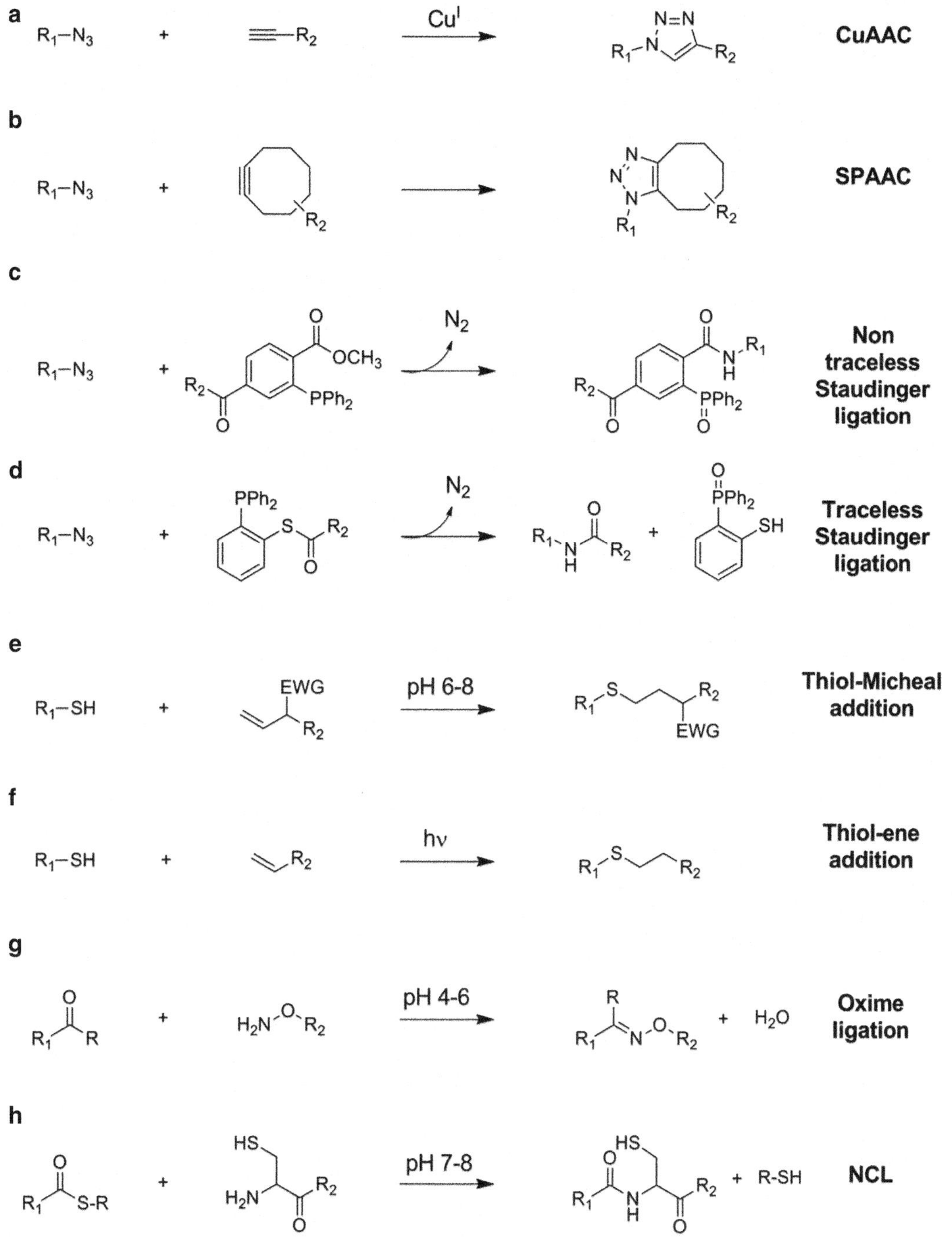

Scheme 1 Reaction schemes for common click-type reactions (**a**) Copper catalyzed azide-alkyne-cycloaddition (CuAAC) (**b**) Strain promoted azide-alkyne-cycloaddition (SPAAC) (**c**) Non-traceless Staudinger reaction ligation (**d**) Traceless Staudinger ligation (**e**) Thiol-Micheal addition (**f**) Thiol-ene addition (**g**) Oxime ligation (**h**) Native chemical ligation (NCL)

intramolecular strain of the cyclooctyne used as the azido counterpart; thus, no metal catalyst is required. However, the high reactivity of the cyclooctyne makes it more feasible to cross-react with other functional groups, such as cysteine thiols, leading to aspecific binding. Also, this technique is limited by the costs and the difficulties in synthesizing the cyclooctyne building blocks to be incorporated in the peptide ligand. Nevertheless, many cyclooctyne-based probes have been developed, whose reactivity and properties are modulated by electronic and steric effects [23].

For instance, the azadibenzocyclooctyne (ADIBO) shows a good compromise between reactivity and synthetic accessibility and, interestingly, was claimed to outperform conventional CuAAC in the immobilization of acetylene-functionalized glass slides with azido-functionalized peptides [24]. Pfeifer and collaborators reported on the use of ADIBO-activated slide surfaces for the preparation of high-density fluorescently labeled peptide microarrays [24]. Excellent immobilization kinetics, good spot homogeneities, and reproducible signal intensities were obtained. Also, interestingly, the specific immobilization of bovine serum albumin (BSA) and dextran via SPAAC led to reduced nonspecific binding of fluorescently labeled IgG to the microarray surface in comparison with other techniques. SPAAC-mediated microarray surface functionalization with peptides was also exploited by Chaikof and coworkers [25]. Immobilization occurred in a fast (<15 min), selective, and tunable fashion. Notably, the generation of a physiologically stable linker methodology also allowed the authors to perform peptide decoration of mammalian cells without compromising their integrity.

2.2 Staudinger Ligation

Azides are chemically inert and relatively stable functional groups under standard biological conditions and only rarely appear in natural biological environments. Not surprisingly, azides have been used in several conjugation strategies, including the Staudinger ligation, which occurs between an azide and a phosphine compound to yield a native amide bond [11, 26]. Staudinger ligation can be divided into two subclasses, traceless and non-traceless, depending on whether the phosphine oxide generated during the reaction is contained or not in the ligated product (Scheme 1c, d). Noteworthy, this reaction can be performed in aqueous buffer with no need for metal catalyst. The reaction kinetics, as like as the final yields, are mainly determined by the structure of the phosphine taking part in the reaction. However, generally, reaction kinetics are considerably slower than CuAAC. The reaction is also amenable to some side reactions, mainly oxidation of the phosphine, which may result in lowered yields. However, the bioorthogonal nature of both the azide and the phosphine functions has resulted in the Staudinger ligation finding numerous applications in various immobilization strategies [27, 28].

For example, Kohn and coworkers provided new insights on the substrate specificity of protein tyrosine phosphatases (PTPs) through the generation of a phosphotyrosine (pTyr)-peptide microarray on a phosphane-modified glass slide obtained by peptide ligand immobilization via non-traceless Staudinger ligation [29]. Similarly, azide-tagged N-Ras proteins were selectively immobilized onto phosphane-modified glass surfaces, and protein functional activity was retained [30, 31]. Likewise, the Raines group reported on the immobilization of the S-protein with good efficiency and almost fully intact functionality [32]. Several others reported on the use of Staudinger ligation for selective immobilization methods, highlighting the considerable potential of this reaction for bioconjugation purposes.

2.3 Thiol-Michael Addition

The Michael addition involves the addition of a nucleophile, also called a "Michael donor," to an activated electrophilic olefin, the "Michael acceptor," resulting in a "Michael adduct" (Scheme 1e). While typical Michael donors refer to enolates, a wide range of chemical functionalities possess sufficient nucleophilicity to perform as Michael donors [33]. In the biological context, both amines and thiols are exploitable as non-enolate nucleophiles. Cysteine thiols are usually more nucleophilic than amines, and the attack of the free sulfhydryl group to the activated alkene to afford thioethers generally proceeds with fast reaction kinetics and high conversion. Moreover, smooth reaction conditions, i.e., weakly basic aqueous buffer at room temperature, are optimal for bioconjugation purposes and enable a fair selectivity with respect to the potentially competing free amine groups. Due to the plethora of electron-withdrawing activating groups that enable the thiol-Michael addition, the reaction kinetics are highly dependent on the nature of the Michael acceptor [33]. Recently, acceptor reactivity has been ranked, with maleimide, vinyl sulfone, and acrylates performing the best [34]. Accordingly, bioprobes conjugation mediated by these functionalities has been extensively exploited to generate functionalized sensor surfaces [35–37]. A case example is given by the work reported by Fu et al. [38]. An array of β-galactosidase (β-Gal)-anchoring peptides was generated by means of simple modification of aminated microwells with bifunctional linker SMCC, followed by rapid and covalent immobilization of peptides through thiol-maleimide addition. Subsequent β-Gal anchoring afforded an immobilized enzyme that exhibited considerably higher activity with respect to otherwise immobilized β-Gal. In addition, peptide-modified surfaces were found to positively affect the thermal stability of the bound enzyme, like the stability under storage conditions. Taken all together, authors estimated a 20-fold increase in enzyme activity for β-Gal on peptide-modified surface, highlighting how deep is the impact an appropriate selection of the immobilization strategy may have in obtained results.

2.4 Thiol-ene Addition

Since the introduction of the archetypal concept of click chemistry by Sharpless and coworkers, many fast and chemoselective transformations have emerged as attractive click-type processes. Among these, the thermal- or UV-initiated addition of a thiol to an alkene through a radical mechanism (Scheme 1f), commonly termed thiol-ene reaction, has earned the click status in view of its favorable peculiarities [39]. High efficiency, broad orthogonality, and compatibility with physiologic conditions make thiol-ene addition a feasible candidate for selective conjugation. Additionally, the robust thioether linkage which is formed upon thiol-ene addition displays remarkable stability in a wide range of environmental conditions. However, careful tuning of the reaction parameters is required to limit undesired side reactions, e.g., alkene polymerization, and to ensure excellent reaction yields [11]. Intriguingly, the modular nature of the reaction in virtue of the photoinitiation feature allows spatial and temporal control of supports functionalization. For example, a landmark contribution to photochemical surface patterning has been reported from the Waldmann group [40]. Photochemical coupling of a set of olefin-tagged biomolecules to a thiol-modified surface allowed the fabrication of a microarray with precise control on protein immobilization even in the sub-micrometer scale. Functional integrity of immobilized molecules was also assessed and found to be comparable to that one in solution phase. Hawker and collaborators exploited thiol-ene chemistry to produce multifunctional microarrays embedded at the surface of poly(ethylene glycol)-based hydrogels [41]. Mild reaction conditions (UV irradiation at 365 nm for 2 min) allowed the incorporation of a wide range of orthogonal chemical handles exploitable for click-mediated selective conjugation of biomolecules, including functional peptides to direct cell adhesion. The combination of polymer direct functionalization with orthogonal postfunctionalization allowed the generation of different platforms for multiple display of functionally distinct biomolecules for different investigation purposes.

2.5 Oxime Ligation

The generation of an oxime bond between two biomolecules through the condensation of a carbonyl group (aldehyde or ketone) with an aminooxy group is commonly referred as oxime ligation (Scheme 1g). This reaction is particularly attractive due to its true click character: high conversion efficiency, chemoselectivity toward other functional groups, and mild reaction conditions in aqueous media are indeed distinctive features of the oxime ligation [42]. In addition, water is the only side product formed in the process, and no metal catalyst is required. Despite this, oxime ligation has not benefited from the same broad popularity of other click strategies such as CuAAC. This is likely due to the traditional limitations associated with the synthesis of aldehyde- and aminooxy-functionalized biomolecules. However, recent advances have enabled oxime ligation to gain a progressively leading role for bioconjugation purposes [11, 43, 44]. One appealing feature of oxime ligation lies in its reversibility. Indeed, unlike other covalent linkages, the oxime bond is reversible as a function of pH. Conveniently, the operational range of pH 4–8 largely avoids undesired hydrolysis of the oxime bond and guarantees the stability of the conjugate under standard biological conditions. Oxime bond hydrolysis can instead be triggered at lower or higher pH values. Notably, the pH responsiveness of the oxime bond enables the generation of dynamic biomaterials which can be adopted for controlled capture-and-release strategies. Patterned peptide-presenting surfaces are essential tools in the investigation of cell behavior in biomaterials and tissue engineering. Functionalized surfaces with patterned topography were realized by capping either the aminooxy or aldehyde group with a photo-labile group. Upon unmasking of the reactive groups with site-specific UV irradiation, ligand immobilization can be performed via oxime ligation to functionalize the surface with controlled spatial resolution. This approach found application in the Dumy, Yousaf, and Barner-Kowollik groups [45–47]. Maynard and collaborators have exploited a combination of oxime ligation and CuAAC to immobilize different proteins on a functionalized surface. The same authors reported a different strategy to surface patterned functionalization by electron-beam lithography. Aminooxy groups on the surface were then coupled to ketone-functionalized RGD peptides, which were shown to retain functional integrity [48].

3 Native Chemical Ligation

Native chemical ligation consists in the condensation of two free peptide fragments to yield a new construct linked by a native amide bond (Scheme 1h). The reaction occurs between a peptide thioester and another peptide fragment bearing an N-terminal cysteine and proceeds through an initial (reversible) transthioesterification step followed by an irreversible S,N-acyl shift which originates the new native amide bond. Although not always regarded as a prototypic click reaction, NCL chemoselectivity and efficiency, along with mild operational conditions, are surely appealing features in the peptide chemistry arena. Remarkably, NCL likely represents the gold-standard technique for the total chemical synthesis of proteins [49]. To overcome synthetic limitations associated with the synthesis of peptide thioesters, new strategies entailing thioester surrogates and precursors have been developed, like unnatural thiolated amino acids which have been synthesized to expand the potential of NCL [50]. NCL reaction kinetics are highly dependent on the substrates participating in the reaction; however, the time required for nearly quantitative product formation can be considerably short (<1 h). Despite NCL has recently experienced a growing application in the preparation of peptide bioconjugates, its potential for the functionalization of surfaces for analytical applications remains widely unexplored. Only few examples have indeed been reported to date. Among these, Helms et al. realized a cysteine-functionalized biosensor surface for conjugation with peptide thioesters [7]. Specific and complete immobilization of a decapeptide occurred with fast kinetics, and the peptide ability to engage in specific binding to its target was preserved. Interestingly, a peculiar feature of cysteine-functionalized sensors is that functional groups maintain their reactivity for extended periods of time. As a consequence, ligand density can be tuned stepwise at any time in sensor's life by using short ligation pulses, and the same functionalized surface can be used for separate experiments. Conversely, peptides bearing an N-terminal cysteine can be chemoselectively immobilized onto thioester-functionalized slides, as, for example, reported by the Yao group [51]. More recently, Dendane et al. exploited this kind of approach for the site-specific and chemoselective immobilization of peptides on hydrogen-terminated silicon nanowires [52].

4 Conclusions and Future Perspectives

Microarray-based screening is progressively gaining central importance in the dissection of a number of biological processes. The development of innovative analytical platforms strongly relies on new synthetic methods to allow controlled, oriented, and robust bioprobe immobilization on sensor surfaces. In this context, the last decade has experienced a growing application of the so-called click reactions for the sophisticated functionalization of sensors. Among these, some of the most popular have been discussed in the present work. Notwithstanding, the click repertoire already includes a more extensive range of chemoselective transformations, such as Diels-Alder reactions, which have been exploited for the generation of peptide and protein bioconjugates. Additional reactions that match the click-type criteria are likely to be discovered in the near future. Moreover, the potential of already existing click-type reactions is presumably going to be fully unrevealed thanks to new synthetic breakthroughs that may overcome current limitations. The combination of different and orthogonal strategies will, also, contribute to widen the available analytical toolbox for new devices development. Overall, a growing synthetic flexibility, a more and more refined control on immobilization parameters, and an increasing range of possible applications are likely to characterize the microarray field over the upcoming years.

References

1. Butte A (2002) The use and analysis of micro-array data. Nat Rev Drug Discov 1:951–960
2. Hoheisel JD (2006) Microarray technology: beyond transcript profiling and genotype analysis. Nat Rev Genet 7:200–210
3. Cretich M, Damin F, Pirri G, Chiari M (2006) Protein and peptide arrays: recent trends and new directions. Biomol Eng 23:77–88
4. Kim D, Herr AE (2013) Protein immobilization techniques for microfluidic assays. Biomicrofluidics 7:1–47
5. Heise C, Bier FF (2005) Immobilization of DNA on microarrays. Top Curr Chem 261:1–25
6. Nimse SB, Song K, Sonawane MD, Sayyed DR, Kim T (2014) Immobilization techniques for microarray: challenges and applications. Sensors 14:22208–22229
7. Helms B, Van Baal I, Merkx M, Meijer EW (2007) Site-specific protein and peptide immobilization on a biosensor surface by pulsed native chemical ligation. Chembiochem 8:1790–1794
8. Köhn M (2009) Immobilization strategies for small molecule, peptide and protein microarrays. J Pept Sci 15:393–397
9. Foong YM, Fu J, Yao SQ, Uttamchandani M (2012) Current advances in peptide and small molecule microarray technologies. Curr Opin Chem Biol 16:234–242
10. Kolb HC, Finn MG, Sharpless KB (2001) Click chemistry: diverse chemical function from a few good reactions. Angew Chem Int Ed Engl 40:2004–2021
11. Tang W, Becker ML (2014) "Click" reactions: a versatile toolbox for the synthesis of peptide-conjugates. Chem Soc Rev 43:7013–7039
12. Rostovtsev VV, Green LG, Fokin VV, Sharpless KB (2002) A stepwise huisgen cycloaddition process: copper(I)-catalyzed regioselective "ligation" of azides and terminal alkynes. Angew Chem Int Ed Engl 41:2596–2599
13. Meldal M, Tornøe CW (2008) Cu-catalyzed azide-alkyne cycloaddition. Chem Rev 108:2952–3015
14. Gori A, Wang C-IA, Harvey PJ, Rosengren KJ, Bhola RF, Gelmi ML, Longhi R, Christie MJ, Lewis RJ, Alewood PF, Brust A (2014) Stabilization of the cysteine-rich conotoxin MrIA by using a 1,2,3-triazole as a disulfide bond mimetic. Angew Chem Int Ed 54:1361–1364
15. Lin PC, Ueng SH, Tseng MC, Ko JL, Huang KT, Yu SC, Adak AK, Chen YJ, Lin CC (2006) Site-specific protein modification through CuI-catalyzed 1,2,3-triazole formation and its implementation in protein microarray fabrication. Angew Chem Int Ed 45:4286–4290
16. ZZilio C, Bernardi A, Palmioli A, Salina M, Tagliabue G, Buscaglia M, Consonni R, Chiari M (2015) New "clickable" polymeric coating for glycan microarrays. Sensor Actuator B: Chemical 215:412–420
17. Lind JU, Acikgöz C, Daugaard AE, Andresen TL, Hvilsted S, Textor M, Larsen NB (2012) Micropatterning of functional conductive polymers with multiple surface chemistries in register. Langmuir 28:6502–6511
18. Koepsel JT, Murphy WL (2009) Patterning discrete stem cell culture environments via localized self-assembled monolayer replacement. Langmuir 25:12825–12834
19. Hansen TS, Lind JU, Daugaard AE, Hvilsted S, Andresen TL, Larsen NB (2010) Complex surface concentration gradients by stenciled "electro click chemistry". Langmuir 26:16171–16177
20. Gallant ND, Lavery KA, Amis EJ, Becker ML (2007) Universal gradient substrates for "click" biofunctionalization. Adv Mater 19:965–969
21. Uttamapinant C, Tangpeerachaikul A, Grecian S, Clarke S, Singh U, Slade P, Gee KR, Ting AY (2012) Fast, cell-compatible click chemistry with copper-chelating azides for biomolecular labeling. Angew Chem Int Ed Engl 51:5852–5856
22. Jewett JC, Bertozzi CR (2010) Cu-free click cycloaddition reactions in chemical biology. Chem Soc Rev 39:1272–1279
23. Codelli JA, Baskin JM, Agard NJ, Bertozzi CR (2008) Second-generation difluorinated cyclooctynes for copper-free click chemistry. J Am Chem Soc 130:11486–11493
24. Prim D, Rebeaud F, Cosandey V, Marti R, Passeraub P, Pfeifer ME (2013) ADIBO-based "click" chemistry for diagnostic peptide microarray fabrication: Physicochemical and assay characteristics. Molecules 18:9833–9849
25. Krishnamurthy VR, Wilson JT, Cui W, Song X, Cummings RD, Chaikof EL (2011) Chemoselective immobilization of peptides on abiotic and cell surfaces at controlled Densities. Langmuir 26:7675–7678
26. Van Berkel SS, Van Eldijk MB, Van Hest JCM (2011) Staudinger ligation as a method for bioconjugation. Angew Chemie Int Ed 50:8806–8827
27. Kalia J, Abbott NL, Raines RT (2007) General method for site-specific protein immobilization by Staudinger ligation. Bioconjug Chem 18:1064–1069
28. Lin P-C, Weinrich D, Waldmann H (2010) Protein biochips: oriented surface immobilization of proteins. Macromol Chem Phys 211:136–144

29. Köhn M, Gutierrez-Rodriguez M, Jonkheijm P, Wetzel S, Wacker R, Schroeder H, Prinz H, Niemeyer CM, Breinbauer R, Szedlacsek SE, Waldmann H (2007) A microarray strategy for mapping the substrate specificity of protein tyrosine phosphatase. Angew Chem Int Ed 46:7700–7703

30. Watzke A, Gutierrez-Rodriguez M, Köhn M, Wacker R, Schroeder H, Breinbauer R, Kuhlmann J, Alexandrov K, Niemeyer CM, Goody RS, Waldmann H (2006) A generic building block for C- and N-terminal protein-labeling and protein-immobilization. Bioorg Med Chem 14:6288–6306

31. Watzke A, Köhn M, Gutierrez-Rodriguez M, Wacker R, Schröder H, Breinbauer R, Kuhlmann J, Alexandrov K, Niemeyer CM, Goody RS, Waldmann H (2006) Site-selective protein immobilization by Staudinger ligation. Angew Chem Int Ed 45:1408–1412

32. Nilsson BL, Hondal RJ, Soellner MB, Raines RT (2003) Protein assembly by orthogonal chemical ligation methods. J Am Chem Soc 125:5268–5269

33. Mather BD, Viswanathan K, Miller KM, Long TE (2006) Michael addition reactions in macromolecular design for emerging technologies. Prog Polym Sci 31:487–531

34. Nair DP, Podgórski M, Chatani S, Gong T, Xi W, Fenoli CR, Bowman CN (2014) The thiol-Michael addition click reaction: a powerful and widely used tool in materials chemistry. Chem Mater 26:724–744

35. Li J, Hu XK, Lipson RH (2013) On-chip enrichment and analysis of peptide subsets using a maleimide-functionalized fluorous affinity biochip and nanostructure initiator mass spectrometry. Anal Chem 85:5499–5505

36. Gao G, Yu K, Kindrachuk J, Brooks DE, Hancock REW, Kizhakkedathu JN (2011) Antibacterial surfaces based on polymer brushes: investigation on the influence of brush properties on antimicrobial peptide immobilization and antimicrobial activity. Biomacromolecules 12:3715–3727

37. Houseman BT, Gawalt ES, Mrksich M (2003) Maleimide-functionalized self-assembled monolayers for the preparation of peptide and carbohydrate biochips. Langmuir 19:1522–1531

38. Fu J, Reinhold J, Woodbury NW (2011) Peptide-modified surfaces for enzyme immobilization. PLoS One 6:2–7

39. Dondoni A (2008) The emergence of thiol-ene coupling as a click process for materials and bioorganic chemistry. Angew Chem Int Ed 47:8995–8997

40. Jonkheijm P, Weinrich D, Köhn M, Engelkamp H, Christianen PCM, Kuhlmann J, Maan JC, Nüsse D, Schroeder H, Wacker R, Breinbauer R, Niemeyer CM, Waldmann H (2008) Photochemical surface patterning by the thiol-ene reaction. Angew Chem Int Ed 47:4421–4424

41. Gupta N, Lin BF, Campos LM, Dimitriou MD, Hikita ST, Treat ND, Tirrell MV, Clegg DO, Kramer EJ, Hawker CJ (2010) A versatile approach to high-throughput microarrays using thiol-ene chemistry. Nat Chem 2:138–145

42. Ulrich S, Boturyn D, Marra A, Renaudet O, Dumy P (2014) Oxime ligation: a chemoselective click-type reaction for accessing multifunctional biomolecular constructs. Chem Eur J 20:34–41

43. Jiménez-Castells C, de la Torre BG, Gutiérrez Gallego R, Andreu D (2007) Optimized synthesis of aminooxy-peptides as glycoprobe precursors for surface-based sugar-protein interaction studies. Bioorg Med Chem Lett 17:5155–5158

44. Moulin A, Martinez J, Fehrentz J-A (2007) Synthesis of peptide aldehydes. J Pept Sci 13:1–15

45. Dendane N, Hoang A, Guillard L, Defrancq E, Vinet F, Dumy P (2007) Efficient surface patterning of oligonucleotides inside a glass capillary through oxime bond formation. Bioconjug Chem 18:671–676

46. Park S, Yousaf MN (2008) An interfacial oxime reaction to immobilize ligands and cells in patterns and gradients to photoactive surfaces. Langmuir 24:6201–6207

47. Pauloehrl T, Delaittre G, Bruns M, Meißler M, Börner HG, Bastmeyer M, Barner-Kowollik C (2012) (Bio)molecular surface patterning by phototriggered oxime ligation. Angew Chem Int Ed 51:9181–9184

48. Kolodziej CM, Kim SH, Broyer RM, Saxer SS, Decker CG, Maynard HD (2012) Combination of integrin-binding peptide and growth factor promotes cell adhesion on electron-beam-fabricated patterns. J Am Chem Soc 134:247–255

49. Dawson PE, Kent SB (2000) Synthesis of native proteins by chemical ligation. Annu Rev Biochem 69:923–960

50. Noisier AF, Albericio F (2014) Advances in ligation techniques for peptide and protein synthesis. Amino Acids Pept Protein 39:1–20

51. Lesaicherre ML, Uttamchandani M, Chen GYJ, Yao SQ (2002) Developing site-specific immobilization strategies of peptides in a microarray. Bioorg Med Chem Lett 12:2079–2083

52. Dendane N, Melnyk O, Xu T, Grandidier B, Boukherroub R, Stiévenard D, Coffinier Y (2012) Direct characterization of native chemical ligation of peptides on silicon nanowires. Langmuir 28:13336–13344

Chapter 12

Manufacturing of Peptide Microarrays Based on Catalyst-Free Click Chemistry

Denis Prim and Marc E. Pfeifer

Abstract

Immobilization of peptides to a solid surface is frequently an important first step before they can be probed with a variety of biological samples in a heterogeneous assay format for research and clinical diagnostic purposes. Peptides can be derivatized in many ways to subsequently covalently attach them to an activated solid surface such as epoxy-functionalized glass slides. Here, we describe a clean, efficient, and reproducible fabrication process based on catalyst-free click chemistry compatible with the construction of low- to high-density peptide microarrays.

Key words Peptide microarrays, Immunoassays, c-Myc peptide, Click chemistry, Catalyst-free, Strain-promoted alkyne-azide cycloaddition (SPAAC), Azadibenzocyclooctyne (ADIBO)

1 Introduction

Strain-promoted alkyne-azide cycloaddition (SPAAC) allows facile coupling of a variety of molecules to solid surfaces without need of additional catalysts or reagents [1]. Synthetically accessible and commercially available modified cyclooctyne probes such as particularly the azadibenzocyclooctyne (ADIBO) [2–4] show rapid reaction kinetics with azide-tagged peptides (Fig. 1). Reactivities of different peptides, however, were shown to vary with also solvent composition and concentration playing an important role [4]. Maximum plateau signal intensities of fluorophore-labeled peptides in our investigations were observed within 10 min [4] which appears compatible with peptide microarray fabrication requirements. Prior automated deposition or more extensive manual preparation of peptide microarrays, it is therefore highly recommended to verify the reaction kinetics of selected peptides beforehand and to identify ideal spotting concentrations and water/organic solvent ratios.

Good intra-spot fluorescence intensity homogeneities and inter-spot fluorescence intensity reproducibility are indicative for

Marina Cretich and Marcella Chiari (eds.), *Peptide Microarrays: Methods and Protocols*, Methods in Molecular Biology, vol. 1352, DOI 10.1007/978-1-4939-3037-1_12, © Springer Science+Business Media New York 2016

uniform solid-phase functionalization and even binding of interaction partners. Spots produced when depositing nanoliter volumes were symmetrical (circular) and had diameters between 250 and 300 μm [4]. Under the conditions described smearing or bleeding of spots was negligible or indiscernible. Manual pipetting of 1 μL of solutions containing azide-modified peptides, for instance, to conduct preliminary studies is also feasible but requires more careful handling in terms of positional accuracy and drying conditions. Temperature control and use of a humidity chamber may help avoid formation of donut-shaped microarray spots due to nonuniform solvent evaporation (Fig. 2). But it also must be mentioned here that slight spot inhomogeneities do not necessarily signify poor and unexploitable data. Indeed, assay results obtained when working with peptides recognized by anti-BRCA1-associated RING domain protein 1 (BARD1) or anti-c-Myc antibodies were reproducible and conclusive ([4, 5] and *unpublished results*).

Another advantage of the reactivity of ADIBO-activated glass slides is the possibility to cap nonfunctionalized areas (Fig. 1) of the glass slide with polyethylene glycol (PEG), dextran, BSA, and other types of molecules to strongly reduce nonspecific binding [3, 4] when working with biological samples or other complex matrices. Indeed, elevated and varying background signals are frequently responsible for poor signal-to-noise ratios (SNR) and limited assay sensitivities.

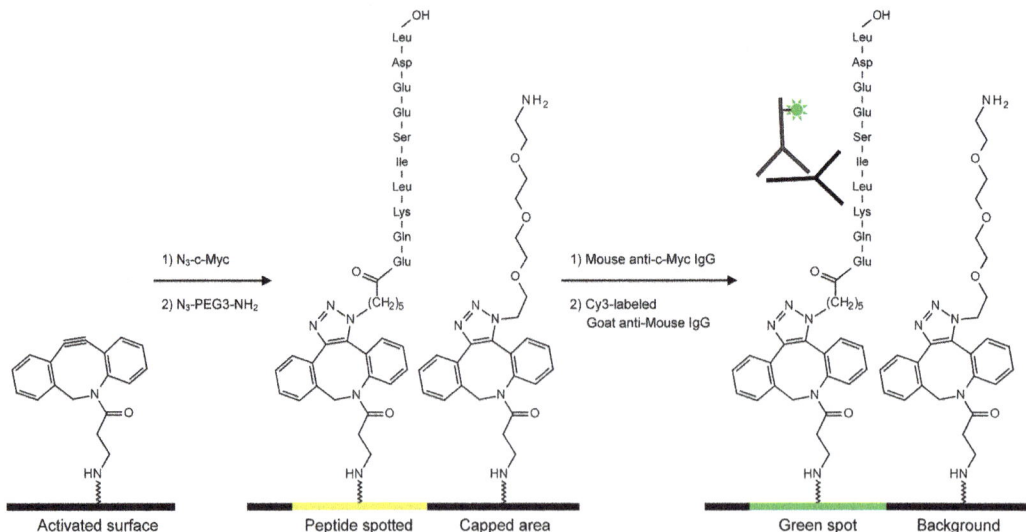

Fig. 1 Covalent immobilization of azide-derivatized peptides to azadibenzocyclooctyne (ADIBO)-activated glass slides via a strain-promoted alkyne-azide cycloaddition (SPAAC) and capping of non-spotted areas with, for instance, a polyethylene glycol (PEG)-type molecule is shown. Subsequently c-Myc peptide baits capture mouse anti-c-Myc antibodies that can be revealed with a Cy3-labeled detection antibody giving green fluorescent spots

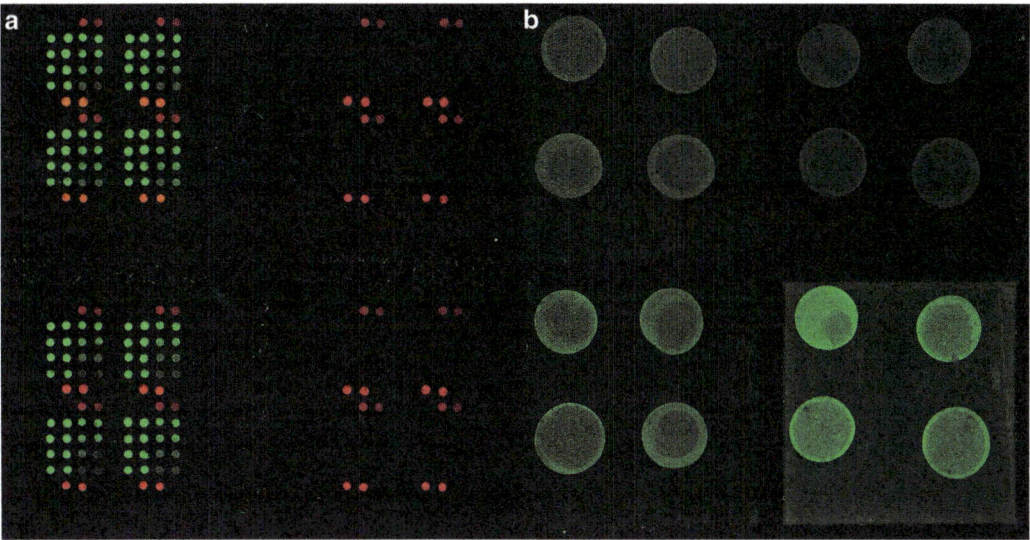

Fig. 2 (**a**) Image of 4×4 sub-arrays depicting an anti-c-Myc peptide antibody titration experiment with 16 spots representing peptide concentrations ranging from 50 to 0.4 μM. The *top-left* four sub-arrays were probed with an anti-c-Myc antibody concentration of 100 ng/mL, *bottom left* with 50 ng/mL, *top right* with 1.56 ng/mL, and *bottom right* with 0.79 ng/mL. Volumes of 1 nL of N_3-c-Myc peptide were deposited with an automated spotting system. (**b**) Immunoassay experiment with anti-c-Myc antibody dissolved in 1 % mouse serum. Volumes of 1 μL of N_3-c-Myc peptide at a concentration of 200 μM were deposited manually with a pipette. Variation in background signals is due to use of different blocking reagents to reduce nonspecific binding (unpublished results)

Below we describe an immunoassay performed with a c-Myc peptide microarray, a primary anti-c-Myc antibody, and a Cy3-labeled detection antibody. Assay results are depicted with spots created either with 1 nL or 1 μL volumes of peptide solution dispensed with an automated system and manually with a pipette, respectively (Fig. 2). Preliminary titration experiments with varying peptide and primary antibody concentrations may be useful to identify an optimal assay dynamic range for quantification purposes.

2   Materials

All solutions for slide functionalization have to be prepared with high-grade anhydrous organic solvents (unless indicated otherwise). Solutions have to be prepared just before use and should be stored under dry atmosphere. It is recommended to filter buffers through 0.45 μm pore size filters.

It is important to check excitation/emission wavelength compatibility of fluorescent dyes with your microarray reader capabilities. Cy3 and Cy5 dyes used in this method are generally compatible with modern microarray reader specifications.

The materials and methods described below illustrate the slide preparation, peptide modification, immobilization, and immuno-assay process steps utilizing the c-Myc (EQKLISEEDL) model peptide.

2.1 Slides

A variety of glass slides is compatible with the method proposed. Here we used glass slides of 75×25 mm dimensions from Corning (Corning, NY, USA) (*see* **Note 1**).

2.2 Slide Functionalization Components

1. Slide cleaning solution ("piranha solution"): Mix 20 mL of hydrogen peroxide 30 % (H_2O_2 30 %) and 60 mL of sulfuric acid (H_2SO_4) $\geq$96 % directly in a beaker chilled with ice (*see* **Note 2**).

2. Silanization solution: Add 1 mL of (3-glycidyloxypropyl)tri-methoxysilane (3-GPTS) and 1 mL of diisopropylethylamine (DIEA) to 100 mL of anhydrous toluene directly in a beaker and stir (*see* **Note 3**).

3. ADIBO solution: Add 100 mg of azadibenzocyclooctyne (ADIBO) amine (Jena Bioscience, Jena, Germany) and 2 mL of diisopropylethylamine (DIEA) to 200 mL of anhydrous DMF in a glass flask and stir (*see* **Note 4**).

4. Slide functionalization control solution: Dissolve 1 mg of N_3-Cy5 (Jena Bioscience, Jena, Germany) in 1 mL of DMSO. Dilute the solution to 20 ng/mL by dissolving 1 μL of N_3-Cy5 1 mg/mL to 50 mL of DMSO 50 % (*see* **Note 5**).

2.3 Azide Modification Components

1. Peptide azide modification: 6-azido-hexanoic acid can be coupled during solid-phase peptide synthesis (SPPS) via a manual procedure or with a peptide synthesizer (*see* **Note 6**).

2. Protein azide modification: An azide labeling reagent kit containing, for instance, NHS-PEG4-N_3 from Piercenet (Piercenet, Rockford, IL, USA) can be used to modify NH_2 groups on proteins or other amine-containing molecules to introduce the azide moiety (*see* **Note 7**).

2.4 Peptide Spotting Components

1. Peptide solution: Dilute peptides at a concentration ranging from 50 to 0.4 μM in DMSO 50 % (8 concentrations at 50, 25, 12.5, 6.25, 3.13, 1.56, 0.78, and 0.40 μM) (*see* **Note 8**).

2. Control solution: Dilute N_3-Cy5 to a concentration of 1.1 μM in DMSO 50 % (*see* **Note 9**).

3. Slide capping solution: Amino-PEG3-azide (Jena Bioscience, Jena, Germany) is directly diluted in DMSO 50 % to obtain a concentration of 3 μM (*see* **Note 10**). Instead of the amino-PEG3-azide, the azide derivative of BSA (prepared as described in Subheading 3.3) can be used for capping purposes.

4. Slide washing solution: Prepare 45 mL of DMSO 50 % in a Falcon tube of 50 mL (*see* **Note 11**).

5. Manual spotting: Use a 0.1–10 μL micropipette set at 1 μL.

2.5 Immunoassay Components

1. Primary antibody: Mouse monoclonal IgG anti-c-Myc (Sigma-Aldrich, St. Louis, MO, USA) is diluted to vary in concentration from 200 to 0.2 ng/mL in the dilution buffer (11 points: 200, 100, 50, 25, 12.5, 6.25, 3.13, 1.56, 0.78, 0.39, and 0.2 ng/mL).

2. Detection antibody: Cy3-labeled goat polyclonal anti-mouse IgG (H + L) (KPL, Gaithersburg, MD, USA) is diluted to 2 μg/mL in the dilution buffer.

2.6 Buffer Solutions

1. Phosphate-buffered saline 10× (PBS 10×): Dissolve 2.0 g KCl, 4.2 g Na_2HPO_4, 2.4 g KH_2PO_4, and 80.0 g NaCl in 1 L of deionized water, and check and adjust pH to 7.4. Commercially available PBS solutions exist. Dilute the solution 10× (100 mL with 900 mL of water) before use (PBS 1×).

2. Washing buffer: Dilute 50 μL of Tween 20 to 1 L of PBS 1×.

3. Blocking buffer: Add 0.5 g of bovine serum albumin (BSA) to 10 mL of PBS 1×.

4. Dilution buffer: Add 0.5 g of bovine serum albumin (BSA) to 50 mL of PBS 1× (*see* **Note 12**).

3 Methods

3.1 Slide Functionalization

1. Prepare the cleaning "piranha solution" directly in the beaker and let the solution cool down before slowly inserting the glass slides. Check that all slides are completely covered by liquid (*see* **Notes 13** and **14**).

2. After 30 min remove slides and wash with a large amount of deionized water three times.

3. Let the slides dry in an oven for 2 h at 100 °C.

4. At the end of the drying process, prepare the "silanization solution" directly in a clean and dry beaker. Keep this solution as much as possible in a dry place (*see* **Note 15**).

5. Once slides are dry and cold, place them in the liquid and let them react for 12 h at room temperature and in a dry place (*see* **Note 16**).

6. Clean slides with two times methanol and then two times acetone before drying them under N_2.

7. Place the slides in an oven at 120 °C for 1 h.

8. At the end of the drying process prepare the "ADIBO solution" and transfer approximately 100 mL in a clean and dry beaker.

9. Once slides are dry and have reached room temperature, place them in the "ADIBO solution" and let them react overnight ($\geq$12 h) at room temperature in a dry place protected from light.

10. Clean slides with methanol and acetone as previously and dry them under argon (*see* **Note 17**).

3.2 Slide Functionalization Control

1. Prepare the "slide functionalization control solution" in 50 mL Falcon tube (*see* **Note 18**).

2. Choose two to three slides from each produced batch of ADIBO-functionalized slides and immerse them, one at the time, in the Falcon tube.

3. Let them react during 10 min and remove the slides from the solution and wash them with a bath of methanol and then acetone.

4. Finally wash them in deionized water before drying them under N_2 flow.

5. Read the Cy5 dye fluorescence intensity with your microarray reader.

6. The fluorescence signal variation across the slide and between the tested slides should be lower than 20 % (*see* **Note 19**).

3.3 BSA Protein Azide Derivative

1. Prepare a 12 mg/mL solution of BSA (Roche Diagnostics, Rotkreuz, Switzerland) in PBS 1×.

2. Mix 320 μL of BSA 12 mg/mL with 5 μL of NHS-PEG4-N_3 and 680 μL of water in an Eppendorf.

3. Keep at 25 °C for 1 h under constant stirring at 500 rpm in a thermoshaker.

4. Transfer the reaction mixture to a 3 mL "Slide-A-Lyzer" dialysis cassette 10 K MWCO from Piercenet or any similar product and perform the dialysis overnight at room temperature.

7. Recover the solution and lyophilize overnight to obtain a white powder (*see* **Note 20**).

3.4 Peptide Spotting

1. With a micropipette dispense peptide solution of 1 μL on the surface of the slides (*see* **Note 21**).
   Or

1. With an automated spotter (and according to the instructions of use), fill the reservoir with the peptide solution and start the spotting procedure.

2. Allow the peptide to react during 30 min to 1 h according to the drying speed of the spots and reaction kinetics of your peptides (*see* **Note 22**).

3. Immerse the slide in the "slide capping solution," mix vigorously, and let it react during 5 min.

4. Wash the slide with the slide washing solution and finally with deionized water.

5. Dry the slide under N_2 flow (*see* **Note 23**).

3.5 Immunoassay

1. Take a spotted slide and place it in a clean Nexterion IC-16. Be careful to place the spotted face up (*see* **Notes 24** and **25**).

2. Add 100 µL of BSA 5 % to each well and place an adhesive tape to seal wells (*see* **Note 26**).

3. After 30 min, remove the solution and wash three times with 200 µL of "washing buffer."

4. Add 100 µL of each primary antibody concentration to each of the eleven wells. Keep the 12th well for the blank by adding 100 µL of the dilution buffer. Place an adhesive tape to seal wells.

5. After 30 min, remove the solution and wash three times with 200 µL of "washing buffer."

6. Add 100 µL of the detection antibody to each well. Do not forget to add it to the blank too. Place an adhesive tape to seal wells.

7. After 30 min, remove the solution and wash three times with 200 µL of "washing buffer."

8. Clean quickly the slide with deionized water (*see* **Note 27**) and dry it under a flow of N_2.

9. Scan the slide (*see* Fig. 2) and extract fluorescence intensity readings according to your microarray reader configuration to measure at a Cy3 dye-compatible wavelength (*see* **Note 28**).

4   Notes

1. It is advantageous to use slides frosted at one end to help you write information on it (only with carbon pen).

2. Slide cleaning solution (piranha solution) is a strong oxidative solution and therefore should be handled carefully. Prepare the minimum amount required and neutralize it before waste disposal. Keep the solution chilled on ice.

3. The epoxide group reacts quickly with water.

4. This solution is stable for several weeks if kept at 5 °C.

5. ADIBO-functionalized slides may be assessed by letting the surface react with a fluorescent dye containing an azide group, for instance, N_3-Cy5. The homogeneity of the fluorescent surface is a good indicator of the quality of the functionalization process.

6. A molecule containing an azido group such as 6-azido-hexanoic acid may be added at the final step of peptide synthesis to modify the N-terminus of the peptide.

7. A convenient method based on *N*-hydroxysuccinimide ester (NHS ester) coupling reaction allows quick and effective coupling of the azide moiety to a free amino group. For peptides or proteins with multiple free amino groups, it is difficult or impossible to know where the modification will occur. The molar ratio between NHS-N_3 reagent and amino group containing substrate allows some control on the number of expected modifications to occur on each protein or peptide.

8. As the reaction kinetics and the binding efficiencies are strongly dependent on the physical properties of your protein or peptide, preliminary studies have to be conducted to determine the best concentrations, solvent compositions, and necessary reaction times.

9. When a large amount of spots are printed on a slide, the use of "controls" is useful to align spots and to check the quality of the printing process without having to perform the full immunoassay. Controls may be small molecules such as N_3-Cy5 or fluorescently labeled peptides or proteins.

10. After spotting of peptides, large areas of the slide still present reactive ADIBO groups. To avoid any "bleeding" of spots, a blocking step to cap all the unreacted areas of the slide is therefore necessary.

11. Cleaning the slide with DMSO 50 % will allow removing of remaining none chemically bound molecules. It is possible to replace this step by methanol and acetone washes.

12. It is possible to prepare more BSA 5 % and dilute it five times with PBS 1× to obtain this solution.

13. To functionalize slides, the use of a staining trough, for instance, the one from Assistent (P/N, 1205, Hecht Assistent, Sondheim/Rhön, Germany), is convenient (*see* Fig. 3).

14. The hydrogen peroxide in the piranha solution induces the presence of bubbles. Shake the slide holder to remove as much as possible bubbles from the surface of the slides.

15. You may keep the silanization solution in a desiccator to reduce contact with humidity or cover the beaker with Parafilm.

16. The epoxy-functionalized slides (obtained after **step 6**) are reactive with water, so it is important to avoid contact with water or your slides will be deactivated.

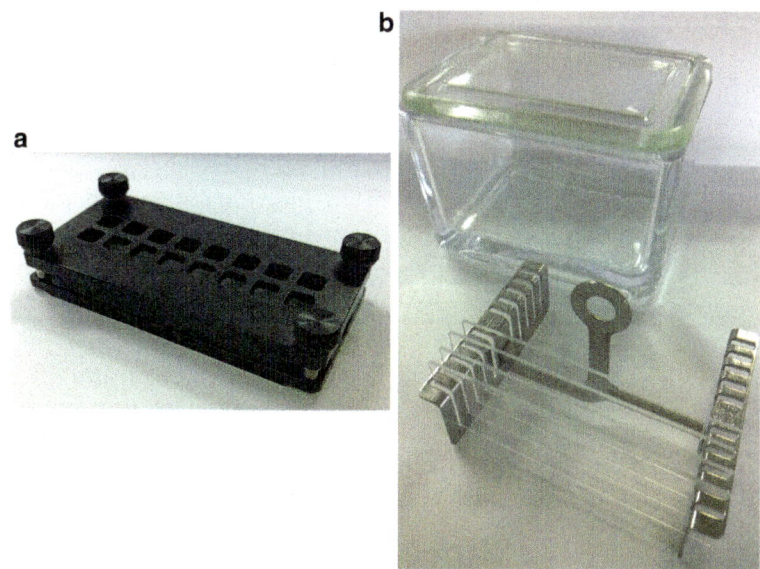

Fig. 3 (a) Nexterion IC-16 hybridization chamber composed of 16 compartments to perform separate immunoassays, i.e., test individual sub-arrays. **(b)** Assistent staining trough glassware and slide holder used to conduct the different slide functionalization steps. Note: no agitation or stirring was necessary

17. ADIBO-functionalized slides are not water sensitive and are therefore compatible with water wash. You can store slides at room temperature for a short term or longer at –20 °C.

18. N_3-Cy5 is poorly soluble in water. It is possible to use sulfo-Cy5-N_3 to work with an aqueous solution; however, better results and shorter reaction time were observed with N_3-Cy5.

19. Software tools typically installed on microarray scanners allow usually simple statistical analyses to determine average fluorescence intensity and standard deviation values for lines across slides as well as for rectangular (or circular) areas.

20. The modified BSA obtained under these conditions contains approximately 13 moieties of PEG4-N_3. The optimal substrate-reagent ratio has to be determined for each substrate molecule.

21. The best way to easily spot your peptides to the right place of the slide is by doing a sketch of your slide with the well and putting the slide on top of it during the spotting. If spots spread too much, change the concentration of H_2O/DMSO, or add a surfactant to the solution or glycerol. Reducing the spotted volume from 1 to 0.5 μL may also be an option. Those parameters may need to be optimized for each situation.

22. If spots are drying too fast, which is often the case when working with nanoliter quantities, control of temperature and humidity will allow a more reproducible drying process. Condensation of water should be avoided.

23. Printed slides can be stored for several days without degradation; however, the stability may depend on the nature of the peptides immobilized.

24. This part is described using the c-Myc model peptide immobilized on the glass surface. Each immunoassay has to be optimized individually and therefore solvent, reaction time, and parameters described here may have to be adapted for other immunoassays. In this case we will use 12 wells of the Nexterion IC-16 (*see* **Note 25**), one for each primary antibody concentration and one for the blank.

25. To hold the slide in place during the immunoassay experiment, we use Nexterion hybridization chamber IC-16 as it will provide up to 16 wells per slide and is reusable after cleaning. The chamber format is common and spotters are usually compatible with this format (*see* Fig. 3).

26. It is possible to incubate your slides during the immunoassay at 37 °C.

27. We recommend you to quickly wash the slide with deionized water before drying it for reading. In fact if you dry after the "washing buffer" step, you will have salt deposits on the surface resulting in poor scan quality.

28. If you have spotted your peptides on the surface manually, it may be tricky to use the automated spot localization tool as the spot size and shape may not be regular. In that case, you may have to adjust manually the spot location with a slightly smaller radius and use one or two additional areas to determine the background signal.

Acknowledgment

This work was supported by the Institute of Life Technologies of the University of Applied Sciences and Arts Western Switzerland (HES-SO), Valais/Wallis, in Sion.

References

1. Jewett JC, Bertozzi CR (2010) Cu-free click cycloaddition reactions in chemical biology. Chem Soc Rev 39:1272–1279

2. Debets MF et al (2010) Azide: a unique dipole for metal-free bioorthogonal ligations. Chembiochem 11:1168–1184

3. Kuzmin A et al (2010) Surface functionalization using catalyst-free azide-alkyne cycloaddition. Bioconjug Chem 21:2076–2085

4. Prim D et al (2013) ADIBO-based "click" chemistry for diagnostic peptide micro-array fabrication: physicochemical and assay characteristics. Molecules 18:9833–9849

5. Cosandey V et al (2012) Construction of a peptide microarray for auto-antibody detection. Chimia 66:803–806

Chapter 13

Clickable Polymeric Coating for Oriented Peptide Immobilization

Laura Sola, Alessandro Gori, Marina Cretich, Chiara Finetti, Caterina Zilio, and Marcella Chiari

Abstract

A new methodology for the fabrication of an high-performance peptide microarray is reported, combining the higher sensitivity of a layered Si–SiO$_2$ substrate with the oriented immobilization of peptides using a N,N-dimethylacrylamide-based polymeric coating that contains alkyne monomers as functional groups. This *clickable* polymer allows the oriented attachment of azido-modified peptides via a copper-mediated azide/alkyne cycloaddition. A similar coating that does not contain the alkyne functionality has been used as comparison, to demonstrate the importance of a proper orientation for facilitating the probe recognition and interaction with the target antibody.

Key words Peptide microarray, Click chemistry, Oriented immobilization, Polymeric coating

1 Introduction

During the last years, peptide microarrays have gained a pivotal role in the development of biosensors. Furthermore, due to their large chemical diversity and facile synthesis, peptides represent a promising class of specific capture molecules in the development of solid-state biosensors [1]. Among the several issues that might affect the performance of these kinds of biosensors, the most important is the coupling efficiency of the analyte to the surface. The most common methods used for the immobilization of a bio-molecule on a surface are based on nonspecific random adsorption or covalent binding between natural available functional groups on protein molecules and complementary coupling groups on the surface.

Physical or non-covalent methods include interactions between the probe and the surface such as hydrophobic, electrostatic, and van der Waals forces [2]. For example, the naturally strong biotin–streptavidin interaction is one of the most exploited non-covalent

Marina Cretich and Marcella Chiari (eds.), *Peptide Microarrays: Methods and Protocols*, Methods in Molecular Biology, vol. 1352, DOI 10.1007/978-1-4939-3037-1_13, © Springer Science+Business Media New York 2016

attachments. However, non-covalent binding may cause changes in the protein microenvironment, protein denaturation, thus leading to loss of functional activity. Furthermore, they are weak interactions, sensitive to variations of temperature, pH, and ionic strength of the environment.

Covalent immobilization takes advantage mainly from the natural abundance of the amine groups in proteins that are very reactive toward a wide range of functional moieties (e.g., active esters, aldehyde, epoxide, etc.) present on a properly modified surface.

For example, lysine residues are frequently used for protein immobilization considering their nucleophile behavior, while cysteines are most commonly used to bind proteins or peptides onto a gold or thiol-modified support [2]. Even though these systems allow a more stable and reproducible attachment, they require drastic conditions for the immobilization process which may lead to loss of binding activity as well. Furthermore, they reduce the probe conformational flexibility and do not offer a precise orientation of the probe onto the surface: in fact, a proper probe orientation increases the exposure of functional domains allowing a better interaction with the target in solution. As a consequence, a good immobilization strategy that preserves the native conformation of probes while providing optimal orientation is needed. In particular, the immobilization strategy should be fast, specific, and high yielding: fast reactions are preferable, since reactions with surfaces are significantly slower than the corresponding ones in solution and long reaction times expose the antibody to possible denaturation. To this end, chemoselective and biorthogonal reactions, such as Staudinger ligation [3], photoactivatable immobilization [4], and copper-mediated cycloaddition [5], have found a wide employment. In particular, the so-called click reactions, including reactions between thiols and maleimides as well as cyclization of azides and alkynes, are suitable to bind chemically modified biomolecules (proteins, antibodies, peptides, etc.) on properly functionalized surfaces.

Here we propose a new method to immobilize peptides in an oriented way through the well-known copper-mediated 1,3-dipolar Huisgen cycloaddition [6] using a N,N-dimethylacrylamide-based copolymer to functionalize a silicon oxide support.

In particular, we have synthesized a novel polymer, named poly(DMA-PMA-MAPS), obtained from the copolymerization of N,N-dimethylacrylamide (DMA), 3-trimethylsilanyl-prop-2-yn methacrylate (PMA), and 3(trimethoxysilyl)-propylmethacrylate (MAPS) (Fig. 1). The polymer consists of (1) a segment of polydimethylacrylamide that interacts with the surface by weak, non-covalent interactions such as hydrogen bonds, van der Waals, and hydrophobic forces, (2) a pending silane hydrolysable monomer that promotes condensation of the polymer with surface silanols or between contiguous chains, and (3) a chemically active monomer which, in the specific case, is an alkyne group. The

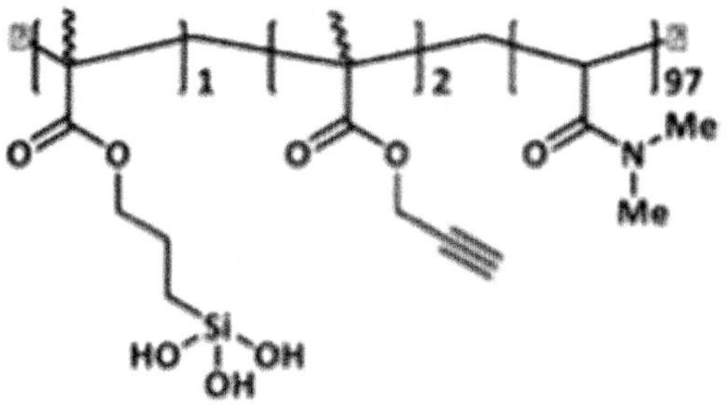

Fig. 1 Chemical structure of the copolymer poly(DMA-PMA-MAPS) used for the formation of the alkyne 3D coating

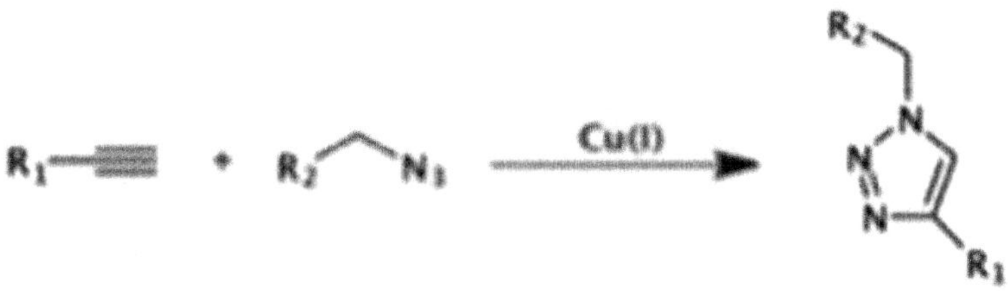

Fig. 2 Copper-mediated 1,3-dipolar Huisgen cycloaddition

polymer reported herein is similar to another polymer developed by Pirri and collaborators [7] to form a coating on glass slides by a combination of physi- and chemisorption that has been widely employed in protein and DNA microarray analysis; this polymer, introduced in 2004, contains N-acryloyloxysuccinimide as functional group which easily reacts with amines.

The novelty of this work [8] consists in the presence of an alkyne functional monomer that replaces the succinimide ester of the parent polymer. In fact, in the panorama of *clickable* reactions, the copper-catalyzed 1,3-dipolar cycloaddition between azide and alkyne (Fig. 2) has gained considerable attention thanks to the dramatic acceleration rate obtained with the catalyst and the beneficial effects of water used as solvent. This reaction does not require protecting the most common functionalities of biomolecules and proceeds with almost complete conversion and selectivity with a surprising indifference to solvent and pH. Even though the use of copper salts with ascorbate might be problematic in bio-conjugation applications, tris[(1-benzyl-1H-1,2,3-triazol-4-yl)methyl]amine, THPTA (Fig. 3), has been shown to effectively enhance the copper-catalyzed cycloaddition without damaging biological scaffolds [9].

Thanks to the regioselectivity of this cycloaddition, the reduction of biomolecular activity due to denaturation and/or poor accessibility to the binding site caused by random orientation on solid surface could be avoided by incorporating a triazole-modified amino acid during the peptide synthesis.

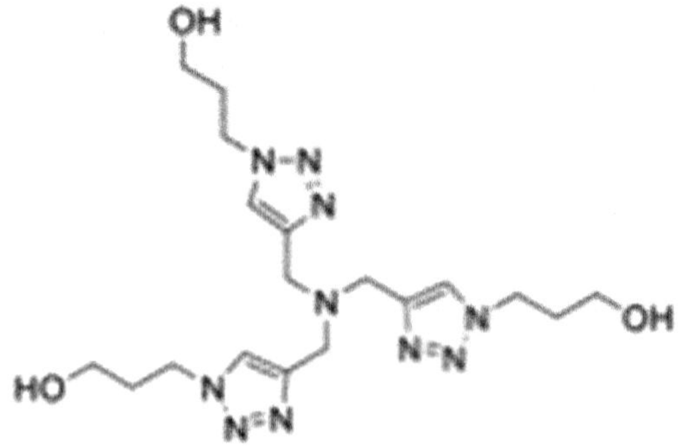

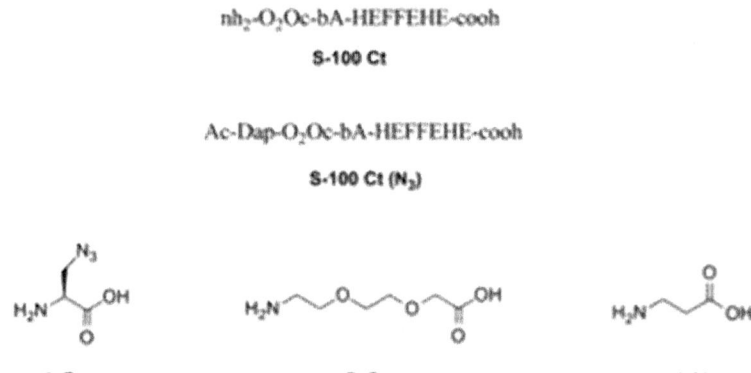

Fig. 3 Chemical structure of the THPTA ligand

nh₂-O₂Oc-bA-HEFFEHE-cooh

S-100 Ct

Ac-Dap-O₂Oc-bA-HEFFEHE-cooh

S-100 Ct (N₃)

L-Dap   **O₂Oc**   **bAla**

Fig. 4 S100 Ct and S100 Ct N_3 sequences. Standard one-letter code is used for natural amino acids (E = glutamic acid, H = histidine, F = phenylalanine, Ac = acetylated). Representative structures for unnatural building blocks are provided. Beta-alanine (bAla) is used as spacer. O_2Oc is used as polar flexible spacer to prevent direct peptide adsorption to the solid surface and facilitate peptide exposure. Azido alanine (Dap) is used to introduce the azido function exploitable for Cu-catalyzed click chemistry

In particular, we have modified a peptide which has been computationally designed and demonstrated to be one of the epitopes of the protein S100B, a recognized marker for neurological disorders [10]. The peptide has been prepared following a classic solid-state synthesis, anchoring of the C-terminal amino acid to a resin through a linker (2-chlorotritylchloride, 2-CTC). The azido-modified peptide has been obtained using a nonnatural azido alanine, as shown in Fig. 4.

After the immobilization on a layered Si-SiO₂ substrate, modified with the new proposed copolymer, it was possible to demonstrate the sensitivity provided by an oriented immobilization of the capturing probe in a microarray-based immunoassay in comparison to a non-oriented attachment (Fig. 5). The choice of

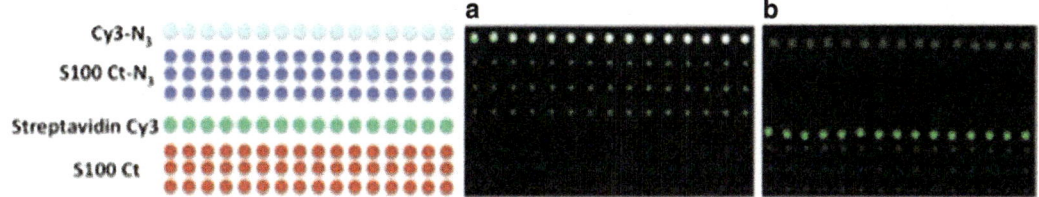

Fig. 5 Spotting scheme and fluorescence images of the arrays obtained on (**a**) a slide coated with poly(DMA-PMA-MAPS) which contains the alkyne functionality and (**b**) on a slide coated with poly(DMA-NAS-MAPS) which contains *N*-acryloyloxysuccinimide active ester as functional group. Azido-modified [13] cyanine3 (Cy3-N3) and cyanine 3-streptavidin have been immobilized as references for facilitating the image analysis. Slides were analyzed using 60 % laser power and 60 % of photomultiplier gain (PMT)

using a Si–SiO$_2$ supports is due to the fluorescence-enhancing properties of this layered material [11], but the coating can be performed also on different supports (glass, nitrocellulose, gold, plastics, etc.).

In particular we have immobilized the azido-modified and the non-modified peptide onto a surface coated with poly(DMA-PMA-MAPS), which contains the alkyne functionality (Fig. 5a) and, as a comparison, on a slide modified with poly(DMA-NAS-MAPS) [7], which contains *N*-acryloyloxysuccinimide active ester as functional group (Fig. 5b). Figure 5a shows the signal obtained, after incubation with fluorescently labeled antibodies against protein S100B, only where the azido-modified peptide (S100-Ct-N$_3$) was immobilized, demonstrating the specificity of the attachment. In fact, almost no signal is obtained where the non-modified S100-Ct peptide is immobilized. On the other hand, on the surface which contains *N*-acryloyloxysuccinimide as active group (Fig. 5b), a fluorescence signal is obtained only where the non-modified peptide has been immobilized on the surface, with a lower fluorescence intensity if compared to the S100-Ct-N$_3$ peptide (Fig. 6), thus demonstrating the importance of a proper orientation for facilitating the probe recognition and interaction with the target antibody.

Another advantage of using this method is the rapidity of the *click reaction*: the immobilization of the azido-modified peptide requires only 30 min in a humid chamber, instead of the overnight incubation required by the nucleophile substitution which takes place on the slides coated with poly(DMA-NAS-MAPS).

2   Materials

2.1   Synthesis of Poly(DMA-PMA-MAPS)

1. *N,N*-Dimethylacrylamide (DMA) filtered on aluminum oxide: fill a small chromatography column with aluminum oxide for 1/3 of its height, then pour 20 mL of DMA on top of the

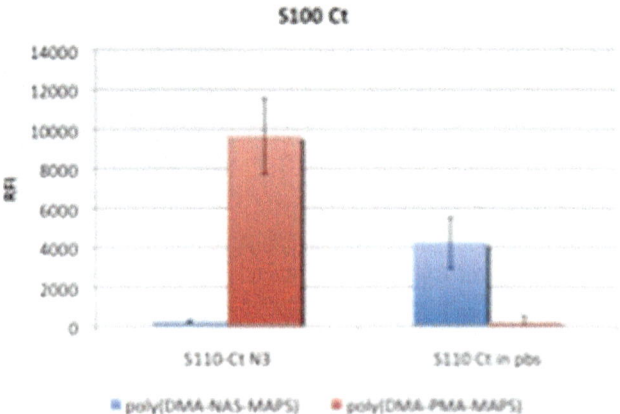

Fig. 6 Fluorescence intensity signals obtained after incubation of the arrays with fluorescently labeled antibodies against S100B

aluminum oxide, and elute it by applying positive pressure from the top of the column (*see* **Note 1**).

2. 3-(Trimethoxysilyl)propyl methacrylate (MAPS) (*see* **Note 2**).

3. 3-Trimethylsilanyl-prop-2-yn methacrylate (PMA) synthesized as reported by Ladmiral et al. [12] (*see* **Note 3**).

4. Anhydrous tetrahydrofuran (THF).

5. α,α′-Azoisobutyronitrile (AIBN): the final concentration of the initiator into the reaction mixture is 2 mM (*see* **Note 4**).

6. Helium connection.

7. Petroleum ether.

8. Potassium carbonate (K_2CO_3).

9. Dialysis tubes, cut-off 14 kDa.

2.2 Synthesis of S100-Ct and S100 Ct N$_3$ Peptide

1. Dichloromethane (DCM).

2. *N,N*-Dimethylformamide (DMF, peptide synthesis grade).

3. Methanol (MeOH).

4. Diethyl ether (Et_2O).

5. Acetonitrile (ACN, HPLC grade).

6. Deionized water (Milli-Q).

7. Standard Fmoc-protected amino acids: Fmoc-L-Glu(O*t*Bu)-OH, Fmoc-L-His(Trt)-OH, and Fmoc-L-Phe-OH.

8. Unnatural Fmoc-protected building blocks: Fmoc-L-Dap(N_3)-OH (For S100 Ct N_3 only), Fmoc-O$_2$Oc-OH, and Fmoc-βAla-OH.

9. 2-Chlorotritylchloride resin (2-CTC), loading: 1.22 meq/g.

10. Acetic anhydride (Ac₂O) 0.3 M in DMF (50 mL): add 1.42 mL Ac₂O to a 50 mL graduated cylinder and add DMF to 50 mL. Transfer to a glass reagent bottle and mix.

11. Ethyl (hydroxyimino)cyanoacetate (Oxyma) 0.5 M in DMF (50 mL): weigh 3.55 g Oxyma and transfer to a 50 mL graduated cylinder, and then add DMF to 40 mL. Mix to complete dissolution. Make up to 50 mL with DMF.

12. *N*,*N*′-Diisopropylcarbodiimide (DIC) 0.5 M in DMF (50 mL): add 3.87 mL DIC to a 50 mL graduated cylinder. Add DMF to 50 mL. Transfer to a glass reagent bottle and mix.

13. *N*,*N*-Diisopropylethylamine (DIEA) 0.3 M in DMF (50 mL): add 2.61 mL DIEA to a 50 mL graduated cylinder and add DMF to 50 mL. Transfer to a glass reagent bottle and mix.

14. Piperidine (PIP) 20 % in DMF (1 L): add 200 mL piperidine to a 1 L graduated cylinder. Make up to 1 L with DMF. Transfer to a 1 L glass bottle and mix (*see* **Note 5**).

15. Thioanisole.

16. Triisopropylsilane (TIS).

17. Trifluoroacetic acid (TFA) cleavage cocktail containing triisopropylsilane (TIS) and thioanisole (TFA/TIS/thioanisole/water = 90/5/2.5/2.5 (vol/vol/vol/vol)): mix 9 mL ice-cold TFA, 0.5 mL TIS, 0.25 mL deionized water, and 0.25 mL thioanisole in a 15 mL disposable polypropylene tube to prepare 10 mL of cleavage cocktail (*see* **Note 6**).

18. Analytical and semipreparative HPLC system (Jasco, Diode array UV–VIS detector).

19. Analytical and semipreparative HPLC reversed-phase columns (C-18).

20. HPLC buffer A (Milli-Q grade water containing 0.1 % TFA (vol/vol)): add 1 mL TFA to 1 L Milli-Q grade water and mix (*see* **Note 5**). Degas with helium for 10 min before use.

21. HPLC buffer B (70 % aqueous ACN containing 0.1 % TFA (vol/vol)): add 300 mL Milli-Q grade water to 700 mL HPLC grade ACN, then add 1 mL of TFA, and mix. The solution can be stored for up to 1 week at room temperature (*see* **Note 5**). Degas with helium for 10 min before use.

2.3 Coating of Silicon Oxide Slides with Poly(DMA-PMA-MAPS) and Poly(DMA-NAS-MAPS)

1. HARRICK Plasma Cleaner, PDC-002 (Ithaca, NY, USA).

2. Ammonium sulfate solution at 40 % saturation level: dissolve 242 g of ammonium sulfate in 1 L of DI water and stir to complete dissolution. Filter the obtained solution on a 0.45 μm membrane under vacuum (*see* **Note 6**).

3. Poly(DMA-PMA-MAPS) synthesized as described in Subheading 3.1.

4. Poly(DMA-NAS-MAPS) synthesized as reported elsewhere [13].

5. Coating solution: 1 % w/v of poly(DMA-PMA-MAPS) in an aqueous solution of 20 % ammonium sulfate (*see* **Note 7**).

6. Vacuum oven.

2.4 Immobilization of the Azido-Modified Probes on the Coated Slides

1. Azido-modified S100 Ct peptide (S100 Ct-N_3), synthesized as reported in Subheading 3.3.

2. Azido-modified cyanine 3 (Cy3-N_3).

3. Sodium phosphate buffer 50 mM pH 7.4: dissolve 1.5 g of sodium phosphate monobasic monohydrate ($NaH_2PO_4 \cdot H_2O$) and 2.5 g of sodium phosphate dibasic dihydrate ($Na_2HPO_4 \cdot 2H_2O$) in 500 mL of DI water.

4. Stock solution containing copper sulfate pentahydrate ($Cu_2SO_4 \cdot 5H_2O$) 2 mM and tris[(1-benzyl-1H-1,2,3-triazol-4-yl)methyl]amine, (THPTA) 8 mM in DI water.

5. Ascorbic acid stock solution 150 mM in sodium phosphate 50 mM pH 7.4.

6. Piezoelectric spotter SciFlex Arrayer S5 (Berlin, Germany).

2.5 Immobilization of the Non-azido-Modified Probes on the Coated Slides

1. Non-azido-modified S100 Ct peptide (S100 Ct), synthesized as reported in Subheading 3.3.

2. Cyanine 3-labeled streptavidin (streptavidin Cy3).

3. Phosphate saline buffer (PBS) 1×.

4. Piezoelectric spotter SciFlex Arrayer S5 (Berlin, Germany).

2.6 Blocking of Poly(DMA-NAS-MAPS) Functional Groups

The active ester of poly(DMA-NAS-MAPS) must be blocked before proceeding with the antibody incubation in order to avoid aspecific binding.

1. Blocking solution composed of ethanolamine 50 mM in tris/HCl 0.1 M pH 9.

2. DI water.

2.7 Slide Incubation and Fluorescence Evaluation

1. Sodium phosphate buffer 50 mM pH 7.4: dissolve 1.5 g of sodium phosphate monobasic monohydrate ($NaH_2PO_4 \cdot H_2O$) and 2.5 g of sodium phosphate dibasic dihydrate ($Na_2HPO_4 \cdot 2H_2O$) in 500 mL of DI water.

2. Antibody anti-S100B raised in rabbits as described elsewhere [9] (Primm srl, Milan, Italy), fluorescently labeled with kit Cy3 monoreactive dye pack from GE Healthcare (Little Chalfont, UK).

3. Incubation buffer composed of: tris/HCl 50 mM pH 7.6, NaCl 150 mM, Tween 20 0.02 %, and BSA 1 % w/v.

4. Washing buffer composed of tris/HCl 50 mM pH 9, NaCl 250 mM, and Tween 20 0.05 %.

5. Confocal laser scanner (ProScanArray, PerkinElmer, Boston, MA).

3 Methods

3.1 Synthesis of Poly(DMA-PMA-MAPS)

The copolymer, made of DMA (with 97 % molar percentage), PMA (2 % molar percentage), and 3-(trimethoxysilyl)propyl methacrylate (MAPS, 1 % molar percentage), is synthesized by free-radical copolymerization. The concentration of the monomer feed in the solvent is 20 % w/v.

The synthesis process is divided into two steps: (a) synthesis of protected polymer which contains 3-trimethylsilanyl-prop-2-yn methacrylate and (b) removal of the protective trimethylsilane groups.

3.1.1 Synthesis of Protected Polymer

1. In a 250 mL three neck round-bottom flask, equipped with condenser, magnetic stirrer, and helium connection, degas 50.00 mL of anhydrous tetrahydrofuran (THF) by purging helium for 15 min (*see* **Note 8**).

2. Add *N,N*-dimethylacrylamide (DMA, 9.42 g, 9.5×10^{-2} mol; *see* **Note 1**), PMA (0.385 g, 1.96×10^3 mol), and the initiator α,α'-azoisobutyronitrile (AIBN, 1.64×10^{-3} g, 1.00×10^{-4} mol; *see* **Note 4**) into the flask, continuing purging helium (*see* **Note 9**).

3. Add 3-(trimethoxysilyl)propyl methacrylate (MAPS, 2.43×10^{-1} g, 9.75×10^{-4} mol; *see* **Note 2**) using a syringe.

4. Remove the gas inlet and heat the solution to 65 °C for 2 h, while stirring, under helium atmosphere (*see* **Note 10**).

5. After the polymerization is completed, dilute 1:1 by adding 20.00 mL of anhydrous THF and stir for 5 min.

6. Precipitate the polymer by slowly dripping (*see* **Note 11**) the reaction mixture into a large excess of petroleum ether (about 1:10 by volume), while stirring. A white precipitate is obtained.

7. Filter the obtained white powder on a Buchner funnel.

8. Dry the copolymer under a vacuum for 1–2 h at room temperature.

9. Store it at –20 °C in a dry environment until the deprotection step.

3.1.2 Removal of the Protective Trimethylsilane Groups

1. Dissolve the obtained polymer in 300 mL of DI water (*see* **Note 12**).

2. Add 380 mg of potassium carbonate (K_2CO_3) to obtain a 9 mM solution, pH 9.

3. Stir the solution at room temperature for 1 h.

4. Dialyze the polymer against water for 3 days (*see* **Note 13**).

5. Lyophilize the polymer (*see* **Note 14**).

3.2 Coating of Silicon Chips with Poly(DMA-PMA-MAPS) or Poly(DMA-NAS- MAPS)

The coating of silicon oxide (*see* **Note 15**) substrates requires two steps: (a) surface pretreatment and (b) adsorption of the copolymer.

3.2.1 Surface Pretreatment

1. The substrates are cleaned and activated by pretreatment with oxygen plasma for 10 min (high radio frequency).

3.2.2 Adsorption of Poly(DMA-PMA-MAPS) or Poly(DMA-NAS-MAPS) to the Substrates

1. Dissolve the coating polymer (poly(DMA-PMA-MAPS) or poly(DMA-NAS-MAPS)) to a final concentration of 1 % w/v in an ammonium sulfate aqueous solution 20 % w/v. Depending on the volume of the chamber and the number of the slides to be coated, weigh the exact amount of copolymer to obtain a water solution at a final concentration of 2 % w/v. When the copolymer is completely dissolved, dilute it 1:1 with the stock solution of ammonium sulfate at a 40 % saturation level. For example, to prepare 10 mL of coating solution, weigh 100 mg of poly(DMA-PMA-MAPS) and add 5 mL of DI water. Stir to complete dissolution and then dilute 1:1 by adding 5 mL of ammonium sulfate solution at 40 % saturation level (*see* **Note 16**).

2. Immerse the substrates into the coating solution for 30 min at room temperature in a plastic chamber (*see* **Note 17**).

3. Wash the substrates vigorously with DI water to remove the excess of the copolymer on the surface.

4. Dry the surfaces with nitrogen stream to avoid stain formation.

5. Cure the slides in a vacuum oven at 80 °C for 15 min (*see* **Note 18**).

6. Store slides at room temperature (*see* **Note 19**).

3.3 Synthesis of S100-Ct and S100-Ct N₃ Peptides

The synthesis of the S100 Ct and S100 Ct-N₃ peptides is divided into four main steps: (a) resin loading, (b) sequence extension, (c) cleavage and analysis, and (d) purification. Carry out all procedures at room temperature (20–25 °C) unless otherwise specified.

3.3.1 Resin Loading

1. Weigh 200 mg 2-CTC resin (1.22 mmol/g) into a peptide synthesis vessel.

2. Add 4 mL DCM to the resin and gently agitate for 5 min. Drain the reaction vessel by applying a gentle nitrogen stream or by vacuum filtration.

3. Add 4 mL DMF to the resin and gently agitate for 10 s. Drain the reaction vessel by applying a gentle nitrogen stream or by vacuum filtration.

4. Repeat **step 3** three more times.

5. Add 4 mL DCM to the resin and gently agitate for 20 min. Drain the reaction vessel by applying a gentle nitrogen stream or by vacuum filtration.

6. Weigh 0.1 mmol (44.4 mg) Fmoc-Glu(tBu)-OH into a 5 mL disposable plastic vial. Add 1 mL DMF and 2 mL DCM. Add 87.1 μL DIEA and gently vortex until complete dissolution.

7. Add resin loading solution from **step 6** to the reaction vessel. Plug reaction vessel and gently agitate for 60 min. Drain the reaction vessel by applying a gentle nitrogen stream or by vacuum filtration.

8. Wash the resin by repeating **step 3** for five times.

9. Add 4 mL 10 % MeOH/DCM (vol/vol) to the resin. Gently agitate for 15 min. Drain the reaction vessel by applying a gentle nitrogen stream or by vacuum filtration.

10. Repeat **step 2** three times.

3.3.2 Sequence Extension

1. Add 4 mL 20 % PIP in DMF to the resin. Plug reaction vessel and gently agitate for 3 min. Drain the reaction vessel by applying a gentle nitrogen stream or by vacuum filtration.

2. Add 4 mL 20 % PIP in DMF to the resin. Plug reaction vessel and gently agitate for 5 min. Drain the reaction vessel by applying a gentle nitrogen stream or by vacuum filtration.

3. Add 4 mL DMF to the resin and gently agitate for 10 s. Drain the reaction vessel by applying a gentle nitrogen stream or by vacuum filtration.

4. Wash the resin by repeating **step 3** for five times.

5. Weigh 0.5 mmol entering Fmoc-protected amino acid in a 3 mL disposable plastic vial. Add 1 mL Oxyma 0.5 M solution and gently mix to assist preliminary dissolution. Add 1 mL DIC 0.5 M solution and gently mix. Keep mixing until complete dissolution of Fmoc-protected amino acid.

6. Add coupling solution from **step 5** to the reaction vessel. Add 1 mL DMF and 1 mL DCM to the reaction vessel. Plug reaction vessel and gently agitate for 60 min. Drain the reaction vessel by applying a gentle nitrogen stream or by vacuum filtration.

7. Wash the resin by repeating **step 3** for five times.

8. Repeat **steps 1–7** iteratively until full peptide sequence (–HEFFEHE) is assembled (*see* **Note 20**).

9. Repeat **steps 1–4**.

10. Couple Fmoc-βAla-OH performing **steps 5–7** followed by **steps 1–4**.

11. Couple Fmoc-O₂Oc-OH performing **steps 5–7** followed by **steps 1–4**.

12. For the synthesis of peptide S100 Ct, proceed directly to **step 15**.

13. For the synthesis of peptide S100 Ct N$_3$, couple Fmoc-Dap(N$_3$)-OH according to **steps 5–7** followed by **steps 1–4** and proceed to **step 14**.

14. Add 1.5 mL Ac$_2$O 0.3 M and 1.5 mL DIEA 0.3 M to the reaction vessel. Gently agitate for 10 min. Drain the reaction vessel by applying a gentle nitrogen stream or by vacuum filtration.

15. Add 4 mL DCM to the resin and gently agitate for 10 s. Drain the reaction vessel by applying a gentle nitrogen stream or by vacuum filtration.

16. Repeat **step 15** for four times.

17. Dry resin under a gentle nitrogen stream or under vacuum (*see* **Note 21**).

3.3.3 TFA Cleavage and Analysis

1. Transfer resin-bound peptide into a fritted peptide synthesis vessel. Insert a magnetic stirrer.

2. Add 5 mL ice-cold TFA cocktail solution to the resin. Gently stir mixture for 2.5 h.

3. Filter cleavage mixture into a 15 mL centrifuge tube. If necessary, apply a gentle pressure by nitrogen stream to favor resin filtration.

4. Rinse filtered resin with 3 mL TFA. Repeat filtration.

5. Repeat **step 4**.

6. Drop filtered cleavage solution into 30 mL ice-cold Et$_2$O under gentle stirring. Make up to 50 mL with ice-cold Et$_2$O. Let the mixture stand at 0 °C for 10 min (*see* **Note 22**).

7. Transfer ether mixture into a 50 mL centrifuge tube. Centrifuge at $560 \times g$ for 2 min. Decant the supernatant from the centrifuge tube carefully.

8. Add 20 mL cold Et$_2$O to the collected residue. Resuspend residue by careful vortexing.

9. Centrifuge at $560 \times g$ for 2 min. Decant the supernatant from the centrifuge tube carefully.

10. Repeat **steps 8–9**.

11. Dry collected product under gentle nitrogen stream for 5 min. Store the dried crude peptide at –20 °C until analysis or purification.

12. Dissolve crude peptide in 50 % HPLC buffer obtained by mixing HPLC buffer A and B mixture.

13. Blow dissolved sample with nitrogen to evaporate residual Et$_2$O.

14. Filter crude peptide solution through a 0.45 mm syringe filter (cellulose acetate membrane); if necessary, apply gentle nitrogen pressure to assist filtration.

15. Inject 10–20 μL into the analytical HPLC system; recommended gradient: 90 % A to 30 % A over 30 min, flow rate 1 mL/min, and UV–VIS detection range 220–340 nm.

3.3.4 Purification

1. Condition semipreparative HPLC column in 100 % HPLC buffer A.

2. Transfer dissolved peptide into a polypropylene syringe equipped with HPLC needle.

3. Inject peptide solution (max 4 mL per time) into the HPLC system. Pump HPLC buffer through the system for 3 min (14 mL/min) to allow full sample loading on the top of HPLC column.

4. Repeat **step 3** until all peptide solution is loaded.

5. Run gradient: 100 % A to 20 % A over 44 min (recommended), flow rate 14 mL/min, and detection wavelength 235 nm, and collect peak in 7 mL fractions.

6. Inject collected peak fractions into the analytical system to assess purity.

7. Combine pure fractions (>95 %) into a round-bottom flask. Concentrate under vacuum and transfer to tared glass vials. Freeze vials and store at –20 °C until freeze-drying step.

8. Freeze-dry samples (*see* **Note 23**).

3.4 Azido-Modified Probe Immobilization on Coated Slides

1. Dissolve the azido-modified peptide (S100 Ct-N_3) in sodium phosphate buffer 50 mM pH 7.4 to obtain a peptide stock solution of 2 mg/mL. Apply the same procedure to prepare the reference solution of Cy3-N_3.

2. Take 10 μL of the peptide (or the Cy3-N_3) stock solution and add 8.2 μL of sodium phosphate buffer pH 7.4.

3. Add 1 μL of the copper sulfate pentahydrate stock solution containing THPTA.

4. Add 0.8 μL of ascorbic acid stock solution.

5. Stir the peptide printing solution.

6. Place the peptide printing solution in a 96-plate well (*see* **Note 24**).

7. Spot the peptide printing solution as shown in Fig. 5, using a piezoelectric spotter.

8. Place the spotted slides in a humid chamber: 30 min for poly(DMA-PMA-MAPS)-coated slides or overnight for poly(DMA-NAS-MAPS)-coated slides.

3.5 Non-azido-Modified Probe Immobilization on Coated Slides

1. Dissolve the non-azido-modified probe (S100 Ct) in PBS to a final concentration of 1 mg/mL. Apply the same procedure to prepare the reference Streptavidin Cy3 solution.

2. Stir the printing solution containing the probe.

3. Place the solution in a 96-plate well (*see* **Note 24**).

4. Spot the peptide printing solution as shown in Fig. 5.

5. Place the spotted slides in a humid chamber: 30 min for poly(DMA-PMA-MAPS)-coated slides or overnight for poly(DMA-NAS-MAPS)-coated slides.

3.6 Blocking of Poly(DMA-NAS-MAPS) Functional Groups

1. After the overnight incubation in the humid chamber, immerse the printed slides (coated with poly(DMA-NAS-MAPS)) in the blocking solution.

2. Leave the slides to stand in the blocking solution for 1 h.

3. Rinse each slide with DI water.

4. Dry the slides with a nitrogen stream.

3.7 Slide Incubation and Fluorescence Evaluation

1. Wash the spotted slides with sodium phosphate buffer for 10 min placing the slides into a container on a shaker (*see* **Note 25**).

2. Dissolve the Cy3-labeled antibody against S100B in the incubation buffer to a final concentration of 1 μg/mL (*see* **Note 26**).

3. Place the printed slides in a petri dish and cover them with the incubation buffer containing the fluorescent antibody.

4. Incubate the slides for 2 h on a shaker at room temperature (*see* **Note 27**).

5. Rinse the slides by immersing them in washing buffer and shaking for 10 min at room temperature (*see* **Note 27**).

6. Dry each slide with a gentle nitrogen stream.

7. Analyze the fluorescence signal using a confocal laser scanner.

4 Notes

1. A larger volume of DMA can be filtered and then stored in a glass vial for no more than a week at 2–4 °C.

2. MAPS is an air-sensitive reagent. It is preferable to use a sealed bottle and take it using a syringe under inert atmosphere.

3. The acryloyl-alkyne monomer is used in the free-radical polymerization protected with a trimethylsilane group (TMS) because of the instability of the triple bond under radical conditions. The TMS moiety is removed after polymerization under mild basic conditions.

4. To avoid weighting small quantities of AIBN, a stock solution can be prepared as well. Prepare a 200 mM solution by

dissolving AIBN into dry THF. Store it at $-20\ °C$ under inert atmosphere.

5. The solution can be stored up to 1 month at room temperature.

6. Store at $2\text{–}4\ °C$.

7. Prepare the coating solution immediately before the coating procedure. Use it within 30 min. Do not store or recycle the coating solution.

8. Use a needle to purge the gas directly into the reaction mixture, with a pressure of about 0.5 bar. The entire system has to be perfectly sealed, so regulate the gas stream to have the maximum flow without having gas leakage.

9. Dissolve small quantity reagents (such as PMA and MAPS) in a small volume of anhydrous THF (1 mL) and add them using a syringe in order to keep the reaction environment dry and inert.

10. Fill a small balloon with helium and place it on top of the condenser, so to keep an inert atmosphere without the need of purging the reaction flask for 2 h.

11. Use a Pasteur pipette to drip the reaction mixture into the petroleum ether. Do not pour the entire crude material all at once.

12. The final concentration of the polymer for the deprotection step is about 3 % w/v.

13. Dialyze the polymer against DI water changing water at least nine times.

14. Store the obtained white powder in dry condition at $-20\ °C$.

15. Other materials such as glass, nitrocellulose, gold, etc. can be coated as well, following the same procedure.

16. *See* Subheading 2.3, **item 2**, in the Materials section to prepare the ammonium sulfate solution at 40 % saturation level.

17. Use preferably a plastic container as glass ones would be coated as well, competing with the surface of the slides that are intended to be coated.

18. Curing step under vacuum is fundamental to assure the formation of covalent bonds between surface silanols and silane reagents.

19. Slides coated with functional groups are best stored in dry conditions in a desiccator, at room temperature and used up to 4 weeks.

20. At the end of every amino acid coupling cycle, the synthesis can be interrupted and the resin-bound mixture can be stored in DMF overnight at room temperature.

21. Store dried resin in a sealed container at –20 °C until TFA cleavage step.

22. The formation of a white precipitate or suspension should be observed at this stage.

23. A white, fluffy solid should be obtained.

24. Centrifuge the plate to remove any air bubble.

25. This step is required only for those slides coated with poly(DMA-PMA-MAPS). The slides coated with poly(DMA-NAS-MAPS) can be used immediately after the blocking step without any further rinsing.

26. Keep it in a dark place to avoid fluorescence quenching.

27. During the incubation, final washing step, and fluorescence analysis, keep the slides in a dark place to avoid fluorescence quenching.

References

1. Fu J, Reinhold J, Woodbury NW (2011) Peptide-modified surfaces for enzyme immobilization. PLoS One 6:e18692

2. Rao SV, Anderson KW, Bachas LG (1998) Oriented immobilization of proteins. Microkim Acta 128:127–143

3. Van Berkel SS, Van Eldijk MB, Van Hest JCM (2011) Staudinger ligation as a method for bioconjugation. Angew Chem Int Ed 50:8806–8827

4. Jung Y, Min Lee J, Kim J, Yoon J, Cho H, Chung BH (2009) Photoactivable antibody binding protein: site-selective and covalent coupling of antibody. Anal Chem 81:936–942

5. Liang L, Astruc D (2011) The copper(I)-catalyzed alkyne-azide cycloaddition (CuAAC) "click" reaction and its applications. An overview. Coordin Chem Rev 255:2933–2945

6. Huisgen R (1989) Kinetics and reaction mechanisms: selected examples from the experience of forty years. Pure Appl Chem 61:613–628

7. Pirri G, Damin F, Chiari M, Bontempi E, Depero LE (2004) Characterization of a polymeric adsorbed coating for DNA microarray glass slides. Anal Chem 76:1352–1358

8. Zilio C, Bernardi A, Palmioli A, Salina M, Tagliabue G, Buscaglia M, Consonni R, Chiari M (2015) New "clickable" polymeric coating for glycan microarrays. Sensor Actuator B 215: 412–420

9. Uttamapinant C, Tangpeerachaikul A, Grecian S, Clarke S, Singh U, Slade P, Gee KR, Ting AY (2012) Fast, cell-compatible click chemistry with copper-chelating azides for biomolecular labeling. Angew Chem Int Ed 51: 5852–5856

10. Peri C, Gagni P, Combi F, Gori A, Chiari M, Longhi R, Cretich M, Colombo G (2013) Rational epitope design for protein targeting. ACS Chem Biol 8:397–404

11. Cretich M, Monroe MR, Reddington A, Zhang X, Daaboul GG, Damin F, Sola L, Unlu SM, Chiari M (2012) Interferometric silicon biochips for label and label-free DNA and protein microarrays. Proteomics 12:2963–2977

12. Ladmiral V, Mantovani G, Clarkson GJ, Cauet S, Irwin JL, Haddleton DM (2006) Synthesis of neoglycopolymers by a combination of "click chemistry" and living radical polymerization. J Am Chem Soc 128:4823–4830

13. Sola L, Chiari M (2012) Modulation of electroosmotic flow in capillary electrophoresis using functional polymer coatings. J Chrom A 1270:324–329

Chapter 14

Oriented Peptide Immobilization on Microspheres

Lisa C. Shriver-Lake, George P. Anderson, and Chris R. Taitt

Abstract

Reproducible immobilization of peptides and proteins on microsphere surfaces is a critical factor for optimal sensitivity and selectivity in bead-based assays. However, peptides with unusually large numbers of lysine residues—whose amines are targeted in the most common microsphere immobilization chemistries—may be particularly challenging to use in bead-based arrays, as they may lose activity through multipoint attachments and incorrect presentation. For this reason, it is imperative to achieve site-directed attachment chemistry, such that a single site of attachment provides reproducibly oriented peptides on the microsphere surface. This can be achieved by inserting a unique targetable residue, such as a cysteine. Here, we present methods for attaching cysteine-containing peptides to standard carboxy-functionalized microsphere surfaces using thiol- rather than amine-directed chemistries. We show that the presence of a cationic detergent (CTAB) and a "passivating" agent such as β-mercaptoethanol facilitates improved bead recovery after peptide immobilization and may enhance functionality of the attached peptides.

Key words Antimicrobial peptide, Bioimmobilization, Bead-based flow cytometry, Luminex

1 Introduction

Since the first description of an antibody array determining the presence and types of cellular antigens [1], protein and peptide microarrays have found a wide variety of applications. Peptide microarrays, in particular, have been applied to diagnostics [2], enzyme profiling [3–5], epitope mapping for antibody characterization [6], vaccine development [7], biomolecular interaction and ligand binding analyses [8, 9], biomarker discovery [10, 11], biomolecular and cellular patterning [12], and biosensing [13–17]. Platforms such as the Luminex xMAP, the Luminex MAGPIX, and the Becton Dickinson FACSArray can accomplish vastly multiplexed analyses using suspension bead arrays, much in the same manner as planar spotted arrays. The Luminex and the FACSArray

Marina Cretich and Marcella Chiari (eds.), *Peptide Microarrays: Methods and Protocols*, Methods in Molecular Biology, vol. 1352, DOI 10.1007/978-1-4939-3037-1_14, © Springer Science+Business Media New York 2016

platforms use mixtures of microspheres, each of which is coated with a different capturing species (e.g., antibody, peptide, nucleic acid probe) or enzyme substrate (e.g., peptide, carbohydrate). Using flow cytometry (xMAP, FACSArray) or fluorescence imaging (MAGPIX), each bead is identified—typically by the quantity of two or more fluorescent dyes within the microspheres—and then the signal intensity of a reporter fluorophore is used to quantify binding or other activity associated with the individual microsphere (e.g., cleavage, catalysis of surface-immobilized peptides/proteins). By combining different sets of microspheres—each with a different surface functionality—users can tailor multiplexed tests to suit their needs.

Unfortunately, while peptide arrays on planar surfaces have been widely applied for a variety of purposes, descriptions of analogous peptide arrays using suspensions of microspheres are much more limited. One potential reason is that the immobilization chemistries used on planar surfaces may be challenging to adapt for attachment of peptides to microspheres, given the limited number of surface functionalities available on Luminex- or FACSArray-compatible microspheres.

Although Luminex and FACSArray microspheres utilize somewhat different immobilization strategies (carbodiimide- and succinimidyl 4-(N-maleimidomethyl)cyclohexane-1-carboxylate [SMCC]-mediated linkages, respectively), linkage of biomolecules to both bead-based platforms is directed against biomolecular amine groups. While these amine-targeted attachment chemistries may be highly effective for antibodies and proteins with relatively low lysine content (~7 %) [18], these same chemistries may not be suitable for cationic peptides with high lysine content. We have observed that the high immobilization efficiencies of amine-directed chemistries can result in inactive or ineffectively displayed peptides on surfaces [19].

As a result, only a limited number of attachment strategies of peptides to suspension array microspheres have been described. Carbodiimide-mediated attachment has been successfully used to attach peptide–aminocellulose conjugates directly to carboxy-functionalized microspheres [20]. While the peptides attached in this manner properly displayed their epitopic sites, this strategy may be limited in its utility for longer, more lysine-rich peptides. In the same study, Heuback also attached peptides in the opposite orientation with N-terminal cysteine residues to microspheres coated with bovine serum albumin via the heterobifunctional cross-linker, sulfo-succinimidyl 4-(p-maleimidophenyl)butyrate [SMPB]; epitope presentation was poor when peptides were immobilized in this orientation. A number of other groups have utilized the high-affinity avidin–biotin interaction to link biotinylated peptides non-covalently to avidin-coated microspheres [21, 22]. This method can be highly effective and result in highly reproducible, oriented immobilization if the biotin is incorporated at a unique location on the peptide. In our own studies using antimicrobial peptides for detection of bacterial cells, surface attachment using avidin–biotin interactions resulted in suboptimal binding efficiency of the immobilized peptides [15]. Furthermore, this strategy will confound any assays utilizing the same avidin–biotin interaction for detection.

In developing bead-assays using arrays of cationic α-helical antimicrobial peptides (AMPs), a significant challenge has been poor recovery of the microspheres in the detection region after peptide immobilization, likely resulting from electrostatic interactions between the negatively charged microspheres and highly cationic peptides. Furthermore, the Luminex and other flow cytometry-based systems can bias detection against aggregates. While bead aggregates may not themselves be lost, signals acquired from bead aggregates are outside of the appropriate detection window, rendering them unusable by the instrumentation; this may result in both inaccurate bead counts and artifactual results in actual assays. An additional challenge has been presentation of the peptides on the microspheres in a reproducible and predictable orientation, such that the binding of bacterial target species is observed. To mitigate these challenges, we have functionalized xMAP microspheres with maleimidyl groups via incorporation of a poly(ethylene glycol) (PEG)-based linker. This intermediate layer serves to provide a "passivating" functionality to prevent nonspecific adsorption and also to display a terminal maleimide that targets a unique cysteine engineered on several AMPs. The resulting peptide-functionalized microspheres can then be used to bind and detect bacteria.

Here, we describe covalent immobilization of the cationic α-helical AMP engineered to possess a C-terminal cysteine onto MagPlex microspheres and use of these peptide microspheres to detect heat-killed *Escherichia coli*. Commercial carboxy-terminated microspheres are first activated with a 1-ethyl-3-(3-dimethylaminopropyl)-carbodiimide(EDC)/ *N*-hydroxysuccinimidyl ester (NHS) and then reacted with maleimide–PEG (600)–amine to yield a maleimide-functionalized surface (Fig. 1). Cysteine-containing peptides are then attached (via the surface maleimides) and can then be used to detect target bacteria in binding assays. Inclusion of a cationic surfactant and additional surface "passivating" species can improve both bead recoveries and binding function of the attached peptides.

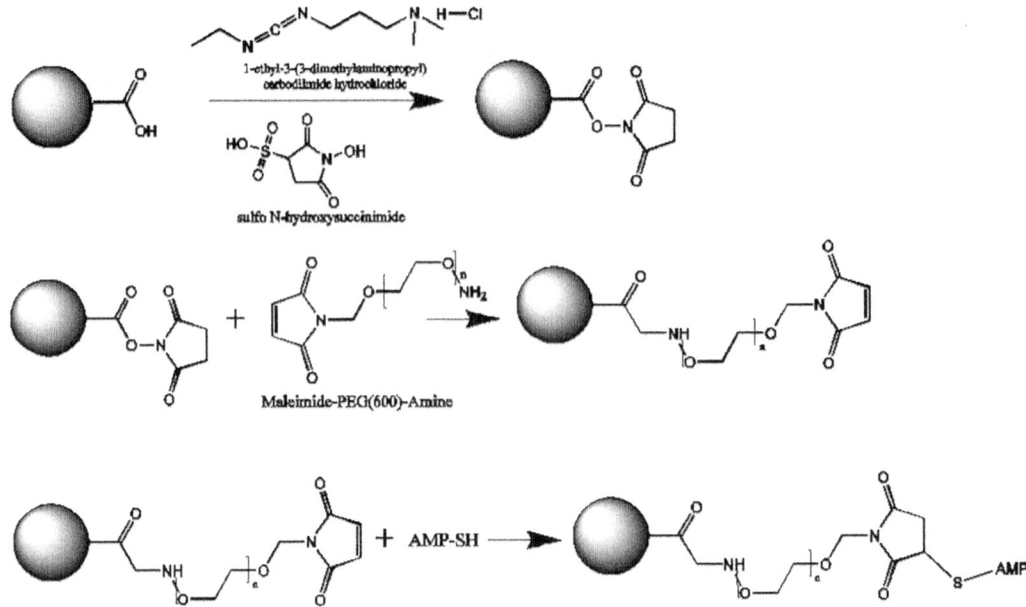

Fig. 1 Schematic of attachment. Washed microspheres are first activated with EDC and sulfo-NHS (top reaction) and then linked to maleimide–PEG (600)–amine (middle reaction), yielding a maleimide-decorated surface. An antimicrobial peptide possessing a unique cysteine (AMP-SH) is then linked to the maleimide-functionalized microsphere (bottom reaction)

2   Materials

2.1   Attachment of Thiol-Containing Molecules to COOH Microspheres

1. Microspheres: COOH-functionalized, xMAP-compatible microspheres (e.g., MagPlex) can be purchased from the following vendors: Luminex (Austin, TX, USA), BioRad (Hercules, CA, USA), Spherotech (Lake Forest, IL, USA), Invitrogen/Life Technologies (Grand Island, NY, USA), and Sigma-Aldrich (St. Louis, MO, USA) (*see* **Note 1**).

2. Magnetic holder: LifeSep 1.5S single-tube magnetic holder (Bangs Laboratories, Fishers, IN, USA) or similar magnetic device.

3. Microfuge tubes—1.8 mL.

4. Phosphate buffer: 0.1 M sodium phosphate buffer, pH 6.

5. EDC/sulfo-NHS solution: Combine equal volumes of EDC solution and the sulfo-NHS solution (below). This solution must be used within 2 h of preparation.

    (a) EDC solution: 50 mg/mL 1-ethyl-3-[3-dimethylamino-propyl]carbodiimide hydrochloride (EDC, Thermo Scientific, Rockford, IL, USA) in phosphate buffer (*see* **Notes 2** and **3**).

(b) Sulfo-NHS solution: 50 mg/mL *N*-hydroxysulfosuc-cinimide (sulfo-NHS, Thermo) in phosphate buffer. Prepare the same volume of solution as used for EDC solution above.

6. Maleimide–PEG (600)–amine solution: 10 mg/mL maleimide–PEG (600)–amine (catalog no. PG2-AMML-600; NANOCS, New York, NY, USA) in phosphate buffer (*see* **Notes 4** and **5**).

7. Tricine/CTAB buffer: 0.1 M Tricine, pH 7.4, with 0.05 % cet-yltrimethylammonium bromide (CTAB; Acros, Thermo).

8. Peptides: Four antimicrobial peptides were custom synthesized with a C-terminal cysteine and were purified to 90 % purity by Biosynthesis, Inc. (Lewisville, TX, USA). The peptide sequences are reported in Table 1. Peptide solutions: 25 μg peptide in 100 μL Tricine/CTAB buffer (*see* **Note 6**).

9. Mercaptoethanol (*see* **Note 7**): β-Mercaptoethanol can be added to peptide solutions to a final molar ratio of 3:1 pep-tide/mercaptoethanol (*see* **Note 8**); if included, incubate for 15 min before use.

2.2 Activity Measurement of Peptide-Functionalized Microspheres

1. Microspheres prepared in Subheading 3.1.

2. Instrumentation: All measurements were carried out on a Luminex 100.

3. Polypropylene round-bottom 96-well plate (Fisher Scientific, Waltham, MA, USA).

4. Microtitre plate shaker: Daigger Micromixer Mx4t (Daigger, Vernon Hills, IL, USA) or a similar plate shaker.

5. Magnetic plate: BioTek Instrument Model 96F (BioTek Instruments, Winooski, VT, USA).

6. PBST: Phosphate-buffered saline (PBS), pH 7.4, with 0.05 % Tween 20.

7. Heat-killed *E. coli* O157:H7 (KPL, Inc., Gaithersburg, MD, USA), diluted in PBST.

Table 1

Amino acid sequences of antimicrobial peptides (AMPs) used in this work

Peptide	Sequence
Melittin-C27	H_2N-GIGAVLKVLTTGLPALISWIKRKRQQC-OH
Histatin-C25	H_2N-DSHAKRHHGYKRKFHEKHHSHRGYC-OH
Magainin-2-C24	H_2N-GIGKFLHSAKKFGKAFVGEIMNSC-OH
Tritrpticin-C14	H_2N-VRRFPWWWPFLRRC-OH

8. Biotin anti-*E. coli* O157:H7 (KPL) antibody solution: 5 μg/mL in PBST (*see* **Note 9**).

9. Streptavidin–R-phycoerythrin (SA–PE) solution: Stock solution from Columbia Biosciences (Frederick, MD, USA). Dilute to 7.5 μg/mL in PBST.

3   Methods

3.1 Attachment of Thiol-Containing Molecules to COOH Microspheres

Attachment of thiol-containing molecules to carboxylated microspheres is accomplished in several phases (Fig. 1). Surface carboxyls on the microspheres are first reacted with the primary amine on maleimide–PEG (600)–amine, leaving a pendant maleimide moiety. The maleimide is then reacted with peptides possessing a unique cysteine.

For the first steps (functionalization with maleimide–PEG (600)–amine), we have adapted the protocol used by Luminex and other commercial sources for microspheres to simplify the physical manipulations and to minimize loss of microspheres during preparation.

3.1.1 Microsphere Preparation: Washing

Suspensions of microspheres are commonly provided in azide or another preservative. To avoid undesired interference or degradation of the preservative(s), the microspheres are washed several times before activation and coupling.

1. Resuspend microspheres in stock bottle by vortexing and brief bath sonication.

2. Pipet 25 μL of microspheres into Eppendorf tubes (~3.1×10^5 microspheres; scale as desired).

3. Using the magnetic holder, remove supernatant (*see* **Note 10**).

4. Resuspend microspheres in 100 μL phosphate buffer in the tube. Vortex to mix (*see* **Note 3**).

5. Using the magnetic holder, remove supernatant.

6. Resuspend microspheres in 100 μL phosphate buffer. Vortex to mix.

7. Using the magnetic holder, remove supernatant.

8. Resuspend microspheres in 60 μL phosphate buffer. Vortex to mix.

3.1.2 Microsphere Activation with EDC/Sulfo-NHS

EDC-mediated coupling of the microsphere carboxyls to the maleimide–PEG (600)–amine is accomplished using a two-step approach. Inclusion of sulfo-NHS in the reaction results in a sulfo-NHS-ester intermediate with greatly increased stability over the o-acylisourea intermediate formed in the absence of NHS [23].

1. Add 15 μL of the EDC/sulfo-NHS solution to the 60 μL microspheres in the Eppendorf tube (**step 8**, above). The final concentrations of each will be ~5 mg/mL (25 mM) (*see* **Notes 2** and **3**).

2. Mix well by briefly vortexing. Incubate for 30 min in the dark at room temperature with constant or intermittent (every 5–10 min) shaking.

3. Using the magnetic holder, remove supernatant (*see* **Note 10**).

4. Resuspend microspheres in 100 μL phosphate buffer. Vortex to mix.

5. Using the magnetic holder, remove supernatant.

6. Resuspend microspheres in 100 μL phosphate buffer in the tube. Vortex to mix.

7. Using the magnetic holder, remove supernatant, leaving the activated microspheres.

3.1.3 Cross-Linking of Activated Microspheres to Maleimide–PEG (600)– Amine

The NHS-ester-active intermediate on the bead surface covalently links to the amine terminus of maleimide–PEG (600)–amine. This reaction results in decoration of the microspheres with pendant maleimide moieties.

1. Add 25 μL maleimide–PEG (600)–amine solution to the pellet of activated microspheres (*see* **Note 3**).

2. Add an additional 50 μL phosphate buffer.

3. Vortex the microspheres to ensure they are well suspended.

4. Incubate for at least 2 h in the dark. Shake the tube to keep the microspheres suspended or, alternatively, mix frequently (every 10–15 min).

5. Using the magnetic holder, remove supernatant (*see* **Note 10**).

6. Resuspend microspheres in 100 μL phosphate buffer in the tube. Vortex to mix.

7. Using the magnetic holder, remove supernatant, leaving the maleimide-functionalized microspheres on the side of the tube.

3.1.4 Cross-Linking of Activated Microspheres to Antimicrobial Peptides

The maleimide-functionalized microspheres covalently link to thiol moieties on cysteine-containing peptides in a highly specific reaction, provided that the reaction is performed at or below neutral pH (*see* **Note 11**). For highly cationic peptides, the presence of a cationic buffer such as Tricine and a cationic surfactant (both in excess) may mitigate bead aggregation (Figs. 2, 3, and 4) (*see* **Note 12**).

1. Add 100 μL peptide solution in Tricine/CTAB with or without mercaptoethanol to microspheres (*see* **Note 8**).

2. Vortex the microspheres to ensure they are well suspended.

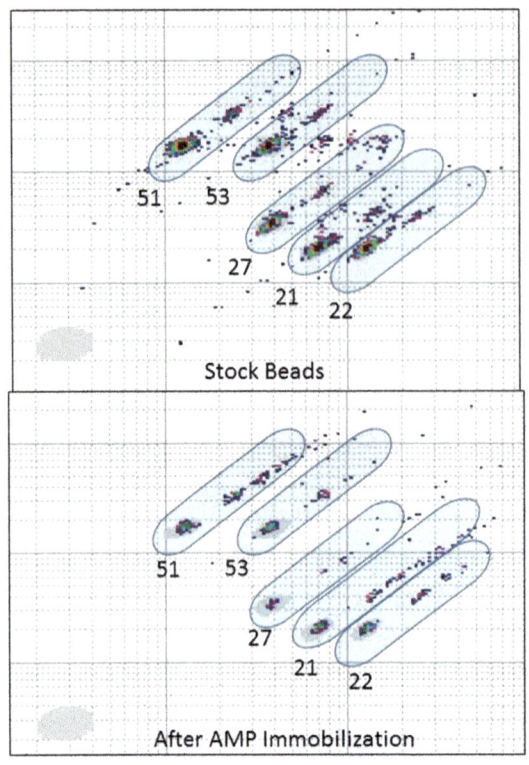

Bead #	Buffer	% Apparent Recovery
21	0.1M Phosphate-buffered saline with 0.5% Tween, pH 7.4	7.4
22	1.0 M Tris, pH 7.4	12.8
27	1.0 M Tris with 0.05% Tween, pH 7.4	2.8
51	1.0 M Tris with 0.05% SDS, pH 7.4	10.6
53	1.0 M Tricine, pH 7.4	17.4

Fig. 2 Bead identification of AMP-coated microspheres by Luminex 100. Microspheres were prepared using different buffers during the final AMP immobilization step (incubation with the peptide solution). The windows used to recognize and identify the individual bead sets are shown as *light gray spots* on the *lower left* end of each shaded oval. Aggregated microspheres extend from the identification spot diagonally to the *upper right*. These aggregates of AMP-coated beads may produce artifactual results in assays, since they are not recognized by Luminex 100 but may have bound analyte. The percent apparent recovery (Table to the right) is indicative of the number of beads recognized by the instrument. In this experiment, bead number 53 has the highest recovery of beads in the identification spot with the lowest number of aggregates

3. Incubate for 1 h in the dark at room temperature. Shake the tube to keep the microspheres suspended or, alternatively, mix frequently (every 10–15 min).

4. After 1 h, move the microspheres to 4 °C and continue the incubation overnight (*see* **Note 13**).

5. The next day, place the microfuge tube in magnetic holder and remove supernatant.

6. Resuspend microspheres in 100 μL phosphate buffer. Vortex to mix.

7. Using the magnetic holder, remove supernatant.

8. Resuspend microspheres in 100 μL phosphate buffer in the tube. Vortex to mix.

9. Using the magnetic holder, remove supernatant, leaving a pellet of microspheres on the side of the tube.

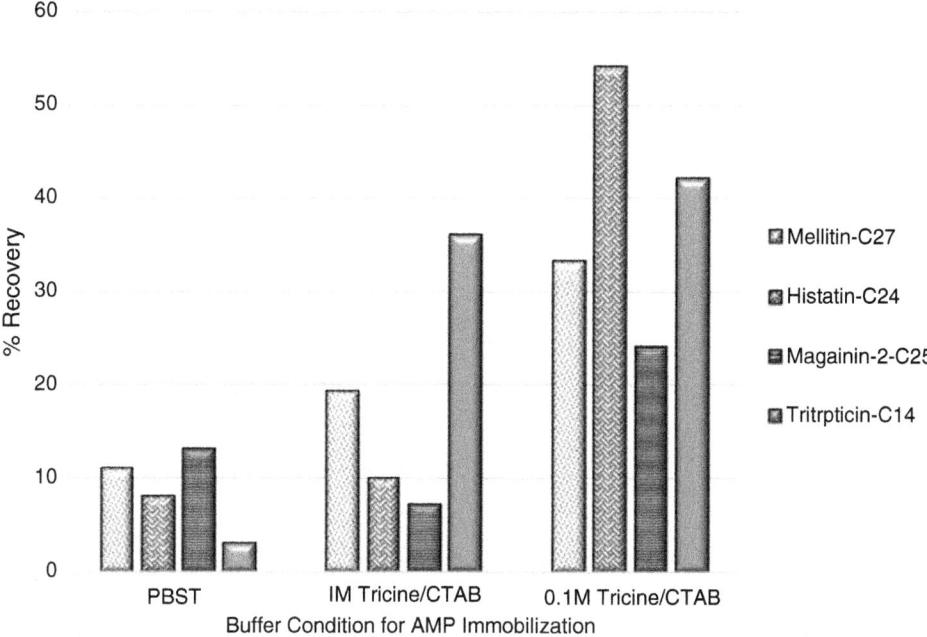

Fig. 3 Recovery of beads with four AMPs immobilized using different buffers. While use of 0.1 M Tricine/CTAB provided the greatest recovery for each of the four AMP-coated microspheres, there are still significant differences between recovery rates for the different AMPs

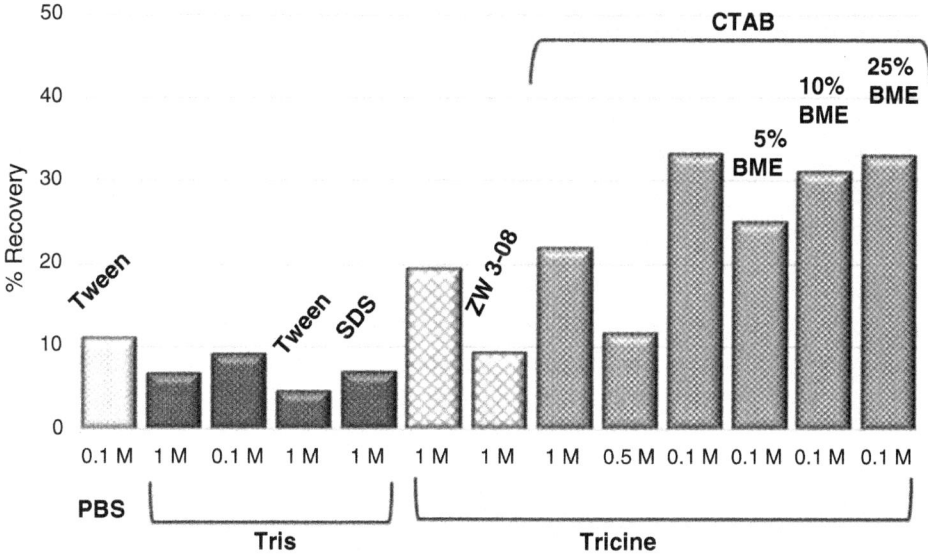

Fig. 4 Recovery of microspheres with immobilized melittin-C27. Melittin-C27 was immobilized onto Luminex microspheres as described in Fig. 1 using a variety of different buffer conditions during the final incubation with peptide. Both Tris and Tricine were used at different concentrations (indicated) in the absence or presence of several surfactants (0.05 % Tween-20, 0.05 % sodium dodecylsulfate (SDS), 0.05 % Zwittergent-3-08 (ZW 3-08), or 0.05 % CTAB). The rightmost three bars represent results from experiments where β-mercaptoethanol (BME) was added to the peptide solution in various molar ratios (indicated)

10. Add 100 μL phosphate buffer in the tube. Vortex to mix. At this point, 2 μL can be removed for determining the number of AMP-coated microspheres that are in the detection region.

11. Store microspheres at 4 °C in the dark for several days to several months depending on the stability of the immobilized biomolecule. Larger storage volumes may be desirable depending on the application.

3.2 Measure Activity of Functionalized Microspheres

Once the microspheres have been functionalized with peptides, we test them for their ability to bind target bacteria, as a means of semi-selective detection and categorization [13, 15–17, 24]. Here, we describe sandwich-format assays using microsphere-immobilized AMPs for capture of *E. coli* O157:H7, with detection of the bound cells by a biotinylated antibody (*see* **Note 14**).

3.2.1 Measure Recovery of Microspheres

Prior to performing any assay, it is important to determine the overall recovery and quality of the microsphere preparation. For this reason, we typically count a small aliquot of the prepared material.

1. Vortex suspensions of the stock microspheres and AMP-coated microspheres.

2. Remove 2 μL of each and place in wells in a 96-well plate (*see* **Note 15**).

3. Bring volume in each well up to 100 μL with the same buffer as the sample.

4. Count plate in the Luminex instrument.

5. Calculate recovery of microspheres in the detection region. Microspheres may be present in the sample as aggregates but will not be counted if they are not in the detection region (*see* Fig. 2).

6. Determine the volume or dilution needed to provide approx. 400 microspheres to each sample (*see* Subheading 3.2.2, **step 2**, below).

3.2.2 Measure Binding of Microspheres

While microsphere recovery is important, the functionality of the prepared microspheres is the most critical factor; even if 100 % recovery is achieved after functionalization, a bead set without the desired activity is essentially useless. Figure 5 shows a sandwich-format assay for detection of *E. coli* O157:H7 using AMP-functionalized microspheres for target capture. Different mixtures of peptide-decorated microspheres can also be used for detection of other bacterial species.

1. Perform serial dilutions of the target cells such that there is 90 μL volume remaining in each well of a 96-well plate (*see* **Note 14**).

2. Add 10 μL of AMP-functionalized bead solution in PBST, such that there are ~400 microspheres of each bead type per well (*see* **Note 16**).

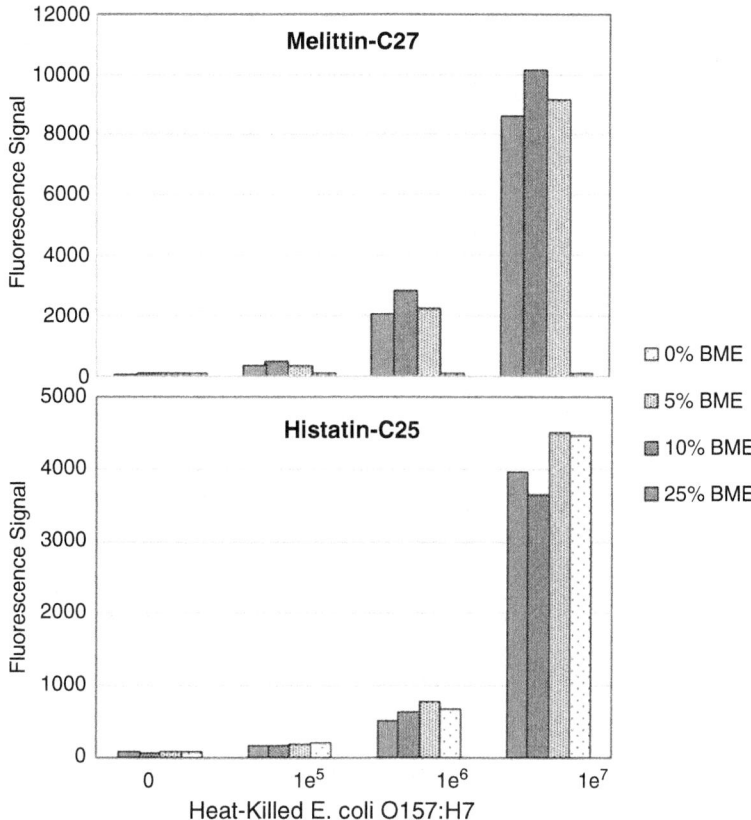

Fig. 5 Sandwich-format assay for the detection of heat-killed *E. coli* O157:H7 employing melittin-C27- and histatin-C25-coated microspheres. Various concentrations of β-mercaptoethanol were included during the final incubation with cysteinyl peptides. While β-mercaptoethanol improved the ability of melittin-decorated microspheres to bind *E. coli*, there was no effect on the histatin-coated beads. These results stress the importance of optimizing immobilization conditions for each peptide

3. Incubate 30 min in the dark at room temperature with shaking (~300 rpm).

4. Place the plate on the BioTek plate magnet. After 1 min, *while keeping the 96-well plate in contact with the magnet*, flip over and back once to remove supernatant (*see* **Note 17**).

5. Add 100 μL PBST to each well.

6. Place the plate on the BioTek plate magnet. After 1 min, while keeping the 96-well plate in contact with the magnet, flip over and back once to remove supernatant.

7. Add 50 μL biotinylated antibodies (*see* **Note 18**).

8. Incubate 30 min, in the dark at room temperature with shaking (~300 rpm).

9. Place the plate on the BioTek plate magnet. After 1 min, while keeping the 96-well plate in contact with the magnet, flip over and back once to remove supernatant.

10. Add 100 μL PBST to each well.

11. Place the plate on the BioTek plate magnet. After 1 min, while keeping the 96-well plate in contact with the magnet, flip over and back once to remove supernatant.

12. Add 50 μL streptavidin–R-phycoerythrin solution.

13. Incubate 30 min, in the dark at room temperature with shaking (~300 rpm).

14. Place the plate on the BioTek plate magnet. After 1 min, while keeping the 96-well plate in contact with the magnet, flip over and back once to remove supernatant.

15. Add 100 μL PBST to each well.

16. Place the plate on the BioTek plate magnet. After 1 min, while keeping the 96-well plate in contact with the magnet, flip over and back once to remove supernatant.

17. Add 100 μL PBST to each well.

18. Count in the Luminex instrument.

4   Notes

1. Microspheres used in Luminex and FACSArray platforms are identified by the quantity of two fluorophores integrated in the microsphere support. Microsphere suspensions should therefore be kept dark as much as possible to avoid unnecessary photobleaching of the integral fluorophores.

2. EDC is highly unstable in the presence of water. To avoid hydrolysis, it should be stored in a dessicator at –20 °C and allowed to warm to room temperature before opening. *Prepare all EDC solutions immediately before use.*

3. Solutions used in the EDC-/sulfo-NHS-mediated coupling step must have no amine-containing components other than maleimide–PEG (600)–amine; amine-containing buffers such as Tris and glycine should not be used. If Tris or another unwanted amine is present, it will be necessary to dialyze or remove by gel filtration prior to reacting with the microspheres.

4. As the maleimide moiety in this compound hydrolyzes over time, it should be prepared just prior to use. Maleimide–PEG (600)–amine is very viscous and cannot be accurately pipetted. However, a small amount can be removed from the stock vial using a spatula and transferred to a tared microfuge tube; once the weight of the maleimide–PEG (600)–amine is quantified,

an appropriate volume of buffer can be added for a final concentration of 10 mg/mL.

5. Anionic buffers such as MES and phosphate can be used up to and including coupling of the maleimide–PEG (600)–amine to the bead surfaces.

6. A total of 25 μg AMP (Subheading 2.1, **step 8**) is used for every 25 μL of the initial stock microspheres ($\sim 3 \times 10^5$ microspheres, Subheading 3.1.1, **step 2**). The volume is brought up to 100 μL.

7. β-Mercaptoethanol is hazardous and has an unpleasant stench. Concentrated solutions of this material should be handled in a hood with appropriate personal protective gear.

8. β-Mercaptoethanol is a reducing agent and can be used to break disulfides resulting from cysteine oxidation during peptide storage. It has the additional benefit that—when added to peptide solutions at concentrations of up to 3:1 peptide/mercaptoethanol (mol:mol)—it may improve both recovery (Fig. 4) and activity (Fig. 5) of peptide-functionalized microspheres.

9. Phycoerythrin-based constructs provide higher signals in the Luminex than Luminex-compatible dyes such as Cy3 and Alexa Fluor 555. However, if direct detection of antibody binding is desired (without the need for an additional streptavidin–R-phycoerythrin step), Cy3- or Alexa Fluor 555-labeled antibodies can be used in place of the biotinylated antibodies.

10. The magnet holds microspheres to the side of the tube in a broad smear. Once the magnet is removed, the microspheres drop to the bottom of the tube.

11. At pH 6–7.5, the reaction rate of maleimide with thiols is 1000-fold greater than its reaction with amines. However, at higher pHs, the maleimide reaction becomes significantly less specific and maleimides may react also with amines [23, 25].

12. The choice of buffer itself (Tris, Tricine, or phosphate) may also be important and will depend on the peptides used. In our own studies with cationic AMPs, we have found significantly greater recovery of microspheres in Tricine but not Tris buffers (Fig. 4).

13. There is little difference between bead recovery and functionality of AMP-coated microspheres incubated for 1 h or overnight. As mentioned in **Note 3**, the reactive groups hydrolyze over time and should not be present after the overnight incubation.

14. The text describes a sandwich-format assay. Alternatively, direct assays (using fluorophore- or biotin-labeled targets) or various competitive assays can also be performed.

15. Microspheres with different identification (dye quantities) parameters can be added to a single well. This allows users to perform multiplexed assays to suit their needs, as well as to minimize the number of separate counting experiments and individual assays performed.

16. When working with large numbers of different bead sets, it is important to ensure that samples do not contain two or more microsphere sets with the same identification parameters.

17. Keeping the plate in contact with the magnetic plate is absolutely critical, as microspheres will be lost otherwise. Alternative methods for separating the supernatant from the microspheres include use of a plate washer developed for use with magnetic microspheres or use of a 96-well filter plate and manually pushing the liquid through.

18. We typically include a control set of microspheres functionalized with anti-chicken IgY antibodies. Cocktails of biotinylated tracer antibodies therefore typically also contain biotinylated chicken IgY.

References

1. Chang T-W (1983) Binding of cells to matrixes of distinct antibodies coated on solid surface. J Immunol Methods 65:217–223

2. Legutki JB, Zhao Z-G, Greving M et al (2014) Scalable high-density peptide arrays for comprehensive health monitoring. Nat Commun 5:4785

3. Rauh D, Fischer F, Gertz M et al (2012) An acetylome peptide microarray reveals specificities and deacetylation substrates for all human sirtuin isoforms. Nat Commun 4:2327

4. Houseman BT, Huh JH, Kron SJ et al (2002) Peptide chips for the quantitative evaluation of protein kinase activity. Nat Biotech 20:270–274

5. Qi H, Wang F, Petrenko VA et al (2014) Peptide microarray with ligands at high density based on symmetrical carrier landscape phage for detection of cellulase. Anal Chem 86:5844–5850

6. Lin J, Bardina L, Shreffler WG et al (2009) Development of a novel peptide microarray for large-scale epitope mapping of food allergens. J Allergy Clin Immunol 124:315–322, 322.e1-3

7. Price JV, Jarrell JA, Furman D et al (2013) Characterization of influenza vaccine immunogenicity using influenza antigen microarrays. PLoS One 8, e64555

8. Villiers M-B, Cortès S, Brakha C et al (2010) Peptide–protein microarrays and surface plasmon resonance detection: biosensors for versatile biomolecular interaction analysis. Biosens Bioelectron 26:1554–1559

9. Pai J, Yoon T, Kim ND et al (2012) High-throughput profiling of peptide–RNA interactions using peptide microarrays. J Am Chem Soc 134:19287–19296

10. Sykes KF, Legutki JB, Stafford P (2013) Immunosignaturing: a critical review. Tr Biotechnol 31:45–51

11. Bleher O, Schindler A, Yin MX et al (2014) Development of a new parallelized, optical biosensor platform for label-free detection of autoimmunity-related antibodies multiplex platforms in diagnostics and bioanalytics. Anal Bioanal Chem 406:3305–3314

12. Deiss F, Matochko WL, Govindasamy N et al (2014) Flow-through synthesis on teflon-patterned paper to produce peptide arrays for cell-based assays. Angew Chem Int Ed Engl 53:6374–6377

13. Kulagina N, Taitt C, Anderson GP et al (2014) Affinity-based detection of biological targets. US 8,658,372 B2

14. Kulagina NV, Anderson GP, Ligler FS et al (2007) Antimicrobial peptides: new recognition molecules for detecting botulinum toxins. Sensors 7:2808–2824

15. Kulagina NV, Lassman ME, Ligler FS et al (2005) Antimicrobial peptides for detection of bacteria in biosensor assays. Anal Chem 77:6504–6508

16. Kulagina NV, Shaffer KM, Ligler FS et al (2007) Antimicrobial peptides as new recognition molecules for screening challenging species. Sens Actuators B Chem 121:150–157

17. Shriver-Lake LC, North SH, Dean SN et al (2013) Antimicrobial peptides for detection and diagnostic assays. In: Piletsky SA, Whitcomb MJ (eds) Designing receptors for the next generation of biosensors. Springer, Heidelberg, pp 85–104

18. Klapper MH (1977) The independent distribution of amino acid near neighbor pairs into polypeptides. Biochem Biophys Res Commun 78:1018–1024

19. North S, Wojciechowski J, Chu V et al (2012) Surface immobilization chemistry influences peptide-based detection of lipopolysaccharide and lipoteichoic acid. J Pept Sci 18:366–372

20. Heubach Y, Planatscher H, Sommersdorf C et al (2013) From spots to beads – PTM-peptide bead arrays for the characterization of anti-histone antibodies. Proteomics 13:1010–1015

21. Bernsteel DJ, Roman DL, Neubig RR (2008) In vitro protein kinase activity measurement by flow cytometry. Anal Biochem 383:180–185

22. Ayoglu B, Szarka E, Huber K et al (2014) Bead arrays for antibody and complement profiling reveal joint contribution of antibody isotypes to C3 deposition. PLoS One 9, e96403

23. Hermanson GT (2008) Bioconjugate techniques. Academic, San Diego

24. Kulagina NV, Shaffer KM, Anderson GP et al (2006) Antimicrobial peptide-based array for Escherichia coli and Salmonella screening. Anal Chim Acta 575:9–15

25. Brewer CF, Reihm JP (1967) Evidence for possible nonspecific reactions between N-ethylmaleimide and proteins. Anal Chem 18:248–255

Chapter 15

A Cell Microarray Format: A Peptide Release System Using a Photo-Cleavable Linker for Cell Toxicity and Cell Uptake Analysis

Kenji Usui, Kin-ya Tomizaki, and Hisakazu Mihara

Abstract

There has been increasing interest in the potential use of microarray technologies to perform systematic and high-throughput cell-based assays. We are currently focused on developing more practical array formats and detection methods that will enable researchers to conduct more detailed analyses in cell microarray studies. In this chapter, we describe the construction of a novel peptide-array format system for analyzing cellular toxicity and cellular uptake. In this system, a peptide is immobilized at the bottom of a conventional 96-well plate using a photo-cleavable linker. The peptide can then be released from the bottom by irradiating the desired wells with UV light, thus allowing the cytotoxicity or cellular uptake of the peptide to be monitored. This system will facilitate the realization of high-throughput cell arrays for cellomics analyses and cell-based phenotypic drug screens.

Key words Cell microarrays, Designed peptide, Photo-cleavable linker, Cytotoxicity, Cellular uptake

1 Introduction

To date, the genome sequences of a number of organisms have been reported. In the post-genome-sequencing era, significant advances in "omics" studies have given rise to a new field known as cellomics, which utilizes cell-based phenotypic assays [1] to study vital cellular phenomena [2, 3]. Therefore, there is increasing demand for systematic cell-based assays to analyze the phenotypic changes induced by chemical stimuli, such as drugs in mammalian cells, via high-throughput screening. Cell microarrays are one of the most promising solutions because they offer more than just the simple miniaturization and mechanization of conventional equipment. To increase the utility of microarray technology, we have constructed peptide microarrays composed of various secondary structures [4–9], which were initially applied to protein analysis [10–14]. We then adapted the design of these peptide libraries for use in cell microarrays [15, 16].

Marina Cretich and Marcella Chiari (eds.), *Peptide Microarrays: Methods and Protocols*, Methods in Molecular Biology, vol. 1352, DOI 10.1007/978-1-4939-3037-1_15, © Springer Science+Business Media New York 2016

As is commonly the case in the development of new protein microarray technologies [6, 9], we are currently focused on developing more practical array formats and detection methods that will enable researchers to conduct more detailed analyses of cellular processes. Consequently, we developed peptide photo-release systems that were combined with a cell counting kit and a plate reader for cytotoxicity analysis [17] and that utilize confocal microscopy for cellular uptake analysis [18]. These systems have a number of advantages compared with existing formats: (1) Peptide release can be temporally controlled because the peptides can be photo-released through irradiation at desired times. (2) Peptide release can be spatially controlled because light can be directed to specific locations with micrometer precision, thereby enabling single-cell assays. (3) Light irradiation is relatively easy and inexpensive compared with other methods of peptide release, including the use of chemolabile linkers [19]. (4) For the cytotoxicity assay, the cytotoxic behavior of a peptide can be monitored after the irradiation and release of the peptide in the desired wells. (5) For the uptake analysis, peptide internalization can be analyzed around the cell adhesion site in detail and in real time because the system utilizes confocal microscopy as the monitoring method.

In this chapter, we describe a method for constructing a novel cell microarray format involving the photo-release of peptides combined with a cell counting kit and a plate reader for cytotoxicity analysis or with confocal microscopy for monitoring peptide permeation (Fig. 1). These systems enabled real-time monitoring of peptide behavior in cells [17, 18]. Additionally, in the case of uptake analysis, the efficiency of peptide permeation in our system was much greater than that of traditional peptide solution assays. We believe that this novel array format will be an effective analysis system for use in both cell engineering and drug discovery research.

2   Materials

2.1 Synthesis of Peptides for Release

1. Fmoc peptide synthesis [20] reagents: NovaSyn TGR resin (Merck Japan, Tokyo, Japan) is used as a Rink amide resin for peptide synthesis. The coupling cocktail contains Fmoc amino acids (Fmoc-AA-OH), 2-(1H-benzotriazole-1-yl)-1,1,3,3-tetramethyluronium hexafluorophosphate (HBTU), 1-hydroxybenzotriazole (HOBt), and diisopropylethylamine (DIEA) in N-methyl pyrrolidone (NMP) (*see* **Note 1**). The Fmoc removal solution consists of 25 % (v/v) piperidine in NMP.

2. Fluorophore-coupling reagents: 5-(and-6)-carboxytetramethylrhodamine (TMR), HBTU, HOBt and DIEA in NMP (*see* **Note 2**).

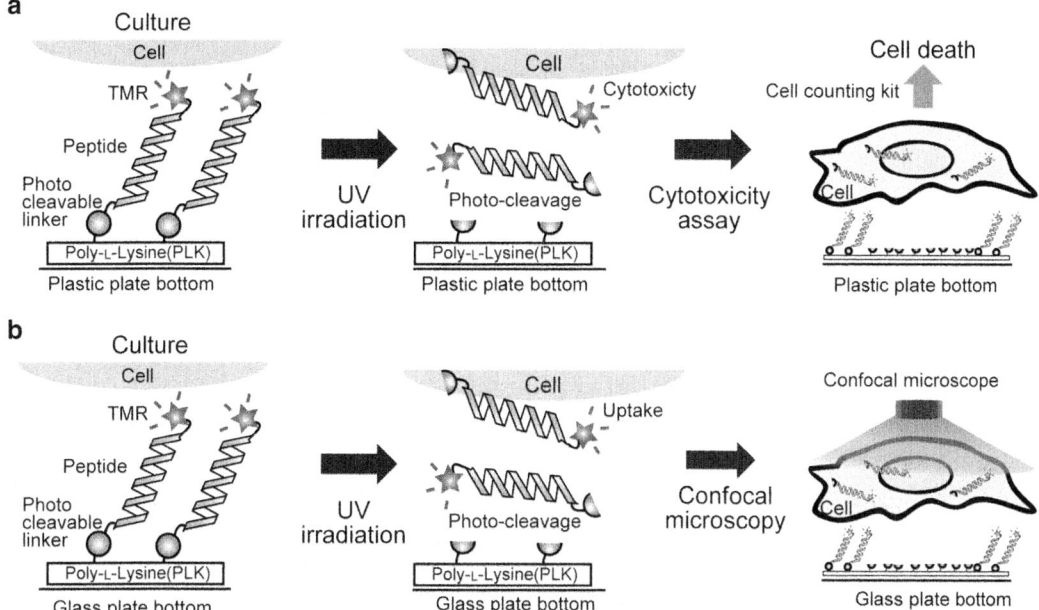

Fig. 1 Illustration of the photo-release systems. (**a**) A novel peptide array format system for cellular toxicity analyses. In this system, a peptide is immobilized to the bottom of a commercially available 96-well plate via a photo-cleavable linker. UV light irradiation of the desired wells releases the peptide from the plate bottom, after which the cytotoxic behavior of the peptide can be monitored using a cell counting kit and a plate reader. This figure was modified and reproduced from [17] with permission from the Nature Publishing Group. (**b**) A novel peptide array format for monitoring cellular uptake. Model peptides were immobilized onto conventional glass plates via a photo-cleavable linker and could be spatiotemporally released using UV irradiation. The incorporation of confocal microscopy allowed for detailed, real-time monitoring of the cellular internalization of the peptides. Modified and reproduced from [18] with permission from the Royal Society of Chemistry

3. Cleavage solution 1: Trifluoroacetic acid/H_2O/triisopropylsilane (95/2.5/2.5, v/v) (the cytotoxicity assay) (*see* **Note 3**).

4. HPLC: HPLC is performed on a Hitachi L7000 or a Shimadzu LC2010C system using a Wakosil 5C18 or a YMC-Pack ODS-A (4.6×150 mm) column for analysis and a YMC ODS A323 (10×250 mm) column for preparative purification with a linear gradient of acetonitrile/0.1 % TFA at a flow rate of 1.0 mL/min for analysis and 3.0 mL/min for preparative separation, respectively.

5. MALDI-TOFMS is measured on a Shimadzu KOMPACT MALDI III with 3,5-dimethoxy-4-hydroxycinnamic acid as a matrix.

6. TMR absorbance is measured on a multiplate reader using microtiter plates in 20 mM Tris–HCl containing 6 M guanidine hydrochloride (pH 7.0).

2.2 Synthesis of Photo-Cleavable Linker

1. Fmoc peptide synthesis [20] reagents: 2-Chlorotrityl chloride resin (Merck) is used for peptide synthesis. Initially, Fmoc-Gly-OH was used as the first coupling amino acid residue with DIEA in dried dichloromethane (DCM)/N, N-dimethylformamide (DMF) (5/4, v/v). Fmoc-2-(2-aminoethoxy)ethoxy acetic acid, Fmoc-4-[4-(1-aminoethyl)-2-methoxy-5-nitrophenoxy]-butanoic acid, and maleimido-propionic acid were used with HBTU, HOBt, and DIEA in NMP (*see* **Note 1**). Fmoc removal solution consists of 20 % piperidine in DCM/DMF (1/1, v/v) or 25 % piperidine in NMP.

2. Cleavage solution 2: Acetic acid/trifluoroethanol/dichloromethane (1/1/8, v/v).

2.3 Peptide Immobilization

1. Activation: N-hydroxysuccinimide (NHS) and 1-ethyl-3-(3-dimethylaminopropyl) carbodiimide (WSC) in DMF.

2. Poly-L-lysine hydrobromide coupling: Poly-L-lysine hydrobromide (PLK, Mw = 150–300 kDa) in HEPES buffer (50 mM 2-[4-(2-hydroxyethyl)-1-piperazinyl]ethanesulfonic acid (HEPES), 250 mM NaCl, pH 7.0) (*see* **Note 4**).

3. Peptide solution: Peptide in HEPES buffer (50 mM 2-[4-(2-hydroxyethyl)-1-piperazinyl]ethanesulfonic acid (HEPES), 250 mM NaCl, pH 7.0).

4. 96-Well culture plate: 96-Well culture plate (Falcon, BD Biosciences, San Jose, CA, USA) or 96-well glass-bottom plate (SF-G-P96, FPI, Kyoto, Japan).

5. Wash buffer: PBS buffer (pH 7.4) (Gibco, Tokyo, Japan).

2.4 Peptide Photo-Release

1. UV light irradiation system: A UV-spot light source (Photocure 200, Hamamatsu Photonics, Hamamatsu, Japan) with a band-pass filter (362 ± 30 nm) and condenser lens (Photocure 200 series, E5147-06).

2.5 Cell Toxicity Assay

1. Cells: Human cervical carcinoma (HeLa) cells, African green monkey kidney (COS7) cells, mouse embryonic fibroblast (10T1/2) cells (*see* **Note 5**).

2. Culture media: Dulbecco's modified Eagle's medium (DMEM) supplemented with 1 % fetal bovine serum (FBS, BioWest, Nuaillé, France) (*see* **Note 5**).

3. Assay media: 5 g DMEM powder and 1.85 g NaHCO$_3$ in 500 mL MilliQ water was filtered through a 0.45 μm syringe filter. Then, 100 U/mL penicillin and 100 μg/mL streptomycin were added to the filtered solution.

4. Assay kit: Cell Counting Kit-8 (Dojindo Laboratories, Kumamoto, Japan).

5. Plate reader: Multiplate reader.

2.6 Cell
Uptake Assay

1. Cells: HeLa cells (*see* **Note 5**).

2. Culture medium: Same as medium described in Subheading 2.5, **item 2**.

3. Assay medium: Same as medium described in Subheading 2.5, **item 3**.

4. Mitotracker assay: Mitotracker (Life technologies Japan, Tokyo, Japan).

5. Confocal microscopy: IX70 microscope with FV300 system (Olympus, Tokyo, Japan).

6. Image Data Software: Fluoroview FV300 Ver. 4.3 (Olympus software).

3   Methods

We developed a novel peptide array format that incorporates a photo-cleavable linker for monitoring the cytotoxicity or cellular uptake of peptides (Fig. 1). In developing this method, we chose a [4-(1-aminoethyl)-2-methoxy-5-nitrophenoxy]-butanoic acid derivative [21–24] as the photo-cleavable linker (Fig. 2a). This compound can be cleaved by UV irradiation at 365 nm, a wavelength that has been proven to be harmless to cells [25, 26]. We also selected the KLA peptide (KLAKLAKKLAKLAKKLAKLAK) as the model peptide [27] (Fig. 2b). After synthesizing the 5-(and-6)-carboxytetramethylrhodamine (TMR)-KLA peptide, we optimized the conditions for monitoring cytotoxicity, using a cell

a

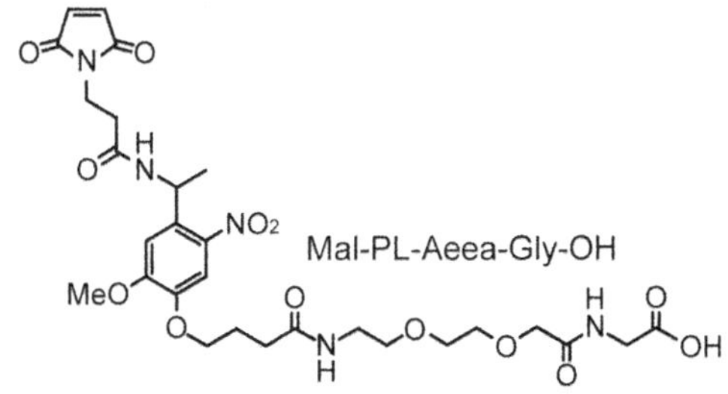

Mal : Maleimidopropionic acid

Mal-PL-Aeea-Gly-OH

PL : 4-[4-(1-aminoethyl)-2- methoxy-5-nitrophenoxy]-butanoic acid

Aeea : 2-(2-aminoethoxy) ethoxy acetic acid

b

TMR-KLA-C peptide

TMR-GKLAKLAKKLAKLAKKLAKLAKGC-NH2   TMR : 5-(6)-carboxytetramethylrhodamine

Fig. 2 (**a**) Structure of the photo-cleavable linker (M-linker). A [4-(1-aminoethyl)-2-methoxy-5-nitrophenoxy]-butanoic acid derivative was chosen as the photo-cleavable linker. This compound can be cleaved by UV irradiation at 365 nm, a wavelength that has been proven to be harmless to cells at the time-scales used in the assay. (**b**) Sequence of the TMR-KLA-C peptide. The peptide had previously shown a high degree of cell penetration and cytotoxicity. We added Gly (as a linker) and Cys (for immobilization) to the C-terminus and Gly and 5-(and-6)-carboxytetramethylrhodamine (TMR) to the N-terminus for detection

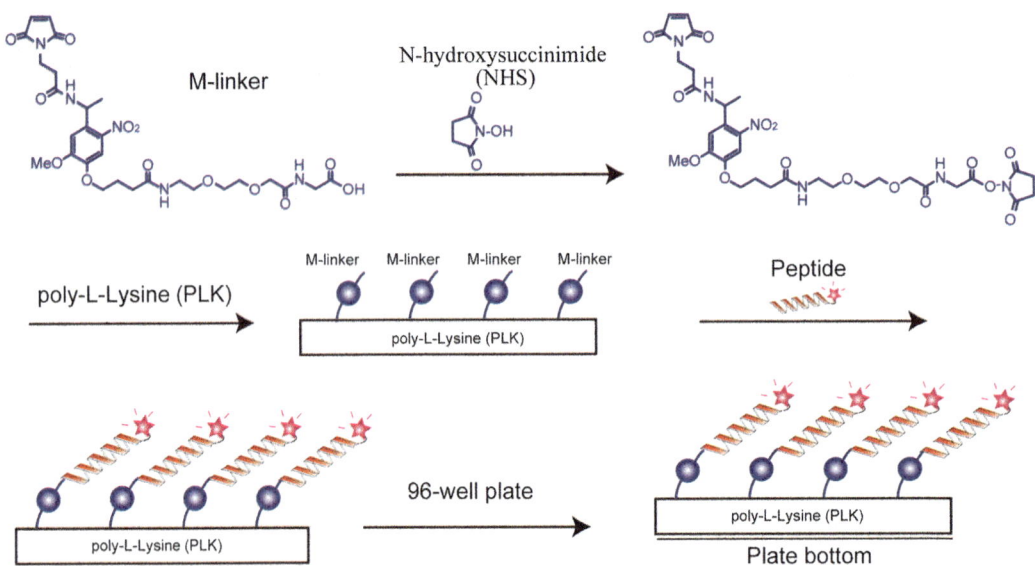

Fig. 3 Outline of the peptide immobilization steps performed in this study. The detailed method is described in Subheading 3.3. This figure was modified and reproduced from [18] with permission from the Royal Society of Chemistry

counting kit and a plate reader, and for cellular uptake, using confocal microscopy. We also optimized the peptide immobilization and photo-release conditions. In this method, we describe these optimized conditions, and details of the available protocols are shown in Subheadings 3.1–3.3.

After the construction of the photo-release array, we checked that the immobilized peptide did not exhibit cytotoxicity and that cells could be cultivated on the peptide-coated plate from the beginning of the experiments. Then, we demonstrated cytotoxicity and cellular uptake monitoring. These protocols are described in detail in Subheadings 3.4–3.6 (Fig. 3).

The results of the cytotoxicity assay were almost identical to those of the free peptide solution assay [17] (Fig. 4a). In the case of the cellular uptake monitoring assay, the amount of internalized peptide increased as the incubation time increased after irradiation [18] (Fig. 4b). Peptide internalization was also found to require approximately 60 min; in contrast, the internalization of Mitotracker required a much shorter time (less than 30 min), with mitochondrial localization occurring in less than 60 min. This result implied that the mechanism of peptide uptake differs from that of Mitotracker uptake. With the improved monitoring capability of our method, uptake, cytotoxicity, and other peptide functions can be analyzed in detail in real time.

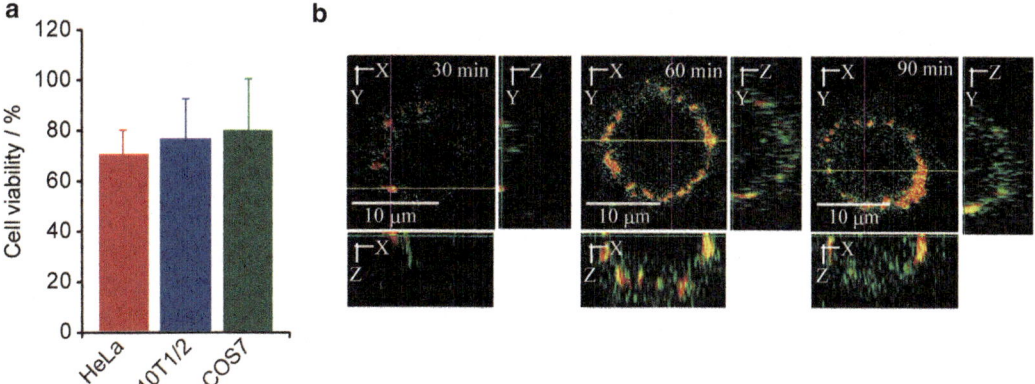

Fig. 4 (**a**) Results of cell toxicity assays using a cell counting kit and a plate reader with the TMR-KLA peptide. Relative viability was calculated by dividing the absorbance of tested cells by that of control cells incubated without immobilized peptide (non-coating) and with light irradiation under normal growth conditions (assuming that 100 % viability = 0 % toxicity). These results were almost identical to those of the free peptide solution assay. This figure was reproduced from [17] with permission from the Nature Publishing Group. (**b**) Cellular uptake results obtained with confocal microscopy. Confocal micrographs (single-cell images) of HeLa cells incubated for various periods of time (30, 60, and 90 min) after plates containing immobilized TMR-KLA peptide (*red*) and Mitotracker (*green*) were irradiated with UV light for 15 min. Side views show confocal projections of cells in the *xz* and *zy* planes. The amount of internalized peptide increased as the incubation time increased after irradiation. This figure was reproduced from [18] with permission from the Royal Society of Chemistry

3.1 Design and Synthesis of Peptides for Release

1. Peptides for release are designed. To date, we have constructed peptide arrays containing peptides with loop [28], α-helical [4, 8], or β-sheet [5] structures, and we have also constructed peptide libraries with peptides containing functional groups such as alkyl chains [5] or sucrose [29]. In developing this photo-release system, we selected the KLA peptide as the model peptide, which had previously shown high degrees of cell penetration and cytotoxicity [27]. We added Gly (as a linker) and Cys (for immobilization) to the C-terminus and Gly and 5-(and-6)-carboxytetramethylrhodamine (TMR) to the N-terminus for detection, resulting in the sequence TMR-GKLAKLAKKLAKLAKKLAKLAKGC (TMR-KLA-C).

2. The selection of an appropriate fluorescent detection method will depend on the assay, and this choice may influence the synthesis strategy. For example, we have already shown that a dual-labeled α-helical peptide library produced efficient fluorescence resonance energy transfer (FRET) responses [4]. However, the synthesis strategy was more complicated than that required for single-labeled peptides.

3. The peptides are synthesized. The precise synthesis strategy will differ depending on which peptides are chosen. Here, we present TMR-KLA-C as an example.

4. The TMR-KLA-C peptide (Fig. 2b) is synthesized on NovaSyn TGR resin (Merck Japan, Tokyo, Japan) by manual synthesis with Fmoc chemistry. Side chain protections are as follows: *t*-butyloxycarbonyl (Boc) for Lys and trityl (Trt) for Cys. Initially, an Fmoc-peptide sequence is synthesized with Fmoc-AA-OH (3 eq.) by the HBTU method (HBTU 3 eq., HOBt 3 eq., DIEA 6 eq.).

5. The fluorophore is attached to the peptides. The TMR moiety is introduced by using TMR (2 eq.) with HBTU (2 eq.), HOBt (2 eq.), and (DIEA, 4 eq.) for 60 min.

6. The peptides are cleaved from the resin, and the side chain protections are removed using cleavage solution 1 at room temperature for 1 h. The peptide is precipitated via the addition of diethylether and collected by centrifugation.

7. The crude peptides are purified by RP-HPLC and characterized by MALDI-TOF MS. The purified samples are then lyophilized.

8. The lyophilized peptide is freshly dissolved before each use. To estimate the concentration of the solution, the TMR absorbance of a diluted peptide solution is measured.

3.2 Synthesis of Photo-Cleavable Linker

1. The photo-cleavable (M-linker) peptide (Fig. 2a) is synthesized on 2-chlorotrityl chloride resin. Initially, Fmoc-Gly-OH (0.5 eq.) is coupled to the resin with DIEA (2.5 eq.) in dried dichloromethane (DCM)/ *N,N*-dimethylformamide (DMF) (5/4, v/v) using a round-bottom flask equipped with calcium chloride for 60 min.

2. Unreacted moieties on the resins are capped with methanol (ca. 5 % of the reaction volume) for 15 min.

3. The resins are transferred to a peptide synthesis tube and washed with DCM, methanol, and DCM/DMF (1/1, v/v). The Fmoc group is removed using 20 % piperidine in DCM/DMF (1/1, v/v) for 15 min. After the removal of Fmoc, the resin is washed with DCM/DMF (1/1, v/v) and diethylether. Fmoc-2-(2-aminoethoxy) ethoxy acetic acid, Fmoc-4-[4-(1-aminoethyl)-2-methoxy-5-nitrophenoxy]-butanoic acid, and maleimidopropionic acid were coupled by the HBTU–HOBt method.

4. The peptide is cleaved from the resin with cleavage solution 2 at room temperature for 2 h. After the removal of the resin, the addition of hexane to the peptide solution and the evaporation of the solvent are repeated three to five times. MilliQ water is then added to the evaporated sample, and the peptide solution was lyophilized.

3.3 Peptide Immobilization

1. An outline of the immobilization procedure is shown in Fig. 3. In this section, the conditions for the construction of the cytotoxicity assay system are described.

2. The C-terminal of the M-linker is activated using N-hydroxysuccinimide (NHS) (10 eq.) and WSC (10 eq.) in 100 μL DMF (2 h).

3. A total of 5 μL of the activated M-linker solution and 10 μL of PLK in HEPES buffer are added to a mixture of 95 μL of DMF and 90 μL of HEPES buffer, and the solution is incubated for 30 min.

4. TMR-KLA-C in HEPES buffer is added to the reaction buffer (final concentrations: [PLK] = 0.10 mg/mL, [M-linker] = 0.14 mM, [TMR-KLA-C] = 0.27 mM), and the reaction solution is incubated for 30 min to produce the peptide-M-linker-PLK conjugate (*see* **Note 6**).

5. Then, 300 μL of HEPES buffer is added to the conjugate solution, and 50 μL of the solution is dispensed into the wells of a 96-well culture plate for immobilization. The samples are incubated for 30 h (*see* **Note 7**).

6. The wells are washed with 200 μL of wash buffer and 200 μL of MilliQ water (ten times each) and dried with N_2 on a clean bench.

3.4 Cell Culture on the Peptide-Immobilized 96-Well Plate and Peptide Photo-Release

1. Cells are seeded at 2×10^3 cells in 100 μL of culture medium per well of the peptide-immobilized 96-well plate.

2. After 1 day of culture, the medium is removed, and the cells are gently washed three times using assay media.

3. UV light irradiation is performed from a height of 20 cm using the UV-spot light source with an appropriate filter (362 ± 30 nm) and a condenser lens at 37 °C for 60 min (cytotoxicity assay) (*see* **Note 8**).

3.5 Cell Toxicity Assay with a Cell Counting Kit and a Plate Reader

1. After UV light irradiation, the cells are incubated for 2 h at 37 °C in a 5 % CO_2 environment.

2. The cells are incubated with 5 μL of the Cell Counting Kit solution for 4 h.

3. The absorbance is measured at 450 nm using the plate reader at 30 °C.

4. The difference in absorbance between the control and peptide-treated samples is calculated to determine the degree of viability associated with each cell.

5. The relative viability is calculated by dividing the absorbance of the tested cells by that of the control cells incubated without immobilized peptide (non-coating) and with light irradiation (60 min) under normal growth conditions (assuming that 100 % viability = 0 % toxicity) (Fig. 4a).

3.6 Cellular Uptake in Conjunction with Confocal Microscopy

1. After UV light irradiation, incubation, and removal of the media, 100 μL of 100 nM Mitotracker in assay media is added to the wells, and the plate is incubated for 30 min at 37 °C.

2. The cells are washed three times with assay media.

3. The cells are observed via confocal microscopy. TMR fluorescence is excited using a 543 nm HeNe(G) laser and recorded using an appropriate filter (>570 nm). Mitotracker fluorescence is excited using a 488 nm Ar laser and recorded using an appropriate filter (>510 nm).

4. The peptide permeation around the cell adhesion site can be analyzed in detail in real time (Fig. 4b).

4 Notes

1. These are standard peptide synthesis reagents. If necessary, HBTU-HOBt can be substituted with stronger reagents, such as HATU, depending on the library peptide sequences.

2. These reagents are examples of fluorophore coupling reagents. The fluorophore, and hence the coupling method, will vary depending on the assay.

3. The identities and amounts of scavengers should be changed if they do not work well for particular peptides.

4. The PLK molecular weight was optimized in [17].

5. These reagents are examples of possible cell types and media, which will vary depending on the experiment.

6. These concentrations are examples of peptide concentrations. The optimal concentrations will depend on the experiment. For the cytotoxicity assay, the concentrations were optimized in [17]. For the cell uptake assay, the concentrations were optimized in [18]. The concentration can be reduced after the detection is optimized.

7. This time period is an example of the immobilization time. The specific time needed is adjustable depending on the experiment. For the cytotoxicity assay, the time was optimized in [17]. For the cell uptake assay, the time was optimized in [18].

8. This time period is an example of the irradiation time. This time is adjustable depending on the experiment. For the cytotoxicity assay, the time was optimized in [17]. For the cell uptake assay, the time was optimized in [18].

Acknowledgments

We thank Mr. T. Kakiyama and Mr. T. Kikuchi (Tokyo Institute of Technology, Yokohama) for valuable discussions and generous support. This study was supported in part by grants from JSPS KAKENHI and NAGASE Science Technology Foundation. K.U. is also grateful to the JSPS KAKENHI Grant Number 26750375 from MEXT and the Grant-in-Aid for Encouragement of Young Scientists from Nakatani Foundation. K.-Y. T. acknowledges the Ryukoku University Science and Technology Fund.

References

1. Stockwell BR (2004) Exploring biology with small organic molecules. Nature 432:846–854

2. Kandpal R, Saviola B, Felton J (2009) The era of 'omics unlimited. Biotechniques 46: 351–355

3. Fernandes TG, Diogo MM, Clark DS, Dordick JS, Cabral JM (2009) High-throughput cellular microarray platforms: applications in drug discovery, toxicology and stem cell research. Trends Biotechnol 27:342–349

4. Usui K, Takahashi M, Nokihara K, Mihara H (2004) Peptide arrays with designed alpha-helical structures for characterization of proteins from FRET fingerprint patterns. Mol Divers 8:209–218

5. Usui K, Ojima T, Takahashi M, Nokihara K, Mihara H (2004) Peptide arrays with designed secondary structures for protein characterization using fluorescent fingerprint patterns. Biopolymers 76:129–139

6. Usui K, Tomizaki K-Y, Ohyama T, Nokihara K, Mihara H (2006) A novel peptide microarray for protein detection and analysis utilizing a dry peptide array system. Mol Biosyst 2:113–121

7. Usui K, Tomizaki K-Y, Mihara H (2006) Protein-fingerprint data mining of a designed alpha-helical peptide array. Mol Biosyst 2:417–420

8. Usui K, Tomizaki K-Y, Mihara H (2007) Screening of alpha-helical peptide ligands controlling a calcineurin-phosphatase activity. Bioorg Med Chem Lett 17:167–171

9. Usui K, Tomizaki K-Y, Mihara H (2009) A designed peptide chip: protein fingerprinting technology with a dry peptide array and statistical data mining. Methods Mol Biol 570: 273–284

10. Kodadek T (2002) Development of protein-detecting microarrays and related devices. Trends Biochem Sci 27:295–300

11. Tomizaki K-Y, Usui K, Mihara H (2005) Protein-detecting microarrays: current accomplishments and requirements. Chembiochem 6:782–799

12. Uttamchandani M, Yao SQ (2008) Peptide microarrays: next generation biochips for detection, diagnostics and high-throughput screening. Curr Pharm Des 14:2428–2438

13. Tomizaki K-Y, Usui K, Mihara H (2009) Proteins: array-based techniques. In: Begley TP (ed) Wiley encyclopedia of chemical biology, John Wiley and Sons, Inc., Hoboken, New Jersey, USA pp 144–158

14. Tomizaki KY, Usui K, Mihara H (2010) Protein-protein interactions and selection: array-based techniques for screening disease-associated biomarkers in predictive/early diagnosis. FEBS J 277:1996–2005

15. Usui K, Kakiyama T, Tomizaki K-Y, Mie M, Kobatake E, Mihara H (2011) Cell fingerprint patterns using designed α-helical peptides to screen for cell-specific toxicity. Bioorg Med Chem Lett 21:6281–6284

16. Usui K, Kikuchi T, Mie M, Kobatake E, Mihara H (2013) Systematic screening of the cellular uptake of designed alpha-helix peptides. Bioorg Med Chem 21:2560–2567

17. Kakiyama T, Usui K, Tomizaki K-Y, Mie M, Kobatake E, Mihara H (2013) A peptide release system using a photo-cleavable linker in a cell array format for cell-toxicity analysis. Polymer J 45:535–539

18. Usui K, Kikuchi T, Tomizaki K-Y, Kakiyama T, Mihara H (2013) Novel array format for monitoring cellular uptake using a photo-cleavable linker for peptide release. Chem Commun 49:6394–6396

19. Hoff A, André T, Fischer R, Voss S, Hulko M, Marquardt U, Wiesmüller K-H, Brock R (2004) Chemolabile cellular microarrays for

screening small molecules and peptides. Mol Divers 8:311–320

20. Chan WC, White PD (2000) Fmoc solid phase peptide synthesis: a practical approach. Oxford University Press, New York

21. Holmes CP, Jones DG (1995) Reagents for combinatorial organic synthesis: development of a new o-nitrobenzyl photolabile linker for solid phase synthesis. J Org Chem 60:2318–2319

22. Whitehouse DL, Savinov SN, Austin DJ (1997) An improved synthesis and selective coupling of a hydroxy based photolabile linker for solid phase organic synthesis. Tetrahedron Lett 38:7851–7852

23. Rinnová M, Nováková M, Kasicka V, Jirácek J (2000) Side reactions during photochemical cleavage of an alpha-methyl-6-nitroveratryl-based photolabile linker. J Pept Sci 6: 355–365

24. Nakayama K, Tachikawa T, Majima T (2008) Protein recording material: photorecord/erasable protein array using a UV-eliminative linker. Langmuir 24:1625–1628

25. Zhang Y, Dong Z, Nomura M, Zhong S, Chen N, Bode AM, Dong Z (2001) Signal transduction pathways involved in phosphorylation and activation of p70S6K following exposure to UVA irradiation. J Biol Chem 276:20913–20923

26. Soughayer JS, Wang Y, Li H, Cheung S-H, Rossi FM, Stanbridge EJ, Sims CE, Allbritton NL (2004) Characterization of TAT-mediated transport of detachable kinase substrates. Biochemistry 43:8528–8540

27. Ellerby HM, Arap W, Ellerby LM, Kain R, Andrusiak R, Rio GD, Krajewski S, Lombardo CR, Rao R, Ruoslahti E, Bredesen DE, Pasqualini R (1999) Anti-cancer activity of targeted pro-apoptotic peptides. Nat Med 5:1032–1038

28. Takahashi M, Nokihara K, Mihara H (2003) Construction of a protein-detection system using a loop peptide library with a fluorescence label. Chem Biol 10:53–60

29. Usui K, Ojima T, Tomizaki K-Y, Mihara H (2005) A designed glycopeptide array for characterization of sugar-binding proteins toward a glycopeptide chip technology. NanoBiotechnology 1:191–200

Part III

Peptide Microarrays for Medical Applications

Peptide Microarrays for Medical Applications in Autoimmunity, Infection, and Cancer

Carsten Grötzinger

Abstract

The diversity of the antigen-specific humoral immune response reflects the interaction of the immune system with pathogens and autoantigens. Peptide microarray analysis opens up new perspectives for the use of antibodies as diagnostic biomarkers and provides unique access to a more differentiated serological diagnosis. This review focusses on latest applications of peptide microarrays for the serologic medical diagnosis of autoimmunity, infectious diseases, and cancer.

Key words Peptide microarray, Autoantibody, Serodiagnostics, Infection, Cancer

1 Introduction

Antibodies that recognize a linear stretch of a polypeptide chain have been shown to bind peptides mimicking these amino acid sequences. The size of such an antibody binding site, a continuous epitope, was found to be usually between five and eight amino acids in length. Short synthetic peptides can therefore be utilized as probes to specifically detect antibodies from complex biological fluids. The ease and speed of production make peptides ideal probes and peptide microarrays an efficient screening platform for the determination of the antibody spectrum in human serum, liquor, saliva, and other body fluids.

Very early during their development, peptide arrays had been utilized for the mapping of antibody binding sites. This has been carried out for a vast variety of monoclonal antibodies. SPOT membranes and other array formats have been used to characterize the antibody repertoire in response to immunization in animals (reviewed in ref. 1). Over the past years, peptide microarray technology has been adopted and proven useful not only for epitope mapping of monoclonal antibodies but also for the detailed analysis of antibody profiles in human serology.

Marina Cretich and Marcella Chiari (eds.), *Peptide Microarrays: Methods and Protocols*, Methods in Molecular Biology, vol. 1352, DOI 10.1007/978-1-4939-3037-1_16, © Springer Science+Business Media New York 2016

2   Autoimmunity

One of the most obvious fields of application for peptide microarray-based immunoprofiling is the delineation of antibody reactivities in autoimmune diseases. Starting in the 1990s, several systematic papers have reported the special utility and explicit advantages of using peptide arrays versus antibody arrays (reviewed in ref. 2). Some of these studies found a higher sensitivity of immunoassays based on synthetic peptides versus other assays utilizing the respective antigen in its full-length native or recombinant form. Of note, it has been shown that in sera of autoimmune disease patients, antibodies reacting with individual peptides of a self-protein but not with the complete protein can be detected. In addition, other antibody subpopulations reacting with both peptides and protein were identified in the same sera. This peculiar phenomenon may be explained by either different sensitivities of the assays or limited exposure of the epitope in a full-length protein under certain assay conditions. Protein array or ELISA formats usually do not involve denaturing steps for the antigen immobilized.

Early on, serological analysis of autoimmune disorder patient sera using peptide macroarray revealed dominant epitopes in protein antigens and provided evidence for antibody specificities that are associated with specific disease or disease subsets. The SPOT method was utilized to create peptide libraries for epitope mapping of the lupus erythematosus autoimmune antigens Sm D1 [3], SSB/La [3, 4], and others. Many of these studies demonstrated the great potential of peptide-based diagnostic tests to serologically discriminate patients with certain autoimmune disorder from those with other rheumatic or inflammatory diseases. However, the expansion and maturation of the humoral immune response known as epitope spreading observed for both systemic and organ-specific autoimmune diseases might impair the sensitivity of diagnostic test based on single peptides, i.e., single-antibody biomarkers only [5, 6]. Peptide microarrays should present as ideal platforms for the use of cocktails of peptides to cover a majority of individuals, not only to detect relevant antibodies but also distinguish fine specificities which might be symptomatic for pathogenesis or prognosis. This is especially relevant as in some cases the relationship of the antibodies' relative concentrations will determine the final clinical manifestation of the autoimmune disease.

One pioneering publication on the use of peptide microarrays for antibody profiling of clinical samples was the study of Robinson et al. on the multiplex characterization of autoantibody responses [7]. In this paper, about 200 distinct putative autoantigens of various nature and origin were spotted on glass slides, including purified or recombinant proteins and nucleic acids. In addition to these antigens, the authors also used 154 overlapping and immunodominant synthetic peptides representing autoantigens like snRPN (small nuclear ribonucleoprotein polypeptide N) and lupus erythematosus Sm proteins as well as PARP (poly ADP-ribose polymerase) and histones. The results demonstrated that subsets of linear peptides were recognized by antibodies from patients with autoimmune diseases and further enabled the epitope mapping of autoantibody responses. The potential clinical impact of such analyses was highlighted by a more recent paper from the same group. Proteomic analysis of autoantibody reactivities using a 1536-feature microarray of 225 peptides and proteins enabled the stratification of patients with early rheumatoid arthritis into clinically relevant disease subsets. Citrullinated epitopes represented by synthetic peptides on the array were found to be present in a group of patients with features predictive of severe disease [8].

Our own group set up a peptide microarray displaying the entire sequence of the autoantigenic human thyrotropin receptor in overlapping synthetic 15-mer peptides [9], and demonstrated the unique capacity of this technology to detect antibodies in a multiplex manner with a simultaneous characterization that breaks down into their single, i.e., monoclonal, specificities. Although peptide microarray technology proved valuable for the delineation of murine monoclonal antibodies against this autoantigen at concentrations as low as of a few nanogram per milliliter [10], it failed to detect autoantibodies in sera of patients with Hashimoto thyroiditis or Graves' disease. This may most likely be attributed to the known conformational character of many autoantibodies against the TSH receptor [11].

Finally, in a recent study by Hecker et al. [12], peptide microarrays representing 24 proteins by more than 3000 overlapping 15-mer peptides were utilized to detect antibodies in sera of ten patients with limited cutaneous systemic sclerosis and ten healthy blood donors. The authors observed marked individual differences as the majority of peptide sequences were bound by antibodies of one serum sample only. In the sera of patients, but not the healthy controls, they identified antibodies to three peptides that have a characteristic sequence motif in common (GP-R/S-RR). These peptides mapped to two known linear epitopes at the N-terminus of centromere protein A (CENPA), confirming the utility of the peptide microarray approach. The same group previously had described the methodology of microarray handling, staining, image acquisition, and data analysis for the elucidation of autoimmune epitopes [13]. In addition, these authors have studied the epitope diversity of a polyclonal antibody model serum against the Sjøgren/systemic lupus erythematosus autoantigen TRIM21 [14], demonstrating the feasibility of patient antibody screening for the purpose of identifying patient subgroups by their characteristic epitope reactivities. In this paper, peptide chip results were corroborated by mass spectrometric analysis, confirming in detail the results of the immunological assay by quantitative biochemical analytics.

3   Infection

3.1   Bacterial Infection

Like with other seroanalytic areas, the SPOT technique had been utilized early on in the analysis of antibody reactions against bacteria to characterize immune reaction upon infection and vaccination. With the advent of high-density peptide microarrays, this method has also been used to enhance the quantity of features to be tested. One of the applications of peptide chips has been the identification of targets from antibody reactions. Gaseitsiwe et al. [15] have constructed a peptide microarray with 61 Mycobacterium tuberculosis presented as linear 15 aa peptides with a 3 aa shift yielding 7446 individual peptide epitopes. It was then used to profile sera of 34 individuals with active pulmonary tuberculosis and 35 healthy individuals. Data showed that 89 out of these 7446 peptides were differentially recognized and turned out to be of high diagnostic value. By analyzing sera from different geographical regions the study also demonstrated that antigen recognition patterns in sera of tuberculosis patients from Armenia and Sweden showed very similar *Mycobacterium tuberculosis* epitopes regardless of the differential genetic background of patients. This study clearly underlines the potential of peptide microarrays not only for identifying reactivity patterns of antibodies, but also to derive information about meaningful targets in bacterial infection for vaccine development.

In a related approach, peptide microarrays were also used to select diagnostic antigens of microbial infection by Echinococcus spp. with a chip presenting 45 peptides [16]. In this paper, the identification of diagnostic targets was the main focus of the study. It demonstrated that a single peptide out of the 45 was not able to discriminate reliably (or provide enough sensitivity) between different Echinococcus strains. As with similar diagnostic challenges, a combination of multiple features may increase sensitivity and specificity of the procedure. Diagnostic differentiation of different infection types was also investigated by Maksimov et al. [17]. In Toxoplasma gondii infection, distinct clinical manifestations of the disease are believed to correlate with different clonal types of the pathogen. By serotyping patients using a chip containing 54 Toxoplasma gondii peptides this study tried to determine the corresponding clonal type of infection. It turned out that type II-specific reactions were dominating and higher in intensity among the patient sera studied, confirming previous genotyping studies. In addition, the study found also type I and type III reactions.

The aforementioned approaches were all focusing on the elucidation of antibody profile. Svarovsky and Gonzales-Moa, however, demonstrated that peptide microarrays may also provide a platform to conduct bacterial binding assays [18]. In their paper, the authors describe the use of a 10,410 random peptide microarray to identify peptides competing with binding of bacteria to lipopolysaccharides.

They identified 54 out of these 10,410 peptides that showed a high inhibition rate. All 54 peptides were found to be physically similar and to contain relatively high proportions of the basic amino acids histidine, lysine, arginine, and of the aromatic amino acidtryptophane. High-density approaches like this one obviously provide an excellent tool to explore the potential of random synthetic peptides to bind defined analytes. This property will also be discussed further on in this review. It should be also mentioned that peptide microarrays have proven successful in screening whole cells against ligands immobilized on the chip and for the discovery of antibacterial peptides (for review, *see* ref. 19).

3.2   Viral Infection   During viral infection, nucleic acids of the pathogen normally can be detected earlier than antibodies. This makes detection by PCR the method of choice for most viral infections. However, for some viruses, especially those that give rise to only transient infections with either low or intermittent virus load, serology can still be more sensitive. In addition, antibody profiles are able to add considerably to PCR results. In hepatitis B virus infection, antibodies against the envelope protein can be found in patients with acute disease, while patients with chronic infections lack such antibodies. Both groups carry antibodies against the capsid protein. In addition, the antibody isotype, e.g., determination of the IgM and IgG subclass, will yield valuable information about the infections status that cannot be gained by PCR alone [20].

Epitope patterns revealed by different patient subsets could potentially be informative in predicting the patient's prognosis or in gaining insight into the immune systems' battle against the infection. The Robinson group pioneered in developing antigen microarrays to profile the breadth, strength and kinetics of epitope-specific antiviral antibody responses in vaccine trials with a simian-human immunodeficiency virus (SHIV) [21]. Four hundred and ten overlapping peptides deduced from four different SHIV antigens as well as 20 protein preparations were assembled in a SHIV proteome array and probed with sera from macaques before and after vaccination, and after challenge with a pathogenic SHIV derivative. In this way, they surveyed the evolution of anti-SHIV immune response and observed a strong convergence of antiviral B-cell specificities with a restricted set of linear epitopes in the presence of the immunogen, while vaccine-induced anti-SHIV responses against a spectrum of non-immunodominant epitopes were lost after challenge. The array profiles rather than single antibody reactivities thereby allowed distinguishing vaccinated from challenged animals, and the breadth and persistence of antiviral antibody responses served as a predictor of mortality.

As mentioned above, high-density peptide microarrays of random peptide sequence may allow identifying antibody reactivity patterns connected to clinical subtypes indicative of disease prognosis and success of therapeutic approaches. The group of Johnston has demonstrated this in a number of various applications ranging from infection to cancer. In virological studies they demonstrated the ability of a 10,410 random peptide microarray to characterize humoral immunity induced by a vaccine and also to predict vaccine efficacy [22, 23]. Although the chip used consisted of completely randomized peptide sequences and was not in any way focused on viral proteins, this approach allowed delineating such prognostic and predictive biomarker patterns in animal and human plasma samples obtained after infection or vaccination. While this random peptide approach works efficiently, more traditional approaches using known defined and protein-derived peptides immobilized on a chip are able to yield similar results with a lower number of features. Price and coworkers [24] have developed influenza virus hemagglutinin peptide microarrays and have characterized sera from 76 subjects taken immediately before and 4 weeks after vaccination with a trivalent influenza vaccine used in the 2008/2009 season. They were able to demonstrate antibody reactivities that correlated significantly with age, H1N1 and B-strain post-vaccine titer. Using bioinformatics modelling, they also could predict vaccine-induced neutralization of two seasonal influenza strains with high accuracy. In a recent paper by Pérez-Bercoff [25], high-density microchips containing 17,174 individual peptides derived from the entire cytomegalovirus proteome were utilized to characterize the immune response after hematopoietic stem cell transplantation. This proved successful, as differential immunoprofiles were recognized over different time points after infection (epitope switch). This approach may also be exploited to identify therapeutic targets in cytomegalovirus and other herpesvirus infections.

Peptide microarrays provide the opportunity to differentiate between individual antibody specificities within a heterogeneous immune response, making them an ideal platform for the serological diagnosis of viral infection. Our group developed a 900 feature peptide microarray for the parallel and differentiated immunodiagnosis of viral infections with closely related herpes viruses and hepatitis c virus genotypes (manuscript in preparation). Infections with these viruses, e.g., herpes simplex virus type 1 and 2, elicit, to a large extend, a type-common immune response with broad cross-reactivity. Thus, it is not possible to serologically discriminate between these viruses with protein-based assays. We therefore selected peptide sequences that resembled type-common as well as type-specific domains of different virus proteins. Screening of 106 characterized patient sera revealed high correlations of 90–100 % in terms of sensitivity and specificity with the reference tests. Moreover, distinct antibody reactions with type-specific peptides allowed us to demonstrate the "serologic genotyping" of hepatitis c virus genotypes and subtypes as well as differentiated diagnosis of herpes virus infections.

4   Autoantibody Signatures in Cancer

For several decades it has been known that in cancer patients circulating antibodies specific for tumor-cell-associated antigens can be detected [26, 27]. These B-cell responses to sporadic cancer are part of the immune reaction observed against tumor cells. Before the advent of peptide microchips, quite a large number of studies have been performed using phage or yeast display on one hand and SPOT technology on the other hand to describe this B-cell response in greater detail. Likewise, protein arrays have been used as tools to discover serum autoantibody markers in cancer [28]. Kilic et al. [29] used microchip technology to immobilize 51 synthetic potentially immunogenic peptides selected on the basis of cancer expression and immunome databases. They then analyzed sera from 18 patients with esophageal adenocarcinoma and 14 with gastroesophageal reflux disease to detect circulating autoantibodies. Using a combination of two of the derived protein markers, a diagnostic sensitivity of 88.9 % and a specificity of 100 % was attained. A more focused approach was chosen by Linnebacher et al. [30] in printing 308 peptides derived from human topoisomerase IIa on a peptide microarray. Sera from two patients with colorectal cancer and two control individuals were then used to detect natural epitope-specific antibodies on this chip. Corroborated by proteome analysis, the authors were able to confirm the presence of antibodies with sequential epitopes of the topoisomerase IIa in patient sera. They also suggest that screening for specific epitope-paratope interactions may assist in setting up novel assays for monitoring personalized therapies, as individual properties of antigen-antibody interactions remain distinguishable.

The aforementioned strategy of using random peptide microarrays has also been introduced for the analysis of cancer autoantibodies in patient sera. In 2012, Hughes et al. [31] have applied this methodology to analyze samples from brain cancer patients and healthy controls. The results demonstrated the power of this approach in not only distinguishing brain cancer from controls, but also important features of the pathobiology of these tumors such as type, grade, and methylation status, the latter being an important biomarker predictive of the response of glioblastoma patients to temozolomide therapy. The same group published a study in 2014 [32] describing a blinded evaluation of 100 samples from 5 cancer types and 20 normal serum samples on their random peptide microchip platform. It showed a 95 % classification accuracy, whereas a further analysis involving more than 1500 historical samples representing different disease pathologies resulted in an average accuracy of 98 %. Similar large-scale studies with nonrandom peptides are so far not available. Yet, given their high diagnostic specificity and sensitivity as well as their potentially universal applicability, random peptide microarrays appear to evolve as a powerful platform to address future serodiagnostic challenges in many clinical areas.

5   Conclusions

The enormous potential of peptide microarrays to detect, differentiate and characterize antibodies in a manner unique to this high-content, high-throughput technology imparts the appropriate diagnostic level to make antibody profiles accessible as a novel bio-markers. It may be either the presence or absence of antibodies directed against specific epitopes or rather a complex pattern of antibodies that could represent a serologic biomarker which is able to predict the severity of a disease and assist in medical decision making. Hence, the use of antibodies as diagnostic biomarkers may be one of the most promising strategies to identify patient subgroups in autoimmunity, infection, and cancer.

References

1. Andresen H, Grotzinger C (2009) Deciphering the antibodyome—peptide arrays for serum antibody biomarker diagnostics. Curr Proteomics 6:1–12

2. Fournel S, Muller S (2003) Synthetic peptides in the diagnosis of systemic autoimmune diseases. Curr Protein Pept Sci 4:261–274

3. Riemekasten G, Marell J, Trebeljahr G, Klein R, Hausdorf G, Haupl T, Schneider-Mergener J, Burmester GR, Hiepe F (1998) A novel epitope on the C-terminus of SmD1 is recognized by the majority of sera from patients with systemic lupus erythematosus. J Clin Invest 102:754–763

4. Haaheim LR, Halse AK, Kvakestad R, Stern B, Normann O, Jonsson R (1996) Serum antibodies from patients with primary Sjögren's syndrome and systemic lupus erythematosus recognize multiple epitopes on the La(SS-B) autoantigen resembling viral protein sequences. Scand J Immunol 43:115–121

5. James JA, Harley JB (1998) B-cell epitope spreading in autoimmunity. Immunol Rev 164:185–200

6. Monneaux F, Muller S (2002) Epitope spreading in systemic lupus erythematosus: identification of triggering peptide sequences. Arthritis Rheum 46:1430–1438

7. Robinson WH, DiGennaro C, Hueber W, Haab BB, Kamachi M, Dean EJ, Fournel S, Fong D, Genovese MC, de Vegvar HE et al (2002) Autoantigen microarrays for multiplex characterization of autoantibody responses. Nat Med 8:295–301

8. Hueber W, Kidd BA, Tomooka BH, Lee BJ, Bruce B, Fries JF, Sonderstrup G, Monach P, Drijfhout JW, van Venrooij WJ et al (2005) Antigen microarray profiling of autoantibodies in rheumatoid arthritis. Arthritis Rheum 52:2645–2655

9. Andresen H, Zarse K, Grotzinger C, Hollidt JM, Ehrentreich-Forster E, Bier FF, Kreuzer OJ (2006) Development of peptide microarrays for epitope mapping of antibodies against the human TSH receptor. J Immunol Methods 315:11–18

10. Andresen H, Grotzinger C, Zarse K, Kreuzer OJ, Ehrentreich-Forster E, Bier FF (2006) Functional peptide microarrays for specific and sensitive antibody diagnostics. Proteomics 6:1376–1384

11. Ando T, Latif R, Davies TF (2005) Thyrotropin receptor antibodies: new insights into their actions and clinical relevance. Best Pract Res Clin Endocrinol Metab 19:33–52

12. Hecker M, Lorenz P, Steinbeck F, Hong L, Riemekasten G, Li Y, Zettl UK, Thiesen HJ (2012) Computational analysis of high-density peptide microarray data with application from systemic sclerosis to multiple sclerosis. Autoimmun Rev 11:180–190

13. Lorenz P, Kreutzer M, Zerweck J, Schutkowski M, Thiesen HJ (2009) Probing the epitope signatures of IgG antibodies in human serum from patients with autoimmune disease. Methods Mol Biol 524:247–258

14. Al-Majdoub M, Koy C, Lorenz P, Thiesen HJ, Glocker MO (2013) Mass spectrometric and peptide chip characterization of an assembled epitope: analysis of a polyclonal antibody model serum directed against the Sjøgren/systemic lupus erythematosus autoantigen TRIM21. J Mass Spectrom 48:651–659

15. Gaseitsiwe S, Valentini D, Mahdavifar S, Magalhaes I, Hoft DF, Zerweck J, Schutkowski M, Andersson J, Reilly M, Maeurer MJ (2008) Pattern recognition in pulmonary tuberculosis defined by high content peptide microarray chip analysis representing 61 proteins from M. tuberculosis. PLoS One 3:e3840

16. List C, Qi W, Maag E, Gottstein B, Muller N, Felger I (2010) Serodiagnosis of Echinococcus spp. infection: explorative selection of diagnostic antigens by peptide microarray. PLoS Negl Trop Dis 4:e771

17. Maksimov P, Zerweck J, Maksimov A, Hotop A, Gross U, Spekker K, Daubener W, Werdermann S, Niederstrasser O, Petri E et al (2012) Analysis of clonal type-specific antibody reactions in Toxoplasma gondii seropositive humans from Germany by peptide-microarray. PLoS One 7:e34212

18. Svarovsky SA, Gonzalez-Moa MJ (2011) High-throughput platform for rapid deployment of antimicrobial agents. ACS Comb Sci 13:634–638

19. Diehnelt CW (2013) Peptide array based discovery of synthetic antimicrobial peptides. Front Microbiol 4:402

20. Storch GA (2000) Diagnostic virology. Clin Infect Dis 31:739–751

21. Neuman de Vegvar HE, Amara RR, Steinman L, Utz PJ, Robinson HL, Robinson WH (2003) Microarray profiling of antibody responses against simian-human immunodeficiency virus: postchallenge convergence of reactivities independent of host histocompatibility type and vaccine regimen. J Virol 77:11125–11138

22. Legutki JB, Magee DM, Stafford P, Johnston SA (2010) A general method for characterization of humoral immunity induced by a vaccine or infection. Vaccine 28:4529–4537

23. Legutki JB, Johnston SA (2013) Immunosignatures can predict vaccine efficacy. Proc Natl Acad Sci U S A 110:18614–18619

24. Price JV, Tangsombatvisit S, Xu G, Yu J, Levy D, Baechler EC, Gozani O, Varma M, Utz PJ, Liu CL (2012) On silico peptide microarrays for high-resolution mapping of antibody epitopes and diverse protein-protein interactions. Nat Med 18:1434–1440

25. Perez-Bercoff L, Valentini D, Gaseitsiwe S, Mahdavifar S, Schutkowski M, Poiret T, Perez-Bercoff A, Ljungman P, Maeurer MJ (2014) Whole CMV proteome pattern recognition analysis after HSCT identifies unique epitope targets associated with the CMV status. PLoS One 9:e89648

26. Kobold S, Lutkens T, Cao Y, Bokemeyer C, Atanackovic D (2010) Autoantibodies against tumor-related antigens: incidence and biologic significance. Hum Immunol 71: 643–651

27. Willimsky G, Blankenstein T (2007) The adaptive immune response to sporadic cancer. Immunol Rev 220:102–112

28. Kijanka G, Murphy D (2009) Protein arrays as tools for serum autoantibody marker discovery in cancer. J Proteomics 72:936–944

29. Kilic A, Schuchert MJ, Luketich JD, Landreneau RJ, Lokshin AE, Bigbee WL, El-Hefnawy T (2008) Use of novel autoantibody and cancer-related protein arrays for the detection of esophageal adenocarcinoma in serum. J Thorac Cardiovasc Surg 136:199–204

30. Linnebacher M, Lorenz P, Koy C, Jahnke A, Born N, Steinbeck F, Wollbold J, Latzkow T, Thiesen HJ, Glocker MO (2012) Clonality characterization of natural epitope-specific antibodies against the tumor-related antigen topoisomerase IIa by peptide chip and proteome analysis: a pilot study with colorectal carcinoma patient samples. Anal Bioanal Chem 403:227–238

31. Hughes AK, Cichacz Z, Scheck A, Coons SW, Johnston SA, Stafford P (2012) Immunosignaturing can detect products from molecular markers in brain cancer. PLoS One 7:e40201

32. Stafford P, Cichacz Z, Woodbury NW, Johnston SA (2014) Immunosignature system for diagnosis of cancer. Proc Natl Acad Sci U S A 111:E3072–E3080

Chapter 17

Synthetic Peptide-Based ELISA and ELISpot Assay for Identifying Autoantibody Epitopes

Judit Pozsgay, Eszter Szarka, Krisztina Huber, Fruzsina Babos, Anna Magyar, Ferenc Hudecz, and Gabriella Sarmay

Abstract

Enzyme-linked immunosorbent assay (ELISA) is an invaluable diagnostic tool to detect serum autoantibody binding to target antigen. To map the autoantigenic epitope(s), overlapping synthetic peptides covering the total sequence of a protein antigen are used. A large set of peptides synthesized on the crown of pins can be tested by Multipin ELISA for fast screening. Next, to validate the results, the candidate epitope peptides are resynthesized by solid-phase synthesis, coupled to ELISA plate directly, or in a biotinylated form, bound to neutravidin-coated surface and the binding of autoantibodies from patients' sera is tested by indirect ELISA. Further, selected epitope peptides can be applied in enzyme-linked immunospot assay to distinguish individual, citrullinated peptide-specific autoreactive B cells in a pre-stimulated culture of patients' lymphocytes.

Key words Anti-citrullinated peptide antibodies, B cell, Citrulline-peptide, Diagnosis, ELISA, Multipin ELISA, ELISpot, Rheumatoid arthritis

1 Introduction

Antibody secretion by autoreactive B cells has a crucial role in the pathogenesis of systemic autoimmune diseases such as rheumatoid arthritis (RA) and systemic lupus erythematosus (SLE), Sjögren's syndrome (SjS), systemic sclerosis (SSc), polymyositis (PM), and dermatomyositis (DM). Rheumatoid arthritis is the most prevalent disease in this group, affecting approximately 1 % of the population [1]. It is characterized by chronic synovitis, pannus formation causing cartilage and bone destruction [2]. The inflamed joint is infiltrated by multiple leukocytes, including T cells, B cells, macrophages, and neutrophils [3–5].

Several types of autoantibodies are evolved in RA such as rheumatoid factor (RF), anti-citrullinated protein antibodies (ACPA), anti-carbamylated protein antibodies (anti-CarP), and anti-peptidyl arginine deiminase 4 (PAD4) antibodies [6–8]. SLE is associated

Marina Cretich and Marcella Chiari (eds.), *Peptide Microarrays: Methods and Protocols*, Methods in Molecular Biology, vol. 1352, DOI 10.1007/978-1-4939-3037-1_17, © Springer Science+Business Media New York 2016

with more than 100 different autoantibodies [9]. The diagnosis of the RA is based on clinical manifestation and the serological detection of autoantibodies in the sera of patients. Approximately 60–70 % of RA patients' sera are positive for autoantibodies against citrullinated proteins [10], which are the most specific markers of RA (94–99 %) [11, 12]. Citrullination is a posttranslational modification of proteins where the guanidino group of an arginine residue is converted to an ureido group, resulting in the loss of one positive charge, a citrulline residue. This conversion is enzymatically mediated by peptidylarginine deiminases (PADs). Several citrullinated autoantigens have been identified in RA, including collagen, filaggrin, vimentin, fibrinogen, and α-enolase [13–19]. Autoantibodies specific for citrulline containing motifs of these proteins (ACPA) are present in sera as well as in synovial fluid of RA patients. ACPA are produced locally in the inflamed tissues containing citrullinated antigens and form immune complexes in vivo; thus local inflammation may promote the autoimmune response [20].

The key to therapeutic success in RA lies in identifying individuals, who will have severe destructive disease as early as possible, so that effective treatment can be initiated before irreversible damages occur. Thus, diagnosis of RA at a very early stage is crucial. ACPA can be present years before disease onset and predict more severe disease progression [21]. The easy accessibility of autoantibodies in patients' sera makes them ideal as diagnostic and prognostic markers of the disease. Detection of ACPA holds promise for earlier and more accurate diagnosis of RA and may provide improved prognostic information. The *2010 ACR/EULAR Rheumatoid Arthritis Classification Criteria* include ACPA testing for early RA diagnosis [22]. ACPA have also been implicated in RA pathogenesis, and thus identification of citrullinated protein/peptide-specific autoantibodies is also necessary for better understanding the etiology of the disease.

Several assays for detecting ACPA were developed, employing mutated citrullinated vimentin (MCV-assay), filaggrin- and fibrinogen-derived peptides (CCP-assays), and viral citrullinated peptides (VCP-assay) [23–25]. Various cyclic epitopes that mimic true conformational epitopes were selected from libraries of citrullinated peptides for the second-generation anti-CCP assay (CCP2). This assay has slightly better performance than anti-CCP1 and it is currently the most widely used diagnostic test. The third-generation anti-CCP3.1 assay has significantly higher sensitivity for ACPA detection in RF negative sera as well as in the whole population as compared to anti-CCP2 test [26].

However, both anti-CCP2 and CCP3.1 tests detect a mixture of autoantibodies with different, mostly non-cross-reacting specificities. In contrast, a panel of short citrullinated peptides should characterize individual peptide-specific antibody profile of patients, and additionally might detect CCP2- and RF-negative patients as well [27]. Moreover, using this peptide panel in an enzyme-linked immunospot (ELISpot) assay one could identify and study individual peptide-specific antibody producing B cells.

Here, we describe the selection of a set of citrullinated peptides to detect ACPA profile and to identify ACPA secreting B cells. The multipin peptide synthesis has long been described and used to map antigenic epitopes of proteins [28]. The adaption of the ELISA protocol is used with the multipin system to scan sets of citrullinated peptides and to define specific epitopes recognized by autoantibodies in RA sera. The data obtained in the multipin system are validated by conventional ELISA. To ensure the even binding of peptides with various lengths to the ELISA plates, biotinylated forms of the peptides [29] are applied on neutravidin-precoated plates. Citrulline- and arginine-containing peptide pairs are used and the ratios of sera reactivity are calculated. Finally, selected citrullinated peptides are tested by a peptide-specific ELISpot system to identify ACPA-secreting B cells.

2 Materials

2.1 Sera Samples

Blood samples must be taken with ethical permission and after the patients signed a written consent. Sera from RA patients, healthy blood donors, and the seronegative disease control group are collected and stored as 200 μl aliquots in numbered tubes at –20 °C (*see* **Notes 1** and **2**).

2.2 Buffers, Reagents, Equipment for Multipin ELISA

1. Phosphate-buffered saline (PBS): 150 mM NaCl, 3.3 mM KCl, and 8.6 mM Na_2HPO_4, in distilled water. Filtrate through 0.45 μm filter.

2. Washing solution: 0.05 % Tween 20 in PBS.

3. Blocking solution: 2 % bovine serum albumin, 0.05 % Tween 20 in PBS.

4. Serum buffer: 2 % bovine serum albumin, 0.05 % Tween 20 in PBS.

5. Antibody diluents: 2 % bovine serum albumin, 0.05 % Tween 20, in PBS.

6. Regeneration buffer: 1 % SDS, 0.1 % β-mercaptoethanol in PBS.

7. A series of citrulline-containing epitope peptides and the respective unmodified counterparts synthesized on MULTIPIN NCP (Chiron Mimotopes Peptide System) non-cleavable kit in duplicates.

8. ELISA plates.

2.3 Additional Reagents for Indirect ELISA

1. Biotinylated citrulline-containing epitope peptides and the respective unmodified counterparts.

2. Neutravidin stock solution: 0.5 mg/ml stock solution in Milli-Q ultrapure water containing 10 % glycerin (pH has to be between 7 and 8) (*see* **Note 3**).

3. Blocking solution: 2 % bovine serum albumin, 150 mM NaCl in PBS. Filtrate through 0.45 μm filter.

4. Serum buffer: 2 % bovine serum albumin + 2 M NaCl in PBS. Filtrate through 0.45 μm filter.

5. Antibody diluent: 2 % bovine serum albumin, 150 mM NaCl in PBS. Filtrate through 0.45 μm filter.

2.4 Development of ELISA

1. To detect ACPA, dilute the polyclonal rabbit anti-human IgG conjugated with horseradish peroxidase (HRP) 2000-fold in the antibody diluent.

2. HRP substrate: 3,3′,5,5′-Tetramethylbenzidine (TMB) Liquid Substrate (*see* **Notes 4** and **5**).

3. Stop reagent: 2 N sulfuric acid, which form a yellow reaction product that may be read at 450 nm by an ELISA reader.

2.5 Solutions, Reagents, Equipment for ELISpot

1. ELISpot PVDF membrane plate (*see* **Note 6**).

2. B cell purified from blood samples of patients by magnetic cell separation technique (Miltenyi Biotec), prestimulated (*see* **Note 7**).

3. Polyclonal cell activator R848 (TLR7, TLR8 agonist).

4. Recombinant human IL-2.

5. PBS: 150 mM NaCl, 3.3 mM KCl, 8.6 mM Na_2HPO_4, distilled H_2O. Filtrate through 0.45 μm filter.

6. Blocking solution: 10 % Fetal calf serum (FCS).

7. Monoclonal capture antibody for total IgG secretion (MT91/145) (Mabtech AB, Nacka Strand, Sweden).

8. Biotinylated anti-human IgG monoclonal detecting antibody.

9. Neutravidin solution, 5 μg/ml.

10. Citrulline containing peptide epitope selected by the previous tests (*see* Subheadings 3.1 and 3.2).

11. Streptavidin-Horseradish Peroxidase conjugate (Mabtech AB, Nacka Strand, Sweden).

12. TMB substrate for ELISpot assay.

3 Methods

3.1 Multipin ELISA

Conventional solid-phase peptide synthesis (Fmoc/tBu strategy) can be carried out on MULTIPIN NCP (Chiron Mimotopes Peptide System) non-cleavable kit according to the manufacturer's protocol [28]. Citrullinated peptides corresponding to various linear sequences of citrullinated proteins and the unmodified counterparts containing arginine are synthesized on the pins to compare their respective reactivity. These peptides on pins are used in a modified indirect ELISA and ACPA is determined in sera of RA patients using horseradish peroxidase labeled anti-human IgG secondary antibodies.

The advantages of the Multipin ELISA are that (a) a large set of peptides can be screened in parallel for epitope optimization and (b) the same set of pins can be reused several times after regeneration.

3.1.1 Blocking and Serum Treatment

1. Add 200 μl per well from the blocking solution in a 96-well ELISA plate, submerge the block of peptide containing pins in it and incubate them for 1 h at room temperature on a shaker.

2. Prepare a new plate adding 200 μl of the 1:1000 times diluted sera per well in duplicates, accordingly. Add serum also to the control wells (no pins, or pins without peptide); this will be your background. Test healthy control sera and sera from CCP-negative, non-RA patients with other autoimmune diseases for disease control as well. Use also a serum which is highly positive for some of the peptides as positive control (*see* **Notes 1** and **2**).

3. Take out the set of pins from the blocking solution and submerge the pins into the second plate containing the diluted sera (1:1000 in serum buffer).

4. Incubate it overnight on a shaker at 4 °C.

3.1.2 Signal Development

1. Discard the plate with the sera samples. Wash the pins three times with c.a. 50 ml of washing solution in a suitable container, by immersing the pins in the buffer for 10 min.

2. Dilute the horseradish peroxidase conjugated rabbit anti-human IgG 2000-fold in the antibody diluent. Add 200 μl to each well of an ELISA plate and submerge the pins in it. Incubate for 1 h at room temperature on a shaker.

3. Discard the plate with anti-human IgG. Wash the pins with c.a. 50 ml of the washing solution three times as at **step 1**.

4. Prepare a new plate, add TMB reagent 100 μl/well (*see* **Note 4**), and incubate the pins in it.

5. After the sufficient color development, remove the pins and stop the reaction by adding 50 μl of 2 N sulfuric acid to each well (*see* **Note 5**).

6. Read the absorbance of each well at 450 nm with an ELISA reader.

3.1.3 Analysis of the Data

1. OD (optical density) ratios (ELISA ratios) are calculated from the means of two parallel samples, divide the OD values of the citrulline containing peptide with the OD values of arginine containing control peptide. Cutoff values for each peptide pairs are calculated from healthy samples' ELISA ratios (mean ± 2× SD (standard deviation)) (*see* **Note 8**).

2. Specificity of the test is calculated as the percentages of healthy samples having an OD ratio below the cutoff value as compared to total number of samples, while sensitivity is calculated as the percentage of RA patient samples giving an OD ratio higher than the cutoff value, as compared to the total number of patient samples.

3. For statistical analysis ELISA ratios are compared between various groups using Kruskal-Wallis test.

4. GraphPad PRISM 4 software are used for statistical analysis. In all tests the P-values <0.05 are considered significant.

3.1.4 Regeneration of the Pins

1. To regenerate the pins, sonicate them for 20 min at 65 °C in 1 l regeneration buffer.

2. Rinse the pins with distillated water.

3. Wash the pins with distillated water for 30 min, in a shaking thermostat at 60 °C.

4. Rinse them with hot (60 °C) methanol for 15 s, than let them dry at room temperature (*see* **Note 9**).

3.1.5 Storage of the Multipin Blocks

Store dried pins at 4 °C.

3.2 Peptide-Based Indirect ELISA

To validate the data obtained by Multipin ELISA, selected peptides are also synthesized in a C- or N-terminally biotinylated form by solid-phase peptide synthesis according to Fmoc/tBu strategy, and tested in a conventional indirect ELISA.

3.2.1 Pre-absorption of the Sera with Neutravidin

1. To avoid aspecific binding of the anti-avidin antibodies in sera to the neutravidin-precoated plates, preincubate the sera samples with 25 μg/ml Neutravidin in serum buffer for 30 min (*see* **Notes 1** and **2**).

2. Use a serum which is highly positive for one of the peptides as positive control.

3.2.2 Coating the Plates

1. Dilute the neutravidin stock solution 100-fold in PBS. Add 100 μl to each well of an ELISA plate with a multichannel pipette. Cover the plate with an adhesive plastic and incubate it overnight at 4 °C.

2. Remove neutravidin and wash the wells with 200 μl washing solution three times. The solutions or washes can be removed by flicking the plate over a sink. The remaining drops are removed by patting the plate on a paper towel.

3. Dilute the peptides to 1 μg/ml final concentration and pipette 100 μl in duplicates to the respective wells. Incubate 1 h at 37 °C in a thermostat.

4. Remove the peptides and wash the wells with 200 μl of the washing solution three times as at **step 2**.

3.2.3 Blocking and Serum Treatment

1. To block the remaining protein binding sites on the plate add 200 μl blocking solution per well, cover the plate, and incubate it for 30 min at 37 °C in a thermostat.

2. Remove the blocking solution by flicking the plate over the sink and pipette 100 μl of 1:100 diluted, neutravidin-pre-treated sera per well in duplicates (*see* Fig. 1 for plate map).

3. Incubate it overnight in a shaker at 4 °C. To the blank wells add only the serum buffer (*see* **Note 10**).

3.2.4 Signal Development

Develop the signals and analyze the data as described in the previous chapter, at p. 5.

3.3 ELISpot Assay

The ELISpot assay is a highly sensitive test that is suitable to detect individual antibody secreting cells. We adapted this method to determine the frequency of citrullinated peptide specific antibody secreting cells in samples of in vitro-stimulated B cells from RA patients.

3.3.1 Cell Separation and Cell Culture

1. Separate B cells with Magnetic Cell Separator (MACS) Human B Cell Isolation Kit II (Miltenyi Biotec Inc. Auburn, USA) according to the manufacturer's instructions.

2. Activate B cells with 1 μg/ml R848 and 10 ng/ml rhIL-2 for 72 h in RPMI medium containing 10 % FCS.

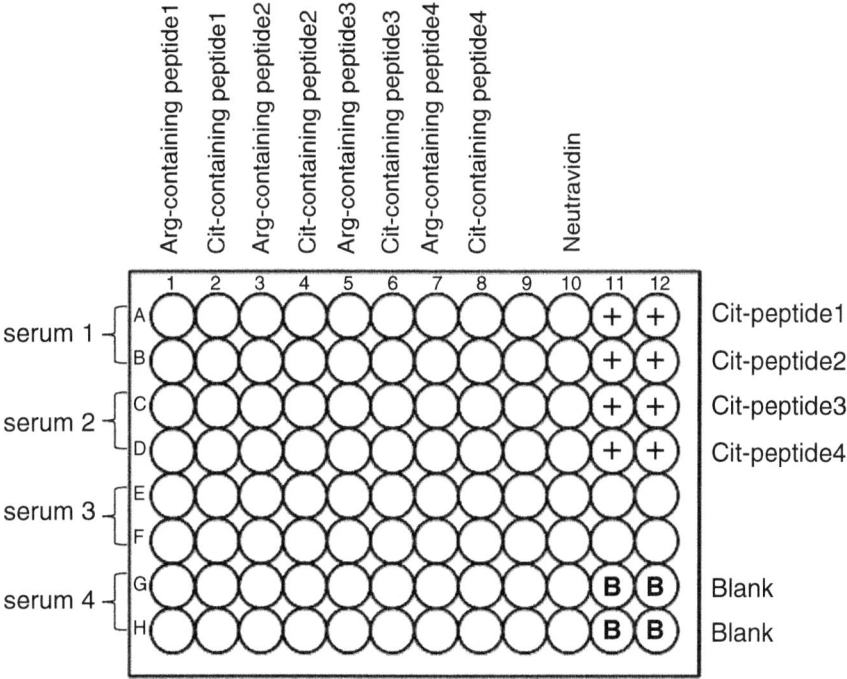

Fig. 1 Map of a typical ELISA plate. *Columns 1–12* are coated with neutravidin. At the second step the peptides and at the third step the diluted sera samples are added. *Column 10* does not contain peptides. For screening many samples, *columns 11–12* in *rows A–D* are needed for control serum, which is highly positive to each citrullinated peptides

3.3.2 Preparation of ELISpot Plate

1. Dilute the neutravidin in sterile PBS to 5 μg/ml and the coating antibodies to 15 μg/ml. For the blank wells use neutravidin coat.

2. Incubate the ELISpot plate with 70 % sterile ethanol, 50 μl/well, to wet the membrane for 2 min (*see* **Notes 11** and **12**).

3. Wash the plate five times with sterile water, 200 μl/well. To avoid leakage, always remove the plate from the plate tray before manually emptying the plate (*see* **Notes 13–15**).

4. Add 100 μl/well neutravidin or capture antibody solution and incubate the plate for 2 h at room temperature.

5. Dilute the biotinylated, citrulline-containing peptides in sterile PBS to 10 μg/ml.

6. Wash plate five times with sterile PBS, 200 μl/well.

7. Incubate the neutravidin-coated wells with the biotinylated citrulline-peptide solution, 100 μl/well, for 1 h at room temperature. Put 100 μl sterile PBS to the antibody coated wells to keep it wet.

3.3.3 Incubation of B Cells on the ELISpot Plate

1. Wash plate five times with sterile PBS, 200 μl/well.

2. For blocking, add 200 μl of the RPMI medium containing 10 % FCS for 30 min at room temperature.

3. Discard the blocking solution from the plate and to the peptide-coated wells add $2–5 \times 10^5$ cells/well in 100 μl RPMI containing 10 % FCS. For the detection of total IgG, as a positive control, put 2.5×10^4 cells to the capture antibody-coated wells.

4. Put the plate in the 37 °C humidified incubator with 5 % CO_2 and incubate for 22 h (*see* **Note 16**).

3.3.4 Detection of Spots

1. To remove the cells, empty the plate and wash five times with PBS, 200 μl/well. You do not need sterile conditions for the detection.

2. Dilute the detection antibody, biotinylated anti-human IgG to 1 μg/ml in PBS containing 0.5 % FCS. Add 100 μl/well and incubate for 2 h at room temperature.

3. Wash plate five times with PBS, 200 μl/well.

4. Dilute the streptavidin-HRP (1:1000) in PBS containing 0.5 % FCS and add 100 μl/well. Incubate for 1 h at room temperature.

5. Wash plate five times with PBS, 200 μl/well.

6. Add 100 μl/well of ready-to-use TMB substrate solution and develop the spots until distinct spots emerge.

7. Stop color development by washing in tap water. Remove the plate from the plate tray and rinse the underside of the membrane.

| Total IgG | Blank | P1-ARG | P1-CIT | P2-ARG | P2-CIT |

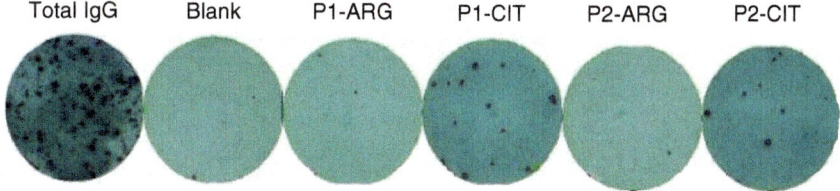

Fig. 2 A representative picture of an ELISpot assay. Each spots represent a single B cell secreting IgG specific for citrulline-containing peptides. *P* peptide, *ARG* arginine-containing peptide, *CIT* citrulline-containing peptide

8. Dry the plate in dark. Inspect and count spots in an ELISpot reader. The frequency of total and peptide-specific IgG-producing cells can be determined using a C.T.L. Immunospot analyzer (CTL-Europe Gmbh, Bonn, Germany) (*see* **Notes 17** and **18**).

   Store developed dried plates in the dark at room temperature to avoid color reduction.

   A typical pattern of spots is shown in Fig. 2.

4   Notes

1. Always use gloves when you work with human samples.
2. Keep human sera at –20 °C in aliquots and the same aliquot is suggested to be refrozen not more than three times.
3. The neutravidin stock solution may be stored at 4 °C for a month.
4. Do not use sodium azide in buffers since it inhibits HRP activity. Always use the TMB substrate at room temperature.
5. Develop the reaction in dark for maximum 15 min to have better signal.
6. The ELISpot protocol works with nitrocellulose membrane containing ELISpot plate as well, but spots are nicer on PVDF membrane.
7. B cells can also be purified by RosetteSep (Stem Cell Technologies) or by fluorescence activated cell sorter (FACS).
8. Sera sample that shows a higher OD ratio than the cutoff value + 2SD is considered to be positive.
9. For testing the regeneration of the pins run the procedure from **step 3** (without adding sera) as described in the first paragraph of this chapter. In case of high OD signals, repeat regeneration step. After five cycles of regeneration checking the efficiency is strongly recommended.

10. Do not shorten the incubation times of the protocol, because the signal may be lost.

11. Do not let the membrane to dry, because the binding capacity will be lower.

12. Put a tray into the sterile box and empty the plate above it.

13. To avoid damage of the PVDF membrane in the wells, do not touch membrane filters with pipette tips.

14. Do not use Tween or other detergent in the buffers because it can cause membrane damage.

15. Be sure that you put back correctly the plate in the plate tray, because if not, the well's fluid can slur.

16. Do not move the plate during the antibody secretion. Open and close the incubator door slowly while the plate is inside.

17. When scanning a plate in the analyzer, make sure the plate is completely inserted into the base.

18. If you coat with different peptide pairs, you will have different backgrounds. At this case you have to compare the number of spots in wells coated with the arginine- and citrulline-containing peptide pairs.

References

1. Abdel-Nasser AM, Rasker JJ, Valkenburg HA (1997) Epidemiological and clinical aspects relating to the variability of rheumatoid arthritis. Semin Arthritis Rheum 27(2):123–140

2. Hill J (2003) Cutting edge: the conversion of arginine to citrulline allows for a high-affinity peptide interaction with the rheumatoid arthritis-associated HLA-DRB1*0401 MHC class II molecule. J Immunol 171(2):538–541

3. Cantaert T et al (2008) B lymphocyte autoimmunity in rheumatoid synovitis is independent of ectopic lymphoid neogenesis. J Immunol 181(1):785–794

4. Yeo L et al (2011) Cytokine mRNA profiling identifies B cells as a major source of RANKL in rheumatoid arthritis. Ann Rheum Dis 70(11):2022–2028

5. Koo J et al (2013) Increased lymphocyte infiltration in rheumatoid arthritis is correlated with an increase in LTi-like cells in synovial fluid. Immune Netw 13(6):240–248

6. Giles JT et al (2014) Association of cross-reactive antibodies targeting peptidyl-arginine deiminase 3 and 4 with rheumatoid arthritis-associated interstitial lung disease. PLoS One 9(6):e98794

7. Trouw LA, Mahler M (2012) Closing the serological gap: promising novel biomarkers for the early diagnosis of rheumatoid arthritis. Autoimmun Rev 12(2):318–322

8. Auger I et al (2012) Autoantibodies to PAD4 and BRAF in rheumatoid arthritis. Autoimmun Rev 11(11):801–803

9. Sherer Y et al (2004) Autoantibody explosion in systemic lupus erythematosus: more than 100 different antibodies found in SLE patients. Semin Arthritis Rheum 34(2):501–537

10. Klareskog L et al (2008) Immunity to citrullinated proteins in rheumatoid arthritis. Annu Rev Immunol 26:651–675

11. Suzuki K et al (2003) High diagnostic performance of ELISA detection of antibodies to citrullinated antigens in rheumatoid arthritis. Scand J Rheumatol 32(4):197–204

12. van Venrooij WJ, Hazes JM, Visser H (2002) Anticitrullinated protein/peptide antibody and its role in the diagnosis and prognosis of early rheumatoid arthritis. Neth J Med 60(10):383–388

13. Simon M (1993) The cytokeratin filament-aggregating protein filaggrin is the target of the so-called "antikeratin antibodies", autoantibodies specific for rheumatoid arthritis. J Clin Invest 92(3):1387–1393

14. Schellekens GA et al (1998) Citrulline is an essential constituent of antigenic determinants

recognized by rheumatoid arthritis-specific autoantibodies. J Clin Invest 101(1):273–281

15. Girbal-Neuhauser E et al (1999) The epitopes targeted by the rheumatoid arthritis-associated antifilaggrin autoantibodies are posttranslationally generated on various sites of (pro)filaggrin by deimination of arginine residues. J Immunol 162(1):585–594

16. Masson-Bessiere C et al (2001) The major synovial targets of the rheumatoid arthritis-specific antifilaggrin autoantibodies are deiminated forms of the alpha- and beta-chains of fibrin. J Immunol 166(6):4177–4184

17. Asaga H, Yamada M, Senshu T (1998) Selective deimination of vimentin in calcium ionophore-induced apoptosis of mouse peritoneal macrophages. Biochem Biophys Res Commun 243(3):641–646

18. Van Steendam K, Tilleman K, Deforce D (2011) The relevance of citrullinated vimentin in the production of antibodies against citrullinated proteins and the pathogenesis of rheumatoid arthritis. Rheumatology (Oxford) 50(5):830–837

19. Lundberg K et al (2008) Antibodies to citrullinated alpha-enolase peptide 1 are specific for rheumatoid arthritis and cross-react with bacterial enolase. Arthritis Rheum 58(10):3009–3019

20. Masson-Bessiere C et al (2000) In the rheumatoid pannus, anti-filaggrin autoantibodies are produced by local plasma cells and constitute a higher proportion of IgG than in synovial fluid and serum. Clin Exp Immunol 119(3):544–552

21. Schellekens GA et al (2000) The diagnostic properties of rheumatoid arthritis antibodies recognizing a cyclic citrullinated peptide. Arthritis Rheum 43(1):155–163

22. Aletaha D et al (2010) 2010 rheumatoid arthritis classification criteria: an American College of Rheumatology/European League Against Rheumatism collaborative initiative. Ann Rheum Dis 69(9):1580–1588

23. Nishimura K et al (2007) Meta-analysis: diagnostic accuracy of anti-cyclic citrullinated peptide antibody and rheumatoid factor for rheumatoid arthritis. Ann Intern Med 146(11):797–808

24. Bang H et al (2007) Mutation and citrullination modifies vimentin to a novel autoantigen for rheumatoid arthritis. Arthritis Rheum 56(8):2503–2511

25. Coenen D et al (2007) Technical and diagnostic performance of 6 assays for the measurement of citrullinated protein/peptide antibodies in the diagnosis of rheumatoid arthritis. Clin Chem 53(3):498–504

26. Swart A et al (2012) Third generation anti-citrullinated peptide antibody assay is a sensitive marker in rheumatoid factor negative rheumatoid arthritis. Clin Chim Acta 414:266–272

27. Szarka E et al (2014) Recognition of new citrulline-containing peptide epitopes by autoantibodies produced in vivo and in vitro by B cells of rheumatoid arthritis patients. Immunology 141(2):181–191

28. Geysen HM et al (1987) Strategies for epitope analysis using peptide synthesis. J Immunol Methods 102(2):259–274

29. Babos F et al (2013) Role of N- or C-terminal biotinylation in autoantibody recognition of citrullin containing filaggrin epitope peptides in rheumatoid arthritis. Bioconjug Chem 24(5):817–827

Chapter 18

IgE and IgG4 Epitope Mapping of Food Allergens with a Peptide Microarray Immunoassay

Javier Martínez-Botas and Belén de la Hoz

Abstract

Peptide microarrays are a powerful tool to identify linear epitopes of food allergens in a high-throughput manner. The main advantages of the microarray-based immunoassay are the possibility to assay thousands of targets simultaneously, the requirement of a low volume of serum, the more robust statistical analysis, and the possibility to test simultaneously several immunoglobulin subclasses. Among them, the last one has a special interest in the field of food allergy, because the development of tolerance to food allergens has been associated with a decrease in IgE and an increase in IgG4 levels against linear epitopes. However, the main limitation to the clinical use of microarray is the automated analysis of the data. Recent studies mapping the linear epitopes of food allergens with peptide microarray immunoassays have identified peptide biomarkers that can be used for early diagnosis of food allergies and to predict their severity or the self-development of tolerance. Using this approach, we have worked on epitope mapping of the two most important food allergens in the Spanish population, cow's milk and chicken eggs. The final aim of these studies is to define subsets of peptides that could be used as biomarkers to improve the diagnosis and prognosis of food allergies. This chapter describes the protocol to produce microarrays using a library of overlapping peptides corresponding to the primary sequences of food allergens and data acquisition and analysis of IgE- and IgG4-binding epitopes.

Key words Peptide microarray, Epitope mapping, Food allergy, Linear epitope, Biomarkers

1 Introduction

Food allergy has become an important health problem in industrialized countries in recent decades. It is generally accepted that food allergy affects 1–2 % of individuals in these populations and 6–8 % of children under 3 years of age [1, 2]. The most common food allergies in children are cow's milk, hen's eggs, peanuts, and fish.

The diagnosis of food allergies is based on a detailed medical history (family history, diet, symptoms), the presence of specific immunoglobulin E (IgE) antibodies directed against proteins in the offending food using a skin prick test or radioallergosorbent test (RAST), and a double-blind placebo-controlled food challenge test (DBPCFC), which demonstrates the relationship

Marina Cretich and Marcella Chiari (eds.), *Peptide Microarrays: Methods and Protocols*, Methods in Molecular Biology, vol. 1352, DOI 10.1007/978-1-4939-3037-1_18, © Springer Science+Business Media New York 2016

between symptoms and the intake of the offending food. The DBPCFC is the gold standard test for confirming a diagnosis. Once the food allergy diagnosis is established, the management of the patient includes the prescription of an offending-food-free diet and regular follow-up visits to reevaluate the patient's allergic status [2]. Patients and their families must be aware that all forms of the offending food must be avoided at all meals, everyday, and must be trained to administer emergency treatment in cases of accidental ingestion, including antihistamines, corticosteroids, and autoinjectable adrenaline for anaphylaxis. In most cases, IgE-mediated cow milk and hen egg allergies are transient and young children outgrow these allergies by 3–5 years of age [3–6]. However, children with peanut or fish allergy do not as readily outgrow their allergies [1].

Although the only current treatment option for persistent food allergy is strict avoidance of the allergen, promising immunotherapeutic treatments have been developed in recent years [7]. Among these, oral immunotherapy (OIT) has been shown to be highly effective in desensitizing individuals to cow's milk, eggs, and peanuts [8–11], although the treatment is not risk-free [9, 12]. OIT involves the administration of gradually increasing quantities of the offending food, starting with very low doses and gradually increasing up to an amount equivalent to the usual daily intake.

IgE plays a central role in the development of immunotolerance, and food-specific IgE antibodies can be used as a marker of persistent allergy and the development of tolerance. High food-specific IgE levels predict clinical reactivity and the persistence of allergy [3, 13], whereas decreasing food-specific IgE levels over time predict developing tolerance [14]. Other specific antibody classes, such as IgG4, have also been associated with the development of tolerance, and high food-specific IgG4 levels during infancy are associated with the earlier development and maintenance of tolerance [13, 15, 16].

Food allergen epitopes are usually classified as either linear or conformational epitopes, based on the recognition of its primary structure (linear sequence of amino acids) or the three-dimensional tertiary structure of the proteins, respectively [17]. In recent years, different food allergens have been identified and the importance of sequential epitopes in food allergies has been established. Sensitization to linear epitopes is much more common in individuals with persistent allergies than in those who have overcome milk [18–23], peanut [24–26], or egg allergy [27, 28]. Linear epitopes are also associated with more-severe reactions to milk [22, 23], peanut [24, 25], lentils [29], fish [30, 31], shrimp [32], and wheat [33]. The development of tolerance to milk is associated with reduced IgE levels and increased IgG4 levels against linear epitopes [22, 34].

Different methods have been used to map the linear epitopes of food allergens (enzyme-linked immunosorbent assay [ELISA], immunoblotting, immunodot blotting, RAST, SPOT® membrane), but peptide microarrays have recently become a powerful tool for identifying linear epitopes in a high-throughput manner. Microarray-based immunoassays allow thousands of peptides to be assayed simultaneously and several immunoglobulin subclasses to be tested simultaneously and require only small volumes of patient samples. However, a major limitation to the clinical use of microarray is the automated analysis of the data [35].

Recent studies mapping the linear epitopes of food allergens with peptide microarray immunoassays have identified peptide biomarkers for the diagnosis and prognosis of food allergies. Savilahti et al. developed a panel of 15 peptides that bind IgE and IgG4 to distinguish at the time of diagnosis those patients with persistent cow's milk allergy from those who will recover early [22]. Lin et al. proposed a decision tree built to classify peanut-allergic and peanut-tolerant individuals based on three peptides [26]. Perez-Gordo et al. identified a 15-amino-acid antigenic region in the Gad m 1 allergen from the Atlantic cod for use as a biomarker of the severity of fish allergy [31]. More recently, Savilahti et al. and de la Hoz et al. proposed two possible approaches to predicting the outcome of OIT for cow's milk allergy with a peptide microarray immunoassay [36, 37].

For the last decade, using the advantages of the microarray technology, we have been involved in epitope mapping the two most important food allergens in the Spanish population, cow's milk and chicken eggs. In this chapter, we describe the methods used for microarray fabrication, sample hybridization, and data analysis.

2 Materials

2.1 Microarray Printing

1. SpotArray 72 spotter (PerkinElmer, Waltham, MA, USA) with SMP6 split pins (Arrayit Corporation, Sunnyvale, CA, USA).

2. Dendron-modified slides, functionalized with N-hydroxyl succinimidyl (NSB27 NHS) (NanoSurface Biosciences POSTECH, Seoul, Korea) (see Note 1).

3. A library of overlapping peptides corresponding to the primary sequences of the food allergen of interest is synthesized commercially, with a purity >70 %, and analyzed with high-performance liquid chromatography and mass spectrometry (GenScript Corporation, Piscataway, NJ, USA) (see Note 2). The custom-made peptides are usually supplied as lyophilized powder, resuspended in phosphate-buffered saline (PBS; bio-Mérieux, Marcy l'Etoile, France) as stock concentrations of 20 mg/mL, and stored at −80 °C until use.

4. Anti-CREB-1 antibody blocking peptide (Santa Cruz Biotechnology, Dallas, TX, USA).

5. Ovomucoid and bovine serum albumin (BSA; Sigma-Aldrich, St. Louis, MO, USA).

6. Protein printing buffer (PPB; Arrayit Corporation).

7. V-bottomed 384-well plates (Greiner Bio-One, Monroe, NC, USA).

8. Aluminum seal tape (Nalge Nunc International, Rochester, NY, USA).

2.2 Microarray Hybridization

1. microBOX™ incubation chamber for glass slides (Quantifoil Instruments GmbH, Jena, Germany) with a slide holder for four glass slides.

2. MixMate orbital shaker (Eppendorf AG, Hamburg, Germany).

3. Hydrophobic Super Pap Pen (Daido Sangyo Co. Ltd., Tokyo, Japan).

4. PBS containing 0.1 % Tween 20 (PBS-T).

5. BSA (Sigma-Aldrich).

6. BlockIt™ blocking buffer (Arrayit Corporation).

7. Secondary antibodies: monoclonal mouse antihuman IgE antibody (clone G7-26) and monoclonal mouse IgG4 antibody (clone G17-4; BD-PharMingen, San Diego, CA, USA).

8. Spike-in control antibody: Alexa Fluor® 647-conjugated anti-CREB-1 rabbit polyclonal antibody (240; Santa Cruz Biotechnology).

9. Alexa Fluor® 546 Monoclonal Antibody Labeling Kit (Invitrogen, Molecular Probes, Eugene, OR, USA).

10. Alexa Fluor® 647 Monoclonal Antibody Labeling Kit (Invitrogen, Molecular Probes).

11. Centricon-30 concentrator (Amicon, Beverly, MA, USA).

12. Centrifuge with a JA14 rotor (Beckman Coulter, Inc., Fullerton, CA).

2.3 Data Acquisition and Analysis

1. ScanArray® Express microarray scanner (PerkinElmer) with two lasers (543 and 633 nm) and the associated software.

2. Arrayit microarray air jet (Arrayit Corporation).

3. Microsoft Excel (Microsoft Corp., Redmond, WA, USA) and TIGR MultiExperiment Viewer (MeV v3.1) (http://www.tm4.org/) software.

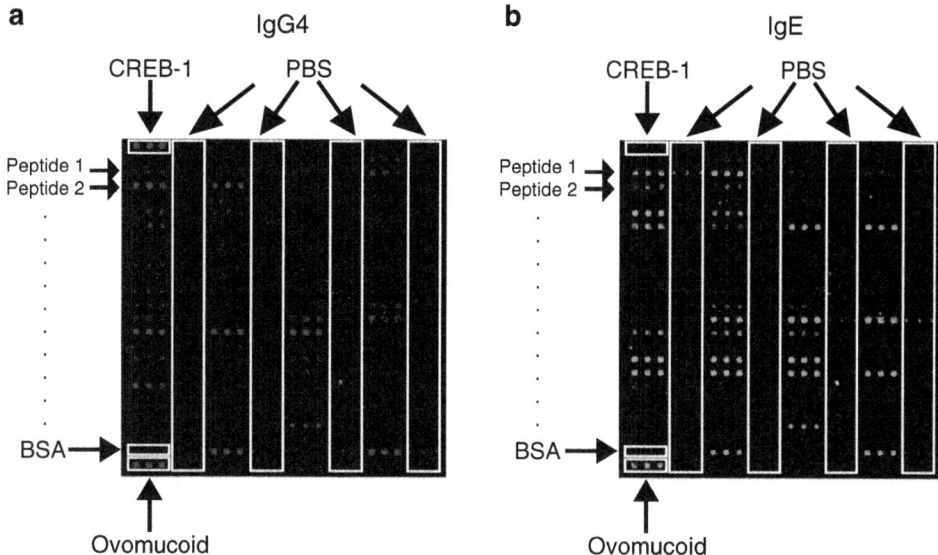

Fig. 1 Example of a peptide microarray. The serum of an egg-allergic patient was incubated with a microarray containing a set of overlapping 20-amino-acid peptides corresponding to the primary sequence of ovomucoid and labeled with Alexa Fluor® 647-conjugated antihuman IgG4 and Alexa Fluor®-546-conjugated antihuman IgE secondary antibodies. CREB-1 blocking peptide was printed as the spike-in control. Ovomucoid and BSA were printed as the positive and negative controls, respectively. PBS was printed to minimized any cross-contamination and for background normalization. Each feature (peptide, proteins, or PBS) was printed in triplicate. The microarray was scanned with a red laser to detect Alexa Fluor® 647 (**a**) and with a green laser to detect Alexa Fluor® 546 (**b**)

3   Methods

3.1   Peptide Microarray Printing

1. To prepare the printing plate, stock solutions of peptides (20 mg/mL) are diluted 1:10 with PBS and transferred (10 μL/well) to V-bottomed 384-well plates, according to the printing protocol (Fig. 1). As the spike-in control, a blocking peptide for the anti-CREB-1 antibody is printed at the beginning of the microarray (100 μg of peptide in 0.5 mL of PBS containing <0.1 % sodium azide and 0.2 % BSA). Ovomucoid and BSA are also printed as the positive and negative controls, respectively. One well containing 10 μL of PBS is included between each peptide or protein to minimize cross-contamination and for background normalization (PBS spots). Once the plate is complete, 10 μL of protein printing buffer (Arrayit Corporation) is added to each well. The final concentration of each peptide is 1 mg/mL. The printing plates are sealed with aluminum seal tape and stored at −80 °C.

2. For printing, a SpotArray 72 spotter is used with SMP6 split pins. Each feature (peptide, protein, or PBS) is usually printed in triplicate. The humidity for printing is adjusted to 60 %. To minimize the cost of the slides and to increase the number of samples that can be processed simultaneously, two microarrays are printed on each slide. After the printing procedure is complete, the microarrays are left overnight in the deck of the printer, and the printer kept closed, to allow the slow equilibration of the humidity with the ambient conditions and the spots to dry slowly. The slides produced are labeled with barcode tags, stored in a box, and sealed in a plastic bag with a desiccant bag (silica gel packet).

3.2 Microarray Hybridization

3.2.1 Antibody Labeling

1. Antihuman IgE secondary antibody is labeled with the Alexa Fluor® 546 Monoclonal Antibody Labeling Kit and antihuman IgG4 secondary antibody is labeled with the Alexa Fluor® 647 Monoclonal Antibody Labeling Kit. Both antibodies are supplied at concentrations of 0.5 mg/mL in 1 mL of aqueous buffered solution containing ≤0.09 % sodium azide. The optimal concentration for labeling with the monoclonal antibody labeling kits is 1 mg/mL and these kits are optimized to label 100 µg of antibody per reaction. The presence of low concentrations of sodium azide (≤3 mM) does not interfere with the conjugation reaction. Therefore, in the first step, the secondary antibodies are concentrated about twofold with a Centricon-30 concentrator, with centrifugation in a JA14 rotor for 15 min at 4 °C.

2. The antibody solution is recovered and 90 µL of antibody solution containing 10 µL of 1 M sodium bicarbonate buffer is transferred to a vial of reactive dye. The vial is capped and gently inverted several times to fully dissolve the dye.

3. The solution is incubated for 1 h at room temperature, with gentle inversion every 10–15 min to mix the two reactants and increase the labeling efficiency.

4. An aliquot (100 µL) of the reaction volume is loaded dropwise into the center of the resin at the top of the spin column and centrifuged for 5 min at $1100 \times g$.

5. The labeled antibody is eluted in approximately 100 µL of PBS (pH 7.2) containing 2 mM sodium azide and stored at 4 °C, protected from light.

3.2.2 Blocking

1. A round area that includes the printed microarray is drawn using the hydrophobic Super Pap Pen (see **Note 3**). The slides are placed in a microBOX™ incubation chamber.

2. To establish constant humidity in the microBOX™ incubation chamber, gauze moistened with deionized water is placed in a basin within the chamber.

3. The slides are loaded into the slide holder and a 1:1 solution of BlockIt™ blocking buffer and PBS-T containing 2 % BSA is placed onto each microarray to block nonspecific binding.

4. The slide holder is loaded into the microBOX™, which is placed on a MixMate orbital shaker (Eppendorf AG) for gentle agitation.

5. The microarrays are incubated for 1 h at room temperature.

6. The blocking solution is removed by aspiration and the samples washed twice with PBS. Using one microBOX™ with two slide holders and four glass slides, 16 samples can be processed simultaneously.

3.2.3   Serum Incubation

1. The patient's serum (50 μL), diluted 1:1 with PBS-T containing 2 % BSA, is placed onto each microarray and then into the microBOX™.

2. The microarrays are incubated overnight on an orbital shaker at 4 °C with gentle agitation.

3. The slides are removed from the microBOX™ and placed into 50 mL tubes (two slides per tube) for washing. The slides must be placed with the nonprinted faces in contact and the printed areas facing outward.

4. The microarrays are washed twice (3 min each) with 40 mL of PBS-T at room temperature on an orbital shaker with gentle agitation.

3.2.4   IgE and IgG4 Detection

1. The microarrays are loaded into the microBOX™ and incubated for 2 h at room temperature with a mixture of antibodies diluted in PBS-T containing 2 % BSA: Alexa Fluor® 647-conjugated anti-CREB-1 rabbit polyclonal antibody, diluted 1:500, antihuman IgE antibody covalently tagged with Alexa® 546 and diluted 1:1000, and antihuman IgG4 antibody covalently tagged with Alexa® 647 and diluted 1:1000 (*see* **Note 4**).

2. The slides are placed into a 50 mL tube and washed twice (3 min each) with 40 mL of PBS-T at room temperature on an orbital shaker with gentle agitation.

3.3   Data Acquisition and Analysis

3.3.1   Microarray Scanning

1. Before scanning, the slides are washed quickly with deionized water and dried under a particle-free air stream produced with the Arrayit microarray air jet.

2. The slides are scanned with ScanArray Express. The Alexa® 546 signal is scanned with a green laser (543 nm) and the Alexa® 647 signal with a red laser (633 nm). The scanner produces green Alexa® 546 and red Alexa® 647 16-bit TIFF image files.

3.3.2 Gridding, Spot Finding, and Quantitation

1. The Alexa® 546 and Alexa® 647 Tiff images are loaded into the ScanArray Express software.

2. The grid information from the gal file generated by the SpotArray 72 spotter software is loaded, and the positive controls at the beginning and end of the microarray (CREB-1 and ovomucoid, respectively) are used for grid alignment (Fig. 1).

3. The adaptive circle algorithm is used for spot finding and quantitation.

4. Before the quantitative data are saved, visual inspection of the spots is required to confirm that all the spots are correctly detected and have the appropriate shape and size.

5. PBS spots with carry-over contamination are removed.

6. After visual inspection, any unreliable spots are flagged and removed from further analysis.

7. The quantitation results are saved in the comma-separated value (CSV) file format.

3.3.3 Data Analysis and Visualization

1. The microarray analysis is performed according to the method of Lin et al. [38, 39]. This method is based on the calculation of a Z-score, which is a measure of the standardized fluorescence intensity for IgE and IgG4 recognition. The Z-score is calculated for each peptide spot (Z_i) according to the formula, $Z_i = [S_i - S_{PBS}]/\text{MAD}(S_{PBS})$, where S for each peptide spot (S_i) or PBS (S_{PBS}) is the median fluorescent signal of the spot divided by the local background and log2 transformed. $\text{MAD}(S_{PBS})$ is the median absolute deviation of all the readouts for PBS spots. The total Z-value for each peptide is the median of the Z-scores for the three replicate spots within the same microarray. Finally, to accommodate overlapping peptides, the weighted average of Z-values is calculated according to the formula: $Z = (0.25 \times Z_{-1}) + (0.5 \times Z_0) + (0.25 \times Z_{+1})$. An individual peptide sample is considered positive if the standardized fluorescence intensity (represented as the weighted average Z-score) exceeds 3. Major specific IgE (sIgE) epitopes are identified based on the IgE/IgG4 Z-score ratio, i.e., when the ratio is >1 in more than 20 % of the patient sera. The Z-score calculation for each data set is performed in Microsoft Excel software using *pivot tables* and links to the original csv data files.

2. Next, a tab-delimited text file containing the weighted average Z-score for each data set (IgE, IgG4, or IgE/IgG4 ratio) is generated in Microsoft Excel. The weighted average Z-scores for each peptide are in rows and the samples in columns. Peptide annotation labels are displayed in the row headers and sample annotation labels are displayed in the column headers (Fig. 2).

Samples

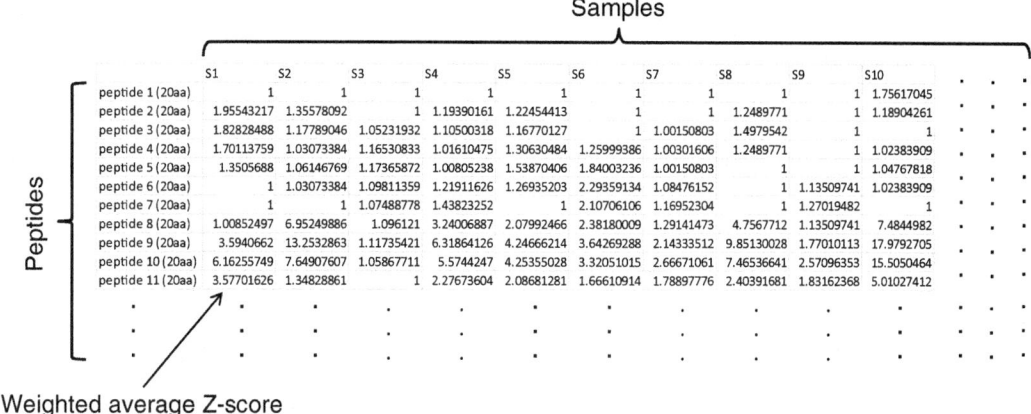

	S1	S2	S3	S4	S5	S6	S7	S8	S9	S10
peptide 1 (20aa)	1	1	1	1	1	1	1	1	1	1.75617045
peptide 2 (20aa)	1.95543217	1.35578092	1	1.19390161	1.22454413	1	1	1.2489771	1	1.18904261
peptide 3 (20aa)	1.82828488	1.17789046	1.05231932	1.10500318	1.16770127	1	1.00150803	1.4979542	1	1
peptide 4 (20aa)	1.70113759	1.03073384	1.16530833	1.01610475	1.30630484	1.25999386	1.00301606	1.2489771	1	1.02383909
peptide 5 (20aa)	1.3505688	1.06146769	1.17365872	1.00805238	1.53870406	1.84003236	1.00150803	1	1	1.04767818
peptide 6 (20aa)	1	1.03073384	1.09811359	1.21911626	1.26935203	2.29359134	1.08476152	1	1.13509741	1.02383909
peptide 7 (20aa)	1	1	1.07488778	1.43823252	1	2.10706106	1.16952304	1	1.27019482	1
peptide 8 (20aa)	1.00852497	6.95249886	1.096121	3.24006887	2.07992466	2.38180009	1.29141473	4.7567712	1.13509741	7.4844982
peptide 9 (20aa)	3.5940662	13.2532863	1.11735421	6.31864126	4.24666214	3.64269288	2.14333512	9.85130028	1.77010113	17.9792705
peptide 10 (20aa)	6.16255749	7.64907607	1.05867711	5.5744247	4.25355028	3.32051015	2.66671061	7.46536641	2.57096353	15.5050464
peptide 11 (20aa)	3.57701626	1.34828861	1	2.27673604	2.08681281	1.66610914	1.78897776	2.40391681	1.83162368	5.01027412

Weighted average Z-score

Fig. 2 Example of a tab-delimited text file containing the weighted average *Z*-score for each data set used for visualization and clustering analysis with the MeV v3.1 software

3. Data visualization and clustering are performed with the TIGR MultiExperiment Viewer (MeV v3.1) software (http://www.tm4.org/). MeV is an open-source software tool used to analyze and visualize processed data from microarray experiments. Although MeV has been developed for gene expression microarrays, many of its features are useful in epitope-mapping peptide microarray analyses (Fig. 3). The display color scheme can be adjusted from a single gradient to a double gradient. A single gradient is used to represent IgE recognition (black to green gradient) or IgG4 recognition (black to red gradient). A double gradient (red–black–green) is used to represent the IgE/IgG4 recognition ratio. Increasing green intensity denotes predominant IgE recognition, increasing red intensity denotes predominant IgG4 recognition, and black represents unrecognized spots or spots that do not differ in their IgE/IgG4 recognition. To visualize the IgE/IgG4 *Z*-score ratio, the data are usually log2 transformed. Log2 (ratio) measures allow data representation, stabilize the variance, compress the range of the data, and increase the normal distribution of the data, which allow the data to be analyzed statistically. The MeV software includes different methods of clustering. A hierarchical clustering analysis can be used to identify differential recognition patterns between allergic patients. It can also be used in several statistical analyses, such as the *t*-test or one-way ANOVA, to identify peptides that are differentially recognized in groups of patients. More complex analyses can be performed using the R programming language, including linear and nonlinear modeling, time-series analysis, classification, clustering, data mining, machine learning analysis, etc. (http://www.r-project.org).

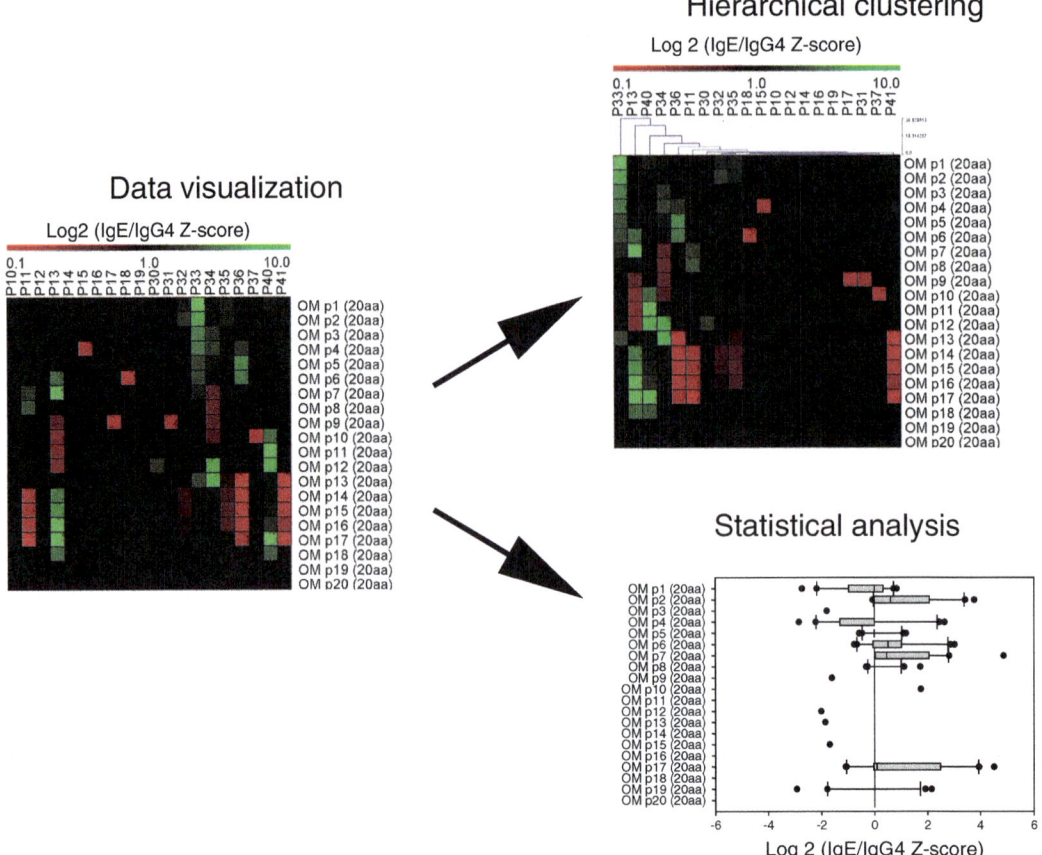

Fig. 3 Visualization and data analysis using the MeV v3.1 software

4   Notes

1. We have tested dendron-modified slide functionalized with *N*-hydroxyl succinimidyl (NSB27 NHS; NanoSurface Biosciences POSTECH) and several slides coated with epoxide groups, and in our experience, the dendron-modified slide gives the best results (Fig. 4). NSB27 NHS slides are coated uniformly with a cone-shaped organic compound, called nano-cone, functionalized with *N*-hydroxyl succinimidyl. Coating the slides with nanocone ensures controlled regular spacing between the proteins or peptides, minimizes the steric hindrance, and enhances the binding kinetics [40]. When the slide surface is coated with this monodispersed molecule, branches of the nanocone form stable chemical bonds with the surface of the substrate, and the functional group at the cone's apex is utilized for the immobilization of a bioactive molecule (Fig. 5). The average spacing on NSB27 NHS slides is 6–7 nm, which is

Serum dilution

1/2   1/5   1/10

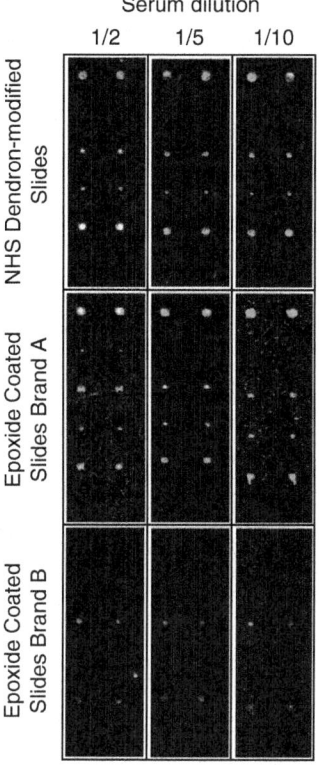

Fig. 4 Performance test for different microarray slides. Different dilutions of serum from an egg-allergic patient were incubated with three different types of slides and labeled with Alexa Fluor® 546-conjugated antihuman IgE secondary antibody

recommended for protein microarrays. NHS-microarray slides contain a highly reactive NHS ester surface, which forms avid covalent bonds with primary amine groups. The NHS slides offer a high loading capacity and extremely low background for the immobilization of proteins (antibodies and antigens).

2. Several peptide lengths have been reported in the literature and the signal intensity generally increases for both IgE and IgG4 as the peptide length increases [41]. In our experience, the best performance is achieved with 20-amino-acid peptides [27]. To design a library of overlapping peptides, it is important to exclude the sequences corresponding to the signal peptide because is not present in the mature protein and is therefore not antigenic.

3. Good performance is also achieved with LifterSlips™ Cover Slips (24x24l-2-4709; Erie Scientific Company, Portsmouth, NH, USA) or CoverWell perfusion chambers (PC1R-0.5; Grace Bio-Labs, Bend, OR, USA).

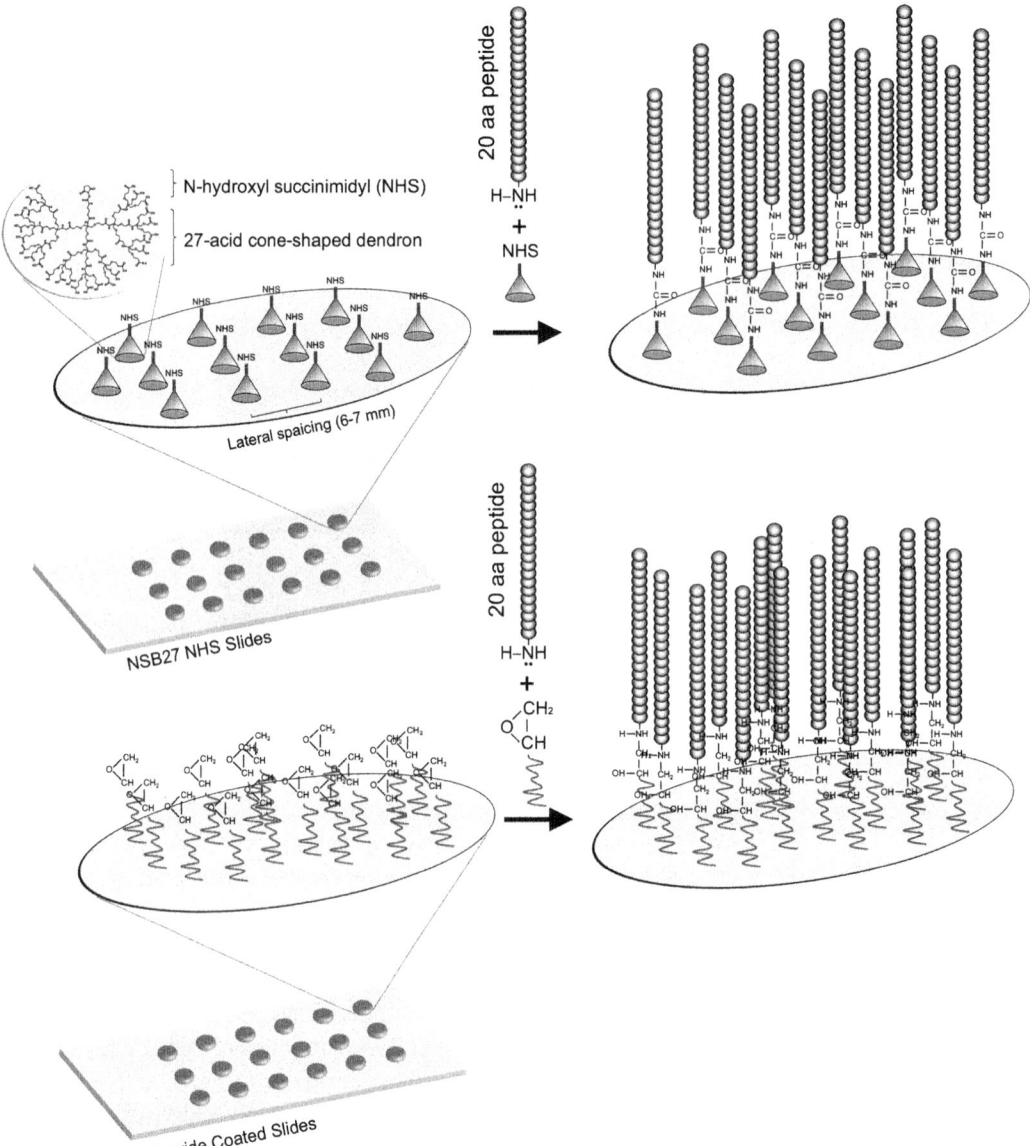

Fig. 5 Schematic illustration of dendron-modified slides functionalized with *N*-hydroxyl succinimidyl (NSB27 NHS slide) and epoxy-coated slides. The chemical structure of the 27-acid cone-shaped dendron is shown. After immobilization and deprotection, the NH_2 group at the cone apex is activated and generates NHS groups, which form avid covalent bonds with the primary amine groups of peptides. The highly reactive epoxide groups on the epoxide-coated slides react specifically with the primary amines of peptides, resulting in covalent bonds

4. The working concentration of each batch of secondary antibody may vary, depending on the labeling efficiency, and must be tested after each labeling reaction using serial dilutions. To improve the sensitivity of IgE recognition, a biotin mouse anti-human IgE (clone G7-26; BD-PharMingen) can be used in combination with Cy™3-streptavidin (PA43001; Amersham GE Healthcare, Little Chalfont, UK).

Acknowledgments

We wish to thank Veronica de Dios and Lorena Crespo for their excellent technical assistance. We also thank Raquel Rodriguez and Joon Won Park for their help and revision of the illustration of dendron-modified slides. This work was supported by grants from the Fondo de Investigación Sanitaria-Instituto de Salud Carlos III and the Sociedad Española de Alergología e Inmunología Clínica.

References

1. Burks AW, Tang M, Sicherer S, Muraro A, Eigenmann PA, Ebisawa M, Fiocchi A, Chiang W, Beyer K, Wood R, Hourihane J, Jones SM, Lack G, Sampson HA (2012) ICON: food allergy. J Allergy Clin Immunol 129(4): 906–920

2. Sicherer SH, Sampson HA (2014) Food allergy: epidemiology, pathogenesis, diagnosis, and treatment. J Allergy Clin Immunol 133(2): 291–307, quiz 308

3. Garcia-Ara MC, Boyano-Martinez MT, Diaz-Pena JM, Martin-Munoz MF, Martin-Esteban M (2004) Cow's milk-specific immunoglobulin E levels as predictors of clinical reactivity in the follow-up of the cow's milk allergy infants. Clin Exp Allergy 34(6):866–870

4. Wood RA, Sicherer SH, Vickery BP, Jones SM, Liu AH, Fleischer DM, Henning AK, Mayer L, Burks AW, Grishin A, Stablein D, Sampson HA (2013) The natural history of milk allergy in an observational cohort. J Allergy Clin Immunol 131(3):805–812

5. Boyano-Martinez T, Garcia-Ara C, Diaz-Pena JM, Martin-Esteban M (2002) Prediction of tolerance on the basis of quantification of egg white-specific IgE antibodies in children with egg allergy. J Allergy Clin Immunol 110(2): 304–309

6. Sicherer SH, Wood RA, Vickery BP, Jones SM, Liu AH, Fleischer DM, Dawson P, Mayer L, Burks AW, Grishin A, Stablein D, Sampson HA (2014) The natural history of egg allergy in an observational cohort. J Allergy Clin Immunol 133(2):492–499

7. Jones SM, Burks AW, Dupont C (2014) State of the art on food allergen immunotherapy: oral, sublingual, and epicutaneous. J Allergy Clin Immunol 133(2):318–323

8. Yeung JP, Kloda LA, McDevitt J, Ben-Shoshan M, Alizadehfar R (2012) Oral immunotherapy for milk allergy. Cochrane Database Syst Rev 11:CD009542

9. Brozek JL, Terracciano L, Hsu J, Kreis J, Compalati E, Santesso N, Fiocchi A, Schunemann HJ (2012) Oral immunotherapy for IgE-mediated cow's milk allergy: a systematic review and meta-analysis. Clin Exp Allergy 42(3):363–374

10. Savilahti EM, Savilahti E (2013) Development of natural tolerance and induced desensitization in cow's milk allergy. Pediatr Allergy Immunol 24(2):114–121

11. Alvaro M, Giner MT, Vazquez M, Lozano J, Dominguez O, Piquer M, Dias M, Jimenez R, Martin MA, Alsina L, Plaza AM (2012) Specific oral desensitization in children with IgE-mediated cow's milk allergy. Evolution in one year. Eur J Pediatr 171(9):1389–1395

12. Nowak-Wegrzyn A, Sampson HA (2011) Future therapies for food allergies. J Allergy Clin Immunol 127(3):558–573, quiz 574–555

13. Savilahti EM, Saarinen KM, Savilahti E (2010) Duration of clinical reactivity in cow's milk allergy is associated with levels of specific immunoglobulin G4 and immunoglobulin A antibodies to beta-lactoglobulin. Clin Exp Allergy 40(2):251–256

14. Shek LP, Soderstrom L, Ahlstedt S, Beyer K, Sampson HA (2004) Determination of food specific IgE levels over time can predict the development of tolerance in cow's milk and hen's egg allergy. J Allergy Clin Immunol 114(2):387–391

15. Tomicic S, Norrman G, Falth-Magnusson K, Jenmalm MC, Devenney I, Bottcher MF (2009) High levels of IgG4 antibodies to foods during infancy are associated with tolerance to corresponding foods later in life. Pediatr Allergy Immunol 20(1):35–41

16. Ruiter B, Knol EF, van Neerven RJ, Garssen J, Bruijnzeel-Koomen CA, Knulst AC, van Hoffen E (2007) Maintenance of tolerance to cow's milk in atopic individuals is characterized by high levels of specific immunoglobulin G4. Clin Exp Allergy 37(7):1103–1110

17. Bannon GA (2004) What makes a food protein an allergen? Curr Allergy Asthma Rep 4(1): 43–46

18. Chatchatee P, Jarvinen KM, Bardina L, Beyer K, Sampson HA (2001) Identification of IgE- and IgG-binding epitopes on alpha(s1)-casein: differences in patients with persistent and transient cow's milk allergy. J Allergy Clin Immunol 107(2):379–383

19. Vila L, Beyer K, Jarvinen KM, Chatchatee P, Bardina L, Sampson HA (2001) Role of conformational and linear epitopes in the achievement of tolerance in cow's milk allergy. Clin Exp Allergy 31(10):1599–1606

20. Jarvinen KM, Beyer K, Vila L, Chatchatee P, Busse PJ, Sampson HA (2002) B-cell epitopes as a screening instrument for persistent cow's milk allergy. J Allergy Clin Immunol 110(2): 293–297

21. Matsumoto N, Okochi M, Matsushima M, Kato R, Takase T, Yoshida Y, Kawase M, Isobe K, Kawabe T, Honda H (2009) Peptide array-based analysis of the specific IgE and IgG4 in cow's milk allergens and its use in allergy evaluation. Peptides 30(10):1840–1847

22. Savilahti EM, Rantanen V, Lin JS, Karinen S, Saarinen KM, Goldis M, Makela MJ, Hautaniemi S, Savilahti E, Sampson HA (2010) Early recovery from cow's milk allergy is associated with decreasing IgE and increasing IgG4 binding to cow's milk epitopes. J Allergy Clin Immunol 125(6):1315–1321, e1319

23. Wang J, Lin J, Bardina L, Goldis M, Nowak-Wegrzyn A, Shreffler WG, Sampson HA (2010) Correlation of IgE/IgG4 milk epitopes and affinity of milk-specific IgE antibodies with different phenotypes of clinical milk allergy. J Allergy Clin Immunol 125(3):695–702, 702. e691–702.e696

24. Shreffler WG, Beyer K, Chu TH, Burks AW, Sampson HA (2004) Microarray immunoassay: association of clinical history, in vitro IgE function, and heterogeneity of allergenic peanut epitopes. J Allergy Clin Immunol 113(4):776–782

25. Flinterman AE, Knol EF, Lencer DA, Bardina L, den Hartog Jager CF, Lin J, Pasmans SG, Bruijnzeel-Koomen CA, Sampson HA, van Hoffen E, Shreffler WG (2008) Peanut epitopes for IgE and IgG4 in peanut-sensitized children in relation to severity of peanut allergy. J Allergy Clin Immunol 121(3):737–743, e710

26. Lin J, Bruni FM, Fu Z, Maloney J, Bardina L, Boner AL, Gimenez G, Sampson HA (2012) A bioinformatics approach to identify patients with symptomatic peanut allergy using peptide microarray immunoassay. J Allergy Clin Immunol 129(5):1321–1328, e1325

27. Martinez-Botas J, Cerecedo I, Zamora J, Vlaicu C, Dieguez MC, Gomez-Coronado D, de Dios V, Terrados S, de la Hoz B (2013) Mapping of the IgE and IgG4 sequential epitopes of ovomucoid with a peptide microarray immunoassay. Int Arch Allergy Immunol 161(1):11–20

28. Jarvinen KM, Beyer K, Vila L, Bardina L, Mishoe M, Sampson HA (2007) Specificity of IgE antibodies to sequential epitopes of hen's egg ovomucoid as a marker for persistence of egg allergy. Allergy 62(7):758–765

29. Vereda A, Andreae DA, Lin J, Shreffler WG, Ibanez MD, Cuesta-Herranz J, Bardina L, Sampson HA (2010) Identification of IgE sequential epitopes of lentil (Len c 1) by means of peptide microarray immunoassay. J Allergy Clin Immunol 126(3):596–601, e591

30. Perez-Gordo M, Lin J, Bardina L, Pastor-Vargas C, Cases B, Vivanco F, Cuesta-Herranz J, Sampson HA (2012) Epitope mapping of Atlantic salmon major allergen by peptide microarray immunoassay. Int Arch Allergy Immunol 157(1):31–40

31. Perez-Gordo M, Pastor-Vargas C, Lin J, Bardina L, Cases B, Ibanez MD, Vivanco F, Cuesta-Herranz J, Sampson HA (2013) Epitope mapping of the major allergen from Atlantic cod in Spanish population reveals different IgE-binding patterns. Mol Nutr Food Res 57(7):1283–1290

32. Ayuso R, Sanchez-Garcia S, Lin J, Fu Z, Ibanez MD, Carrillo T, Blanco C, Goldis M, Bardina L, Sastre J, Sampson HA (2010) Greater epitope recognition of shrimp allergens by children than by adults suggests that shrimp sensitization decreases with age. J Allergy Clin Immunol 125(6):1286–1293, e1283

33. Battais F, Mothes T, Moneret-Vautrin DA, Pineau F, Kanny G, Popineau Y, Bodinier M, Denery-Papini S (2005) Identification of IgE-binding epitopes on gliadins for patients with food allergy to wheat. Allergy 60(6):815–821

34. Cerecedo I, Zamora J, Shreffler WG, Lin J, Bardina L, Dieguez MC, Wang J, Muriel A, de la Hoz B, Sampson HA (2008) Mapping of the IgE and IgG4 sequential epitopes of milk allergens with a peptide microarray-based immunoassay. J Allergy Clin Immunol 122(3):589–594

35. Renard BY, Lower M, Kuhne Y, Reimer U, Rothermel A, Tureci O, Castle JC, Sahin U (2011) rapmad: Robust analysis of peptide microarray data. BMC Bioinformatics 12:324

36. Savilahti EM, Kuitunen M, Valori M, Rantanen V, Bardina L, Gimenez G, Makela MJ, Hautaniemi S, Savilahti E, Sampson HA (2014) Use of IgE and IgG4 epitope binding to predict the outcome of oral immunotherapy in cow's milk allergy. Pediatr Allergy Immunol 25(3):227–235

37. de la Hoz B, Vlaicu C, Cerecedo I, Rodriguez-Alvarez M, Dieguez MC, Fernandez-Rivas M, Martinez-Botas J (2014) Changes In IgE and IgG4 epitopes after milk oral immunotherapy (OIT) [abstract]. J Allergy Clin Immunol 133(Suppl 2):AB49

38. Lin J, Bardina L, Shreffler WG, Andreae DA, Ge Y, Wang J, Bruni FM, Fu Z, Han Y, Sampson HA (2009) Development of a novel peptide microarray for large-scale epitope mapping of food allergens. J Allergy Clin Immunol 124(2):315–322, 322.e1-3

39. Lin J, Bardina L, Shreffler WG (2009) Microarrayed allergen molecules for diagnostics of allergy. Methods Mol Biol 524:259–272

40. Kwak JW, Jeong H, Han SH, Kim Y, Son SM, Mook-Jung I, Hwang D, Park JW (2014) Phosphokinase antibody arrays on dendron-coated surface. PLoS One 9(5), e96456

41. Shreffler WG, Lencer DA, Bardina L, Sampson HA (2005) IgE and IgG4 epitope mapping by microarray immunoassay reveals the diversity of immune response to the peanut allergen, Ara h 2. J Allergy Clin Immunol 116(4):893–899

Chapter 19

IgE Epitope Mapping Using Peptide Microarray Immunoassay

Gustavo Gimenez, Sara Benedé, and Jing Lin

Abstract

IgE epitope mapping of food allergens contributes to a better understanding of the pathogenesis of food allergy and may become an additional tool for food allergy diagnosis/prognosis. Microarray platforms which allow for simultaneous screening of a large number of peptides corresponding to the sequences of food allergens are ideally suited for large-scale IgE epitope mapping. Here we describe the method of performing a reliable and sensitive peptide microarray immunoassay, which was developed in our lab and results in the identification of candidate IgE epitope biomarkers useful in determining allergic disease severity and prognosis, as well as in the prediction of treatment outcomes. A gastric digestion model that can be applied to prescreen peptides and reduce costs in the peptide microarray is also described in this chapter.

Key words IgE, Food allergy, Epitope mapping, Peptide microarray, Gastric digestion, Immunoassay

1 Introduction

Food allergy is usually mediated by immunoglobulin E (IgE) antibodies through their binding to specific IgE epitopes within allergenic proteins. Identification of IgE-binding epitopes in food allergens will contribute to the design of well-tolerated immunotherapeutic agents and a better understanding of the pathogenesis of food allergy. In addition, studies have suggested a role for sequential IgE-binding epitopes as biomarkers for characterizing various phenotypes of food allergy [1].

Due to its ability to evaluate thousands of targets in parallel by using small volume of sample, the peptide microarray immunoassay is ideally suited for large-scale epitope mapping of IgE antibodies [2]. A typical peptide microarray consists of two experimental steps: printing a library of overlapping peptides, corresponding to the amino acid sequence of the target protein(s) onto glass slides coated with a functional layer (i.e., epoxy, aldehyde) where the peptides bind covalently to the functional groups on the slide surface; and

Marina Cretich and Marcella Chiari (eds.), *Peptide Microarrays: Methods and Protocols*, Methods in Molecular Biology, vol. 1352, DOI 10.1007/978-1-4939-3037-1_19, © Springer Science+Business Media New York 2016

immunolabeling of the printed slides with patients' sera. The use of overlapping peptides printed in the microarray format increases the mapping resolution and the accuracy of the test. In the past 10 years, we have developed and optimized a reliable and sensitive peptide microarray immunoassay. Our optimized peptide microarray platform is capable of generating reproducible data with high correlation coefficients ($r > 0.9$) [3], and detecting IgE binding from diluted serum sample with specific IgE levels as low as 0.068 kUA/L [4]. Using our peptide microarray, we have successfully mapped the sequential IgE-binding epitopes for a variety of food allergens, such as allergens from peanut, milk, lentil, and shrimp [5–8], and identified several IgE epitopes as candidate biomarkers for allergy diagnosis/prognosis [4] or tolerance prediction [9].

Synthesizing and printing a large library with hundreds/thousands of overlapping peptides covering entire protein sequences involves a high cost and also limits the number of food allergens that can be tested in one microarray. A selected peptide library covering only a portion of protein sequences with high probability of being epitopes can be an alternative to reduce the expenses and increase the number of allergens/foods that can be tested. In order to induce sensitization and elicit an allergic response, allergenic proteins must retain sufficient structural integrity after digestion [10, 11]. The sequences resistant to gastric digestion have higher chances of being bound by IgE antibodies. Therefore, a simulated protein digestion method mimicking physiological conditions of the stomach during gastric digestion [12] is useful as a prescreening of allergen sequences to select digestion-resistant sequence for peptide microarray.

In this chapter, we describe the method of IgE epitope mapping using peptide microarray immunoassay. This method is not only useful for epitope mapping of IgE antibodies, but can also be applied to other immunoglobulin subclasses (i.e., IgG4) by switching to or adding (for concurrent testing) the corresponding secondary antibody. In addition, as an option to downsize the peptide library for each allergen, we provide the method of performing simulated protein digestion with detailed description of the preparation, processing, and hydrolysis of allergenic proteins.

2   Materials

2.1   Gastric Digestion

1. Phosphatidylcholine (PC) (Larodan, Malmo, Sweden).

2. Simulated gastric fluid (SGF), containing 35 mM NaCl, pH 2.0: SGF is prepared by dissolving 1 g of NaCl in 0.5 L of distilled water (dH_2O) and adjusting pH to 2.0 with 1 N HCl.

3. Porcine pepsin (Sigma-Aldrich, St. Louis, MO, USA): EC 3.4.23.1, 3640 U/mg protein.

4. RP-HPLC-MS/MS access is required for final analysis of gastric digestion.

2.2 Peptide Microarray

1. SuperEpoxy 2 (SME2) glass slides (Arrayit Corporation, Sunnyvale, CA, USA), used for all peptide microarray printings (*see* **Note 1**).

2. A library of overlapping peptides (*see* **Note 2**) (JPT Peptide Technologies GmbH, Berlin, Germany), consisting of 15 amino acids (AA) overlapping by 12 (3-offset), corresponding to the primary sequence of the target proteins, are commercially synthesized either as custom-made or PepStar format (*see* **Note 3**). If enzymatic digestion has been performed for protein sequence pre-screening, only peptides covering the digestion-resistant sequences are commercially synthesized for printing.

3. 384-Well, round-bottom, non-sterile assay plates (Corning Incorporated, Corning, NY, USA) and lids are required for proper storage and printing of the peptides (*see* **Note 4**).

4. NanoPrint 60, equipped with 2×4 Stealth Micro Spotting Pins (SMP3B) and controlled by the NanoPrint Microarray Manager Software (Arrayit Corporation), is used to print peptide libraries (*see* **Note 5**).

5. Hybridization Chamber/Stain Tray (Worldwide Medical Products, Bristol, PA, USA), used for all the incubation steps in the immunolabeling procedure. We recommend purchasing the tray with the black lid, as some of the incubation steps in this assay require protection from light sources.

6. Dako Pap pen (Dako North America Incorporated, Carpinteria, CA, USA).

7. High-throughput wash stations and squeeze bottles (Arrayit Corporation).

8. Dimethyl sulfoxide (DMSO), 99.7 + %, extra dry over molecular sieve (Acros, New Jersey, USA).

9. Phosphate-buffered saline (PBS) stock solution: For peptide microarray washing steps, 1×, 0.1×, and 0.05× solutions are prepared by diluting the stock solution in distilled water 1:10, 1:100, and 1:200, respectively.

10. Phosphate-buffered saline-Tween (PBS-T): 1 mL Tween 20 is diluted with 1× PBS to a final volume of 2 L.

11. Ethylenediaminetetraacetic acid (EDTA), 1 mM in PBS-T: First, a 0.5 M stock solution of EDTA is prepared by dissolving 186.12 g in 800 mL dH$_2$O, adjusting pH to 8.0 by adding approximately 20 g sodium hydroxide (NaOH) while stirring, and adjusting the volume to 1 L with additional dH$_2$O. Second, the 1 mM solution is prepared by diluting 100 µL of the EDTA with PBS-T to a final volume of 50 mL.

12. 15 mM Trizma hydrochloride (Tris–HCl) buffer solution, pH 8.0: 30 mL of 1 M Tris–HCl, pH 8.0, is diluted with dH$_2$O to a final volume of 2 L.

13. Protein printing buffer (PPB), 2× solution (Arrayit Corporation), is used in the working plates to dilute the peptides. This reagent is essential for quality printings and provides efficient coupling and consistent spot morphology.

14. Blocking buffer: 1 % human serum albumin (HSA) in PBS-T is used to block all unbound sites on the slide surface prior to incubation with patient serum samples. It is also the diluent for both serum and detection antibodies.

15. IgE immunolabeling solution: 250 μL of solution containing monoclonal biotin mouse anti-human IgE (BD Biosciences, San Jose, CA, USA) and monoclonal biotin mouse anti-human IgE (ε-chain specific) (Molecular Probes by Life Technologies, Grand Island, NY, USA), each diluted 1:250 in blocking buffer.

16. Dendrimer signal amplification system: 200 μL of solution containing 12 μL UltraAmp Anti-Biotin oyster 550 (Genisphere Incorporated, Hatfield, PA, USA), 4 μL salmon sperm DNA (10 mg/mL) (Agilent Technologies, Santa Clara, CA, USA), and 184 μL UltraAmp Binding Buffer II (Genisphere Incorporated).

17. ScanArray Gx Microarray Scanner (PerkinElmer, Waltham, MA, USA).

---

3 Methods

3.1 Gastric Digestion

1. Phospholipid vesicle solution is prepared by dissolving PC in SGF at a concentration of 9.58 mg/mL. The mixture is sonicated until clear to the eye (~5 min). This reduces the crude lecithin solution into a suspension of single-shelled liposomes.

2. Solutions are then filtered through a 0.22 μm filter to remove any possible contaminants.

3. 39 mg of protein is dissolved in 3 mL of SGF (final concentration of 13 mg/mL). The pH is adjusted to 2.0 with 1 N HCl. Additional SGF is then added to bring the volume to 3.1 mL.

4. Phospholipid vesicle solution (3.6 mL) is then added to the protein mixture (9.58 mg PC/mL SGF) and incubated for 5 min at 37 °C.

5. Following the incubation, 6.030 mL of mixture is transferred to a clean tube and 303 μL of pepsin solution (18,200 U/mL SGF) is added to obtain the physiological ratio of 182 U/mg of protein.

6. Gastric digestion is performed in a shaking incubator for 60 min at 37 °C. The reaction is stopped by increasing pH to 7.5 with NaHCO$_3$. Distilled water is added to the sample until a final volume of 7.800 mL to obtain a final concentration of protein of 4.5 mg/mL. RP-HPLC-MS/MS analysis of the digested samples is performed in order to identify the sequences resistant to digestion.

3.2   Printing of Peptides

1. Working plates containing peptides in printing buffer (3 μL peptide + 3 μL 2× PPB) are stored at –80 °C, and left at room temperature for 10–15 min prior to use. Contents of the working plates are mixed well by placing the plate on a pedestal for 60 s in an ultrasonic bath, and then centrifuged at 130 × g for 3 min at room temperature.

2. Peptides are printed from the working plate onto the SME2 slides in two sets of triplicate spots. Each slide can fit two or three full microarrays, depending on the size of the peptide library or the nature of the experiment (Fig. 1a–c). In addition to the peptides, 20 % of the printed spots must be "blanks" composed of 50 % v/v DMSO in PPB.

3. All printings are carried out under 55–65 % humidity in order to prevent any evaporation from the samples on the working plate. Once completed, humidity is turned off and slides are left to dry overnight in the printer to ensure that peptides are bound to the slides.

4. Following the drying period, the printed slides are scanned for quality check (QC) at low resolution (20 μm) to ensure proper peptide spotting. The microarrays on the printed slides are then outlined with the Dako pap pen and labeled with the printing and slide number (Fig. 1d). Marking the outline of all the arrays in the same sized square shape with a hydrophobic pen ensures proper hybridization of antibodies and amplification molecules to peptides, and also prevents cross-contamination between patient samples on the arrays.

5. All slides are then washed twice, each for 20 min in PBS-T and 0.05× PBS in high-throughput wash stations. This washing step removes excess, unbound peptides, leaving only the covalently bound peptides that are not visible by eye (Fig. 1e). In addition, this pre-wash reduces nonspecific binding and background noise. Slides are then dried in a centrifuge using the same settings as in step "A"—130 × g for 3 min at room temperature. The immunolabeling protocol can now be initiated, or the slides can be stored in vacuum for up to 3 weeks.

3.3   Immunolabeling Protocol

1. Blocking: Slides are placed in the hybridization chamber, 250 μL of blocking buffer is added to each array (*see* **Note 6**), and gentle rotation for 1 h at 31 °C is required. The blocking buffer is then carefully removed by aspiration.

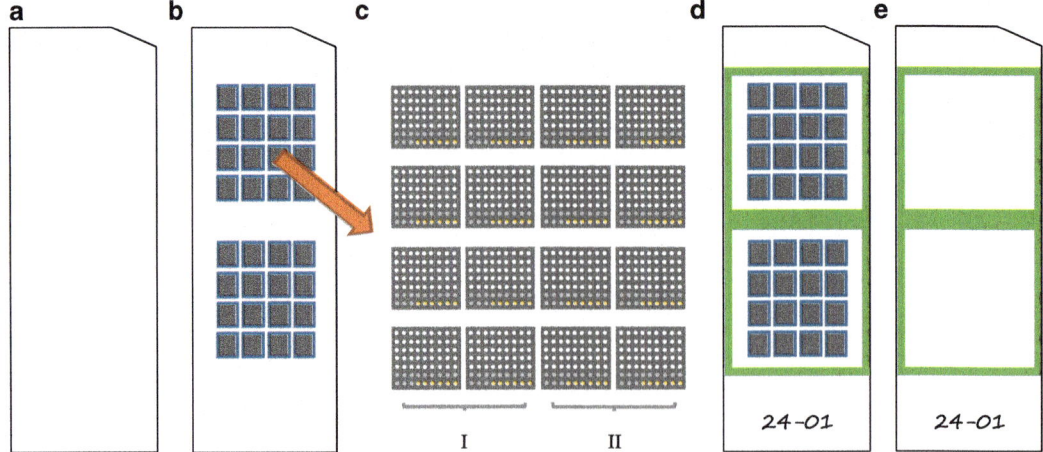

Fig. 1 (**a**) Blank SME-2 slide from Arrayit Corporation, the notch on the *top right* is used as a guide for holding the slide *right* side up. (**b**) Slide with two identical arrays after printing. Depending on the size of the peptide library, changes can be made to fit up to 3 arrays (for 3 individual serum samples) per slide. (**c**) Close-up view of 1 array. The *white* spots represent individual peptides, printed in triplicates. The *light gray* spots indicate "blank" PPB spots used for background normalization. It is important to note that at least 20 % of the array must be composed of PPB spots. The *yellow* spots, usually printed at the *bottom right* corner of each sub-array, represent the fluorescent-labeled BSA spots used as a grid control during alignment. Sub-panel I and sub-panel II are duplicates. Since each peptide is printed in triplicate, the total number of spots per peptide is 6. (**d**) After QC scan, printed slides are marked with a pap pen, enclosing the separate arrays in equal-sized squares for optimal hybridization. The center line must be thick enough to prevent cross-contamination between samples. Each slide is also labeled with the printing and slide number. (**e**) After washing the slides with PBS-T and 0.05× PBS, the arrays will no longer be visible by eye

2. Serum incubation: Patient serum samples (*see* **Note 7**), diluted 1:5 in blocking buffer, are added to the arrays (250 μL/array) and gently rotated overnight at 4 °C. The hybridization chamber must be always protected from light. Following overnight incubation, serum is removed by gently tapping the slides horizontally—this helps prevent cross-contamination between the vertical arrays. The slides are transferred onto a slide rack, dipped two times in PBS-T, and placed in a high-throughput wash station containing fresh PBS-T for 10 min at room temperature. The slides are then placed back into the hybridization chamber and the remaining PBS-T is removed by aspiration.

3. Secondary antibody incubation: For IgE detection, a fresh preparation of the biotinylated secondary antibody cocktail in blocking buffer is required for every experiment. The BD and Life Technologies antibodies are each diluted 1:250. After removing PBS-T from the slides by aspiration, 250 μL of the antibody cocktail is added onto each array, spread across the array outline, and the slides are gently rotated at 31 °C for 4 h

(*see* **Note 8**). The slides are quickly washed with PBS-T from a squeeze bottle, and after clearing the arrays with aspiration, 300 μL of 1 mM EDTA (*see* **Note 9**) in PBS-T is added to each array. This is then removed after rotating for 4 min at room temperature, followed by another quick PBS-T wash. The remaining PBS-T is removed by aspiration.

4. Dendrimer amplification: The next step in the immunolabeling protocol is the addition of dendrimers for signal amplification to increase the sensitivity of the assay. 200 μL of dendrimer buffer is added, spread across the arrays, and is left to sit on the slides for 1 min before it is removed. The dendrimer cocktail (200 μL/array: 12 μL anti-biotin dendrimer, 4 μL salmon sperm DNA, and 184 μL dendrimer buffer) (*see* **Note 10**) are added onto each array, followed by gentle rotation for 3 h at 31 °C, protected from light.

5. Final wash: The final wash aims to remove unbound dendrimer molecules and reduce excess salt deposits on the array surface that may interfere with the subsequent analysis. With the slides still in the hybridization chamber, they are washed quickly with PBS from a squeeze bottle, then dipped ten times in 15 mM Tris–HCl. Slides are dried in a centrifuge, and then dipped five times in 0.1× PBS. After another spin in the centrifuge, the slides are dipped five times in 0.05× PBS, and dried again before they are scanned for analysis. Dried slides may also be stored in vacuum for up to 5 days before scanning without compromising signal intensity.

3.4 Slide Scanning

The processed slides are scanned using a ScanArray®Gx (PerkinElmer, Waltham, MA, USA) microarray scanner. The difference between this scan and the original QC scan after printing is that the scanner settings are changed to a much higher resolution (10 μm) in order to increase sensitivity and detect even low-intensity IgE binding. Scanned images are saved as TIF files, and the fluorescence signals are digitized with ScanArray Express software (Perkin Elmer) using an alignment procedure (*see* **Note 11**). The aligned data for the arrays are then exported as comma-delimited text (.csv) files.

3.5 Data Analysis

(a) A readout (S) for each array element (spot), including the replicates for peptides, controls and blanks (PPB + DMSO), is calculated (as indicated in the equation) by $\log_2$ transformation of the ratio of the median fluorescent signal of the spot to the median fluorescent signal of the local background:

$$S_i = \mathrm{Log}_2\left(\frac{\mathrm{Median\,(Spot)}}{\mathrm{Median\,(Local\ background)}}\right)$$

(b) The S_i for each array element is transformed into a Z score (Z_i) by using the equation below:

$$Z_i = Log_2 \frac{S_i - \text{Median}\left(S_{\text{Blank}}\right)}{\text{MAD}\left(S_{\text{Blank}}\right)}$$

The Median(S_{Blank}) and MAD(S_{Blank}) are the Median and the Median Absolute Deviation (MAD) of all the readouts of the blanks spots, respectively. MAD is a robust version of the standard deviation (*see* **Note 12**). An index Z value of each peptide element is generated from the Median of Z scores of the six replicate spots. An individual peptide sample is considered positive if its index Z exceeds 3, meaning that the signal is above the background noise plus three times the MAD. As MAD is a robust version of the standard deviation, this means that the signal is significantly above the background with p value less than 0.003.

We perform the above transformation using R programming language (http://www.r-project.org/), but it can also be calculated using other software which does not require programming, such as Microsoft Excel. Further data analysis and presentation can be performed using R programming or other commercially available software such as Microsoft Excel, GraphPad Prism, Sigmaplot and TIGR Multiexperiment Viewer (TMeV).

4   Notes

1. SME2 slides by Arrayit Corporation provide high-quality surface chemistry necessary for proper peptide printing, binding, and subsequent immunolabeling. We choose this particular product due to their consistency in performance, 6-month shelf life, and minimal slide-to-slide variation. Once the slides are received, they must be stored in a cool, dry place at room temperature, and used within 6 months of the manufacture date.

2. Using overlapping peptides facilitates both the mapping of epitope sequences and the identification of key amino acids based on IgE binding intensity. In addition, peptides can form secondary structures and epitope sequences of solid-phase bound peptides may not always be exposed and available for binding with IgE antibodies. By using overlapping sequences, we increase the chances of detecting all epitopes.

3. Custom-made peptides are synthesized and aliquoted as lyophilized powders. The purity of these peptides has been measured by HPLC. In general, >70 % purity is required for peptide microarray printing. These peptides are resuspended in DMSO at 2 mg/mL concentration. The PepStar format peptides dif-

fer from custom-made in that the former is immobilized on the glass slides via a flexible linker at the N-terminus, and is provided already in solution in 96- or 384-well stock plates. All peptides (stocks and working plates) must be stored at –80 °C until used.

4. Plates from Corning Incorporated are used as working plates. They have the proper dimensions for accurate printing, and are used to store peptides diluted in printing buffer.

5. Specific printing conditions are required in order to ensure performance and accuracy in the immunoassay. Having a clean workspace, preferably an isolated room to avoid dust buildup and other printing irregularities is very important. Humidity is also an important factor. The humidity level inside the printing chamber should be maintained around 60 % during printing to prevent the evaporation of peptides, and humidity in the room must not exceed 50 %, as this contributes to "pin sticking" and other technical problems with the printing unit. In addition to these conditions, the printer is equipped with a sonicating bath station for washing of the pins before and after printing, as well as in between visits to the working plate. The sonication station is filled with a 10 % ethanol solution, and is used for pin washing for 5 min before and after printing, and for 2 min for in-between visits to the plate. These washes are necessary for optimal cleaning and maintenance of the pins, prevent cross-contamination of peptides, and remove any particles that may interfere with accurate spotting.

6. For this step, and for all subsequent steps involving the addition of buffers, samples, and other reagents to the arrays, the liquids must be spread out across the entire outlined array using a pipette tip. A clean pipette tip is used for each array containing different serum sample to prevent cross-contamination. For buffers and detection cocktails, the same pipette tip can be used. This spreading maximizes the reagent's exposure to the entire array of printed peptides and improves sensitivity. In addition, a small amount of water must be kept in the hybridization chamber throughout the immunolabeling procedure in order to maintain the humidity required to prevent evaporation of the reagents—especially in steps that require incubations at higher temperatures.

7. In addition to individual patient samples, a system of positive and negative controls is required for each peptide microarray experiment to ensure that the assay is working properly. This system can also be used for data normalization between different microarray experiments. Some options include printing fluorescent-labeled proteins in the arrays that can be detected and compared from experiment to experiment, or in the case

of our facility, the use of positive and negative control patient samples in every experiment. The positive control is a serum pool of severely allergic patients for which the IgE binding patterns are well established. The negative control can be a pooled serum sample of non-allergic, non-atopic patients, or individual non-allergic patient's serum sample.

8. Overnight serum incubation at 4 °C and detection antibody incubation for 4 h at 31 °C are the optimal conditions based on our current peptide library and other experimental conditions. Peptides from different companies, with different levels of purity or printed at different concentrations, may show binding at various levels of intensity. Other factors, such as temperature and humidity levels in the laboratory, may also affect binding diversity and intensity. It is strongly recommended for new facilities to test different printing and immunolabeling conditions and find the conditions that give maximum binding sensitivity without compromising specificity or giving excess signals in the negative controls.

9. EDTA serves to bind any metal ions that may be present in order to prevent the activation of nucleases that will hydrolyze and interfere with the dendrimer signal amplification system.

10. Salmon sperm DNA serves two purposes in this assay: It blocks nonspecific binding of dendrimer molecules to the array, and it protects against nucleases that may be present in the buffer.

11. The Microarray Manager software installed with the NanoPrint60 is used to generate a .gal file that, when opened with the ScanArray Express software, creates a grid on the scanned microarray image and is aligned over the printed spots (Fig. 2b). This grid maps where each array element is spotted, including negative control buffer spots and positive grid control spots. In our facility, we print grid control spots composed of a fluorescent-labeled bovine serum albumin (BSA) on the bottom-right corner of each sub-array. These are helpful with the alignment procedure, and ensure that signals from the correct peptides are digitized and analyzed.

12. For a Gaussian distribution, the MAD is the same as the standard deviation. But the distribution of all the readouts of the blank spots is not necessarily Gaussian, the MAD are more robust estimates of the spread than the standard deviation. Given that the data consists of $x_1, x_2, \ldots, xn$, the MAD is defined as $1.48 \times \mathrm{Median}\left(\left|x_i - m\right|\right)$, where $m = \mathrm{Median}(x_i)$. The constant 1.48 is used to ensure that if the data's distribution is really Gaussian, then MAD is equal to the standard deviation.

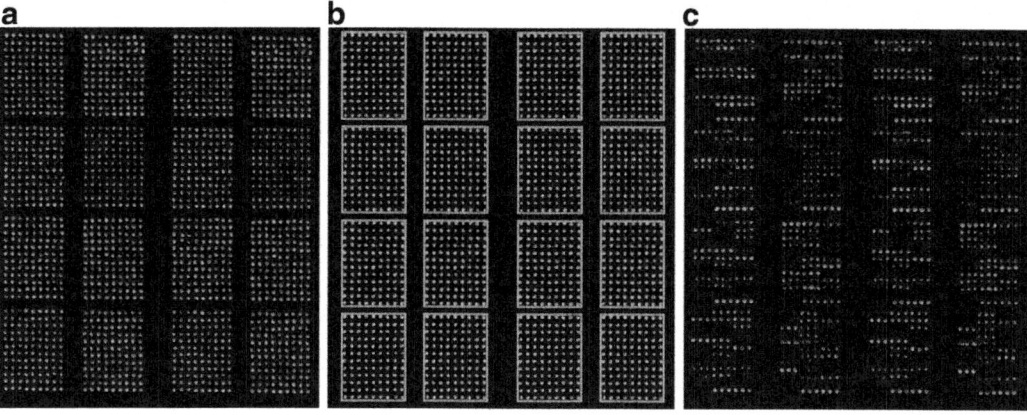

Fig. 2 Scanned images of microarray slides using the ScanArrayGx microarray scanner (PerkinElmer). (**a**) QC scan of the slide after printing to ensure proper printing. In general, spotted peptides are consistent in size and shape, with some minor variations. After washing, all the spots become invisible except the six grid control spots on the *lower right* side of each sub-array. (**b**) A microarray overlaid with the .gal file grid carefully aligned over the spots in order to obtain fluorescence data. (**c**) Final scan of an array immunolabeled with the serum of a severe allergic patient. While binding diversity varies from patient to patient, fluorescent signal from peptides does tend to decrease with lower IgE-level patients, and is blank (with the exception of grid control spots) in negative control samples

References

1. Lin J, Sampson HA (2009) The role of immunoglobulin E-binding epitopes in the characterization of food allergy. Curr Opin Allergy Clin Immunol 9:357–363

2. Shreffler WG, Lencer DA, Bardina L et al (2005) IgE and IgG(4) epitope mapping by microarray immunoassay reveals the diversity of immune response to the peanut allergen, Ara h 2. J Allergy Clin Immunol 116:893–899

3. Lin J, Bardina L, Shreffler WG et al (2009) Development of a novel peptide microarray for large-scale epitope mapping of food allergens. J Allergy Clin Immunol 124:315–322

4. Lin J, Bruni FM, Fu ZY et al (2012) A bioinformatics approach to identify patients with symptomatic peanut allergy using peptide microarray immunoassay. J Allergy Clin Immunol 129:1321–1328

5. Flinterman AE, Knol EF, Lencer DA et al (2008) Peanut epitopes for IgE and IgG4 in peanut-sensitized children in relation to severity of peanut allergy. J Allergy Clin Immunol 121:737–743

6. Wang J, Lin J, Bardina L et al (2010) Correlation of IgE/IgG4 milk epitopes and affinity of milk-specific IgE antibodies with different phenotypes of clinical milk allergy. J Allergy Clin Immunol 125:695–702

7. Vereda A, Andreae DA, Lin J et al (2010) Identification of IgE sequential epitopes of lentil (Len c 1) by means of peptide microarray immunoassay. J Allergy Clin Immunol 126:596–601

8. Ayuso R, Sanchez-Garcia S, Lin J et al (2010) Greater epitope recognition of shrimp allergens by children than by adults suggests that shrimp sensitization decreases with age. J Allergy Clin Immunol 125:1286–1293

9. Savilahti EM, Rantanen V, Lin JS et al (2010) Early recovery from cow's milk allergy is associated with decreasing IgE and increasing IgG4 binding to cow's milk epitopes. J Allergy Clin Immunol 125:1315–1321

10. Polovic N, Blanusa M, Gavrovic-Jankulovic M et al (2007) A matrix effect in pectin-rich fruits hampers digestion of allergen by pepsin in vivo and in vitro. Clin Exp Allergy 37:764–771

11. Ruiter B, Shreffler WG (2012) Innate immunostimulatory properties of allergens and their relevance to food allergy. Semin Immunopathol 34:617–632

12. Benede S, Lopez-Exposito I, Gimenez G et al (2014) Mapping of IgE epitopes in in vitro gastroduodenal digests of beta-lactoglobulin produced with human and simulated fluids. Food Res Int 62:1127–1133

Chapter 20

Spot Synthesis: An Optimized Microarray to Detect IgE Epitopes

Salvatore G. De-Simone, Paloma Napoleão-Pêgo and Thatiane S. De-Simone

Abstract

Peptide microarrays have become increasingly more affordable in recent years with the SPOT technique being one of the most frequently used methods for synthesis and screening of peptides in arrays. Here, a protocol is presented for the identification of the amino acid sites involved in the conversion of human IgG to IgE response during the passive administration of therapeutic, anti-snake venom sera. Similarly, the minimal region of both the IgG and IgE binding epitopes, important for its interaction with ligand, were identified. As the ratio of concentrations for IgG to IgE in human serum is 1:10,000, also presented is a reproductive protocol of chemiluminescence-scanning for the detection of both immunoglobulins.

Key words Microarray, Spot-synthesis, Protein–peptide interactions, Epitope structure, IgE detection, Allergy, Hypersensitivity

1 Introduction

Peptide microarrays have emerged as one the most prominent and revolutionary technologies for multiple purposes. Potential applications range from drug development [1], protein interaction domains [2–5], metal binding [6, 7], nucleic acid binding [8], enzyme–substrate assays [9], and antimicrobial screening [10], but the principal uses are epitope mapping [11–13] and diagnostics [14]. The advances delivered by the technique are miniaturization of microarrays, parallel synthesis, and automation that permit affordable high throughput screening [15]. In the immunology field, its main application has been in the detailed investigation of antigen–antibody interactions at the level of individual amino acid residues [13, 16, 17]. In addition to the identification of peptide epitopes for the respective antibodies, such studies also yield information on the relative specificities of individual positions within the antigenic sequences, as well as the minimal peptide length

Marina Cretich and Marcella Chiari (eds.), *Peptide Microarrays: Methods and Protocols*, Methods in Molecular Biology, vol. 1352, DOI 10.1007/978-1-4939-3037-1_20, © Springer Science+Business Media New York 2016

requirements and cross-reactivity. Investigators, pioneered by Geygen [18], have used peptide arrays to measure the B cell response in outbred animals to study epitope structure. Many investigators have used small overlapping arrays of peptides to define continuous B-cell epitopes of numerous antigens, including various allergens [2, 3, 16, 19]. Because of its ability to assay thousands of targets in parallel by using small volumes of sample, the microarray immunoassay is ideally suited for the determination of individual monoclonal and polyclonal epitope recognition patterns. Furthermore, signal intensity for microarray-based immunoassays is more dependent on affinity than concentration, making it an ideal platform for the detection of epitope-specific IgE immunoglobulins, along with other minor classes and subclasses, present at low concentrations in serum.

Peptide–antibody interactions had been studied extensively using either one-bead-one-peptide [20] and mixture-based combinatorial peptide libraries [21]. Currently, two-dimensional, high-density peptide arrays are used that have their conceptual roots in the strategies for the synthesis of multiple peptides, which were developed during the 1980s. This innovation permitted the simultaneous generation and biological evaluation of up to hundreds of peptides, thus dramatically increasing the efficiency of biomedical research using synthetic peptides. The driving force for this development was a rapid growth in the demand for synthetic peptides from various fields of biomedical research, in particular immunology, where synthetic peptides were increasingly used as immunogens and as tools to study antibody specificities at the level of individual amino acid residues [17, 22].

Two important strategies emerged from this period: The "pin" [18] and "tea bag" [22] methods. These created peptide arrays that represented novel types of combinatorial peptide library. There are many other variants of biological peptide libraries, such as phage display libraries [23], the synthesis of peptides with random sequences based on the split-and-mix strategy [24] and the approach of "a peptide bead" [25]. Another strategy known in the field of combinatorial peptide libraries include positional approximation approach scan [26]. Each strategy had its advantages and disadvantages.

The concept of peptide arrays was first introduced by Fodor et al. [27], who generated peptide arrays on small glass slides. The spatial separation of individual peptides on the glass surface was achieved using photolithographic masks that, in conjunction with the use of photolabile protecting groups, made it possible to selectively address separate areas of the slide. Due to practical limitations, such as insufficient stability of the bond between peptide and glass surface during peptide synthesis, as well as the need to generate large numbers of different photolithographic masks, this method did not prove optimal for the generation of synthetic peptide arrays.

Building on the "pin" method, a conceptually simple and flexible approach for the generation of highly complex peptide arrays or parallel library on planar cellulose supports was developed, the so called synthesis in concentric point or SPOT synthesis [2, 5]. In this method, the peptides are synthesized simultaneously on a solid support (derivatized cellulose sheets in this case) simply by dispensing small droplets onto the surface of a flat porous membrane (defined as an open reactor) where the chemical synthesis occurs to generate arrays of unique molecules (1–10,000 SPOTs). The array format is robust, reliable, inexpensive and now a well-established method for the simultaneous analysis of large number of different peptides. Although the SPOT synthesis is only a semi-quantitative method, it has as a major advantage the capacity to approximate the affinity of the protein–peptide interaction through the measurement of the signal intensity at the spot [28, 29]. Another advantage over other conventional methods is the hydrophilic, mechanically strong physical support that is easily manipulated during synthesis and the screening steps. Furthermore, the cellulose membranes are compatible with various on-support biological screening methods [30].

Here, a protocol based on SPOT-synthesis is presented to identify immunodominant allergenic linear epitopes responsible for the undesirable response that can occur during a treatment with therapeutic antivenom sera. Since the sequences of IgG and IgE epitope mimics are often very different from the known ligands, they would not likely have been identified using conventional chemical mutation methods, in which individual positions of a known sequence are addressed one at a time. From the results, a theoretical analytical strategy to predict B cell epitopes based on scoring and ranking of amino acids sequences also would not have succeeded to determine the IgE based epitopes.

The technical protocols of SPOT synthesis for multiple determinations have been described in detail [31–34, 38] and have been used for the multiple approaches described above. In this chapter, the basic procedure of screening method and scanning detection is presented in relation to SPOT synthesis with an outlined protocol for easy-to-use IgE detection method for an allergenic disease (horse antivenom allergy) where the concentration of the antibodies is in the nano scale (≥ 1 ng/mL) within patient sera.

Clinically, anaphylaxis is an exaggerated response in a sensitized patient with circulating IgE antibodies that can immediately react with the proteins administered in the antivenom. The reaction causes the degranulation of mast cells and basophiles, which consequently leads to a cascading release of vasoactive substances. An anaphylactic reaction can cause multiple signs and symptoms that can culminate in hypotension, shock, and death. Although much is known about the structure–function of this immunological class of proteins, there is limited published evidence about the precise localization of antigenic sites contained within the antibodies themselves [35]. To date, the recognition of horse immunoglobulin's (hIg) by human antibodies and the specific residues involved remain unknown. Therefore, a detailed molecular characterization of the epitopes in hIg that are bound by human antibodies would greatly contribute to our understanding of the observed adverse reactions to therapeutic equine preparations. Here, the horse heavy chain of immunoglobulin G3 (hhcIgG3) was examined since it is the major hIg isotype implicated in the snake venom toxins neutralization [36] and it is the second most prominent antibody in horse serum [37]. The entire sequence of hhcIgG3 was mapped for the epitopes bound by human IgG and IgE along with determining the critical amino acids within the epitope involved with the interactions with human antibodies. Localizing antigenic determinants within the polypeptide chains of protein molecules is essential for subsequent investigations of structural and functional characteristics. By identifying and quantifying components in hIg that contribute to immunological discrimination between self and nonself, antivenoms and antitoxins can be improved by new production techniques engineered specifically to minimize activation of a patient's immune response, but maintain efficacy.

2 Materials

2.1 Equipment

1. Robot Autospot Asp 222 or MultiPep peptide synthesizer with spotting tray (Intavis AG Bioanalytical Instruments, Cologne, Germany).

2. Software Multipep (Intavis) for the generation of peptide lists, pipetting protocols (included in the synthesis kit) and in the operation software of the spotting robot.

3. Standard micropipetting tips (Gilson, Eppendorf).

4. Glassware and polypropylene containers used exclusively in all steps involving organic solvents.

5. Flat reaction/washing troughs with a tightly closing lid made of chemically inert material (glass, teflon, polypropylene) with dimensions slightly larger than the membranes.

6. Eppendorf 1.5 mL plastic tubes.

7. Racks as reservoirs for amino acid solutions.

8. Rocker table with two dispensers for DMF and alcohol adjustable from 5 to 50 mL.

9. Hand-held hair dryer with a non-heating option.

10. Chemical chapel and fume hood.

11. Freezer −80 °C.

12. Membrane: An acid stable cellulose membranes (Intavis) functionalized with amino-polyethyleneglycol (amino-PEG500-UC540, #84300). Other membranes functionalized with β-alanine acquired from Sigma or other companies can be used.

13. Chromatography paper type 3MM (Whatman, Maidstone, UK).

14. Antibodies: Rabbit anti-human-IgG (Abcam plc, USA) and goat anti-human IgE labeled with alkaline phosphatase (KPL, Kirkegaard & Perri Laboratories, USA).

15. Human sera: Two pools of sera were generated to represent either human IgG antibodies or human IgE against horse heavy chain IgG3. The sera with human IgG were stimulated in volunteers that were known not to be sensitive to treatment (NST) by the injection of horse sera-based antivenom. The source of IgE antibodies was from three selected patients, who had been victims of one or more snake bites and who had presented with a severe anaphylactic reaction when treated with a series of three injections of antivenom (polyvalent antibothropic sera). The allergic response was confirmed by clinical history and a positive response to hIg characterized by measuring specific IgE reactivity. A pool consisting of equal volumes of sera from 15 health individuals who had never received any therapeutically injection products produced in horses were used as a negative control.

2.2 Buffers and Solutions

1. *N,N*-dimethylformamide (DMF) is a toxic and flammable bipolar aprotic solvent widely used and highly successful for peptide synthesis. Must be free of contamination by amines and with high degree of purity (*see* **Note 1**).

2. Methanol (MeOH) or ethanol (EtOH) of technical grade (95 %).

3. 1-methyl-2-pyrrolidinone (NMP) should be of the highest purity available. Must be stored in a dark flask containing 5 % molecular sieve (0.4 nm) and free of contamination by amines. Check amines as (*see* **Note 1**) and for purification *see* **Note 2**.

4. *N*-hydroxybenzotriazole anhydrous (HOBt). A coupling agent that should be stored in tightly closed bottles at room temperature in a dry environment.

5. Dichloromethane methylene chloride (DCM).

6. *N,N'*-diisopropylcarbodiimide (DIC) ≥98 %, coupling agent.

7. Dioxane (1,4 dioxane; stabilizer, flammable, suspected carcinogenic).

8. Acetic anhydride (AcAN), capping reagent.

9. Hexahydropyridine (Piperidine) 99.5 % is toxic and should be only handled with gloves under a hood.

10. Dimethylsulfoxide (DMSO).

11. Trifluoroacetic acid (TFA) ≥99.5 % with a degree of purity for synthesis. The acid is very harmful and volatile, which should be handled with gloves under a hood. It is used in the last step, after incorporation of the last amino acid and the treatment with piperidine to remove the amino group termini protection (F-moc).

12. Triisopropylsilane 99 % (TIPS).

13. Bromophenol blue (BPB) indicator, stock solution 10 mg/mL in DMF and kept at ambient temperature (must have an intense orange color that should be discarded when it turns a green color).

14. Mix acetylation: 2 % acetic anhydride in DMF (N,N-dimethyl formamide). For blocking by acetylation of peptides after coupling with F-moc amino acid.

15. Mix piperidine: Piperidine 20 % in DMF. To remove the protective group F-moc, prior to addition of the next amino acid.

16. Fmoc-AA stock solutions. The Fmoc-AA derivatives of all 20 l-amino acids, are available in several companies. The side-chain protecting groups must be Cys (Acm) and Cys (Trt), Asp (OtBu), Glu (OtBu), His (Trt), Lys (Boc), Asn (Trt), Gln (Trt), Arg (Pmc), Ser (tBu), Thr (tBu), Trp (Boc), and Tyr (tBu). Warning: prepare HOBt esters of these derivatives in NMP. Dissolve 1 mmol each of Fmoc-AA in 5 mL NMP containing 0.25 M HOBt to obtain 0.2 M solution of Fmoc-AA stock. Keep these stocks in 10 mL plastic tubes well sealed, frozen in liquid nitrogen and store at −80 °C. For use in the peptide spot synthesis, aliquot for each cycle the amount of each Fmoc-AA required depending on the quantity of each amino acid to be incorporated in the cycle, dilute 1 vol. of Fmoc-AA with 3 vol. of NMP solution. Freeze in liquid nitrogen and store at −80 °C (*see* **Note 3**).

17. Deprotecting Mix: TFA (80 %), DCM (12 %), TIBS (3 %), and water (5 %). The order of mix is important and must be as written. The deprotecting mix removes all of the protecting groups on the side chains of the amino acids incorporated onto the membrane bound peptides (*see* **Note 4**).

18. Tris–HCl 50 mM containing 136 mM NaCl and 2.68 mM KCl, pH 7.0 (TBS).

19. TBS + 3 % casein + 01 % Tween 20, pH 7.0 (TBS-CT).

20. Citrate 48 mM containing 136 mM NaCl and 2.68 mM KCl, pH 7.0 (CBS).

21. Super Signal R West Pico chemiluminescent substrate (Pierce Biotechnology (Rockford, IL, USA)).

3   Methods

3.1   Cellulose Membranes and SPOT-Synthesis

Cellulose is a polysaccharide with free hydroxyl groups. To make it suitable for the synthesis of peptides, it is necessary to modify its surface and change the functionalization from hydroxyl to amino groups. Different approaches have been described for functionalizing the cellulose sheets [39], but using the standard functionalized Amino-PEG500-UC540 cellulose (8×12 cm) obtained from Intavis AG Bioanalytical Instruments, Germany, provides a low background compared to membranes functionalized with amino acids such as β-alanine.

A filter paper (Whatman 3MM) was used to accommodate the membrane in the Spot-synthesis machine. For a spotting volume of 0.1 μL, the distance between the centers of two spots should be at least about 3 mm and at least 8 mm for a 1 μL spotting volume. If not noted differently elsewhere, all washing, incubation, and reaction steps are performed using a rocking shaker. All washing steps should be performed for at least 30 s each, unless mentioned differently.

A major advantage of the SPOT-synthesis method is the opportunity to individually design the spatial arrangement and spot size of membrane-bound compound libraries, as discussed previously. In the immunological assay using the Asp 222 machine and a membrane having dimensions of 19×28 cm, it is possible to accommodate 384 peptides easily (1.5 mm Ø and 50–200 nmol).

3.1.1   Detection Method

Different probing techniques can be used; here the focus is on the description for the detection of antibodies labeled by alkaline phosphatase. This detection can only be performed if the peptides are still attached to the membrane. If not noted differently elsewhere, all washing steps should be performed for at least 5 min each. Solvents or solutions used in washing and incubation steps were gently agitated on a rocker table at room temperature (25 °C). During incubations and washings, the troughs are closed with a lid.

The concentration of the two distinct classes of immunoglobulins in the human sera ranged from 9 to 15 mg/mL for IgG and 320–410 ng/mL for IgE, which equates to 10,000 to one million-fold bias towards IgG than IgE [40]. To account for this bias, the optimal dilution of each pool was empirically determined for detection of positive peptides spots. For IgG, the best detection occurred with a dilution of 1:250 and 1:100 for IgE detection. The difference between bias and the detection dilution suggests that human IgE had a greater affinity for hhcIgG3 than human IgG [41].

1. Generate a list of peptides to be prepared calculating a spot distance of 4 mm (0.1 μL) during array generation and 0.2 μL for elongation cycles.

2. Added 1.0 mL of NMP to each peptide flask and vortex if necessary. The amino acids can be used immediately or stored at 4 °C for 1 week.

3. Calculate the volumes of Fmoc-AA (amino acid) solutions required for each derivative and cycle; consider that a triple coupling procedure may be necessary and that each vial should contain a minimum of 50 μL. For example, in your list of peptides, alanine is required for 26 peptides at cycle 1 and you will use a 17×25 array. Then for A1 you will need $26 \times 0.2 \times 3 = 15.6$ μL of Fmoc-Ala stock solution and you will take 50 μL for this vial. The SPOT software (MultiPep) available can do this calculation for you.

4. Label a set of 1.5 mL plastic tubes with derivative and cycle code (e.g., A1 to N) and distribute the Fmoc-AA stock solutions according to the calculated volumes required. Flash-freeze in liquid nitrogen and store at −80 °C.

1. Activate the amino-PEG functionalized membranes washing in DMF (three times for 2 min) and them one time for 2 min in EtOH (*see* **Note 5**).

2. Dry at room temperature and fix the membrane on the platform of the spot-robot for semi-automated synthesis.

3. Take the set of Fmoc-AA stock aliquots for cycle 1 from the freezer, bring to RT and activate by addition of DIC (4 μL per 100 μL vial; ca. 0.25 M). Leave for 30 min. Then place the vials with the activated Fmoc-AA solutions into the corresponding location in the rack of the spotting robot and start cycle 1. Leave for at least 15 min. Repeat the spotting twice and then let react for 2 h (cover the membranes on the spotter with glass plates). If some spots stay dark blue, you may add additional aliquots. If most spots are yellow to green, then continue (*see* **Note 6**).

4. Wash each membrane with 15 mL acetylation mix (1×30 s and 1×2 min). Then incubate until all remaining blue color has disappeared (3×10 min).

5. Wash each membrane with 15 mL DMF (3×10 min).

6. Add 15 mL piperidine mix and incubate for 5 min.

7. Wash each membrane with 15 mL DMF (3×10 min).

8. Incubate with 15 mL of 1 % BPB stock in DMF. Exchange the solution if traces of remaining piperidine turn the DMF solution into a dark blue solution. Spots should be stained only light blue! Due to the charge specific staining, BPB does not

only bind to N-terminal amino-groups. The side chains and protecting groups of other amino acids can strongly influence the staining intensity. The visible color of the peptides depends on the overall charge and therefore depends on the individual amino acid sequence.

9. Wash each membrane with 20 mL MeOH (3×10 min).

10. Dry with cold air from a hair dryer in between a folder of 3MM sheet.

11. Start at **step 2** for the next elongation cycle.

3.3 Terminal Acetylation

Synthetic peptides mimicking fragments of a longer continuous protein chain should be N-terminally acetylated to avoid an artificial charged terminus.

1. Incubate each membrane with acetylation mix until all remaining blue color has disappeared (15 mL for 30 min).

2. Wash each membrane with 15 mL DMF (3×10 min).

3. Wash each membrane with 15 mL MeOH (3×10 min).

4. Dry with cold air from a hair dryer in between a folder of 3MM.

3.4 Side Chain Deprotection of Membrane-Bound Peptide Arrays

After the peptide assembly is complete, it is necessary to remove all side chain protecting groups from the peptides. This must be performed under a hood as TFA is very harmful.

1. Prepare 40 mL of deprotection mix solution.

2. Place the dried membrane in the reaction trough, add deprotection mix solution, close the trough very tightly and agitate overnight (*see* **Note 7**).

3. Wash each membrane with 15 mL DCM (1×30 seg and 1×4 min).

4. Wash each membrane with 15 mL DMF (1×30 seg and 2×2 min).

5. Wash each membrane with 15 mL MeOH (1×30 seg and 2×2 min).

6. Wash each membrane with 15 mL 1 M acetic acid in water (3×2 min) (*see* **Note 8**).

7. Wash each membrane with 15 mL alcohol (1×30 seg and 2×2 min). The membrane sheets may now be dried with cold air and stored at $-20\ °C$ or further processed as described in the next section.

3.5 Immunological Screening of Spot Membrane

1. Place a membrane in a plastic box and wet it with a few drops of ethanol. This procedure is performed to increase the rehydration of some spot containing hydrophobic peptides (not dry the membrane).

2. Wash the membrane with TBS solution (3×10 min each).

3. Incubate the membrane, with agitation at room temperature for 1–1.5 h or overnight at 8 °C with 15 mL of TBS-CT (freshly prepared) (*see* **Note 9**).

4. Incubated the membranes for 2 h with the primary antibody (human allergenic serum) diluted 1:250 for IgG and 1:150 for IgE in TBS-CT (*see* **Note 10**).

5. Wash the membranes with 10 mL TBS-T (3×10 min each).

6. Incubate with rabbit anti-human IgG (1:5000) or goat anti-human IgE (1:1000) conjugated to alkaline phosphatase diluted in TBS-CT for 1.5 h at room temperature.

7. Wash with TBS twice in CBS solution (10 min each) and add the chemiluminescent substrate (CDD Star Ready-to-use Nitro Block II).

8. After completion of the reaction, the liquid is removed and the humid membrane is covered with cellophane and air bubbles removed with the aid of a rubber roller. At this point the membrane is ready to be scanned.

3.6 Scanning and Measurement of Spot Signal Intensities

Measurements of the spot chemiluminescent signals were stately and a digital image file generated on a MF-ChemiBis 3.2 (DNR Bio-Imaging Systems, Israel) with a resolution of 5 MP. This equipment is capable of obtaining digital images without noise and is controlled automatically by the software TotalLab installed on a home computer finding routine resulting in reproducible signal intensities.

1. Connect the computer and then the MF-ChemiBis 3.2 and start the Gel Capture mode software.

2. Wait the cooling time of the camera (about 10 min) (*see* **Note 11**) and put the humid membrane in the dark drawer CL compartment.

3. Start the reaction by adjusting the position applied in the drawer, sharpness, gain and sensitivity. In the "Light tab", choose the types of filters (place the mouse cursor over the options for more information) and in "Optics tab", start adjusting images with more open iris and choose the option to zoom to close image.

4. Capture the image [about 30 s for IgG (Fig. 1b) and 8–10 min for IgE (Fig. 1c)] and save on the hard drive to define rows and columns in accordance with the number of spots.

5. Adjust spot diameter and set the empty area (spot without peptide) or white.

6. Set the positive (stronger) or spot control (100 %) and subtract the empty areas of the microarray spot area.

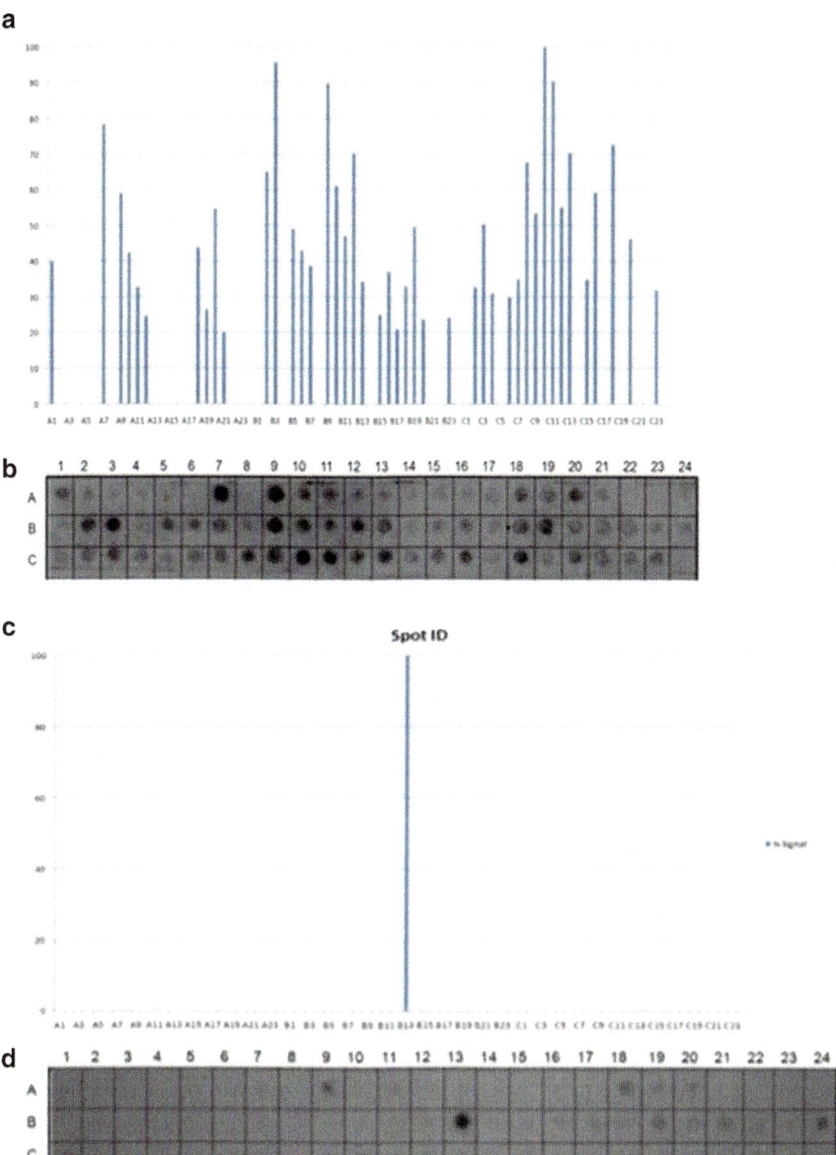

Fig. 1 Analysis of human IgG (**a** and **b**) and IgE (**c** and **d**) reactivity to a cellulose-bound synthetic peptide library representing horse heavy chain IgG3. A membrane spotted with peptides that span the amino acid sequence of hhcIgG3 was probed with a 1:250 (IgG) and 1:100 (IgE) dilution of human sera from volunteers sensitized to horse sera. Bound antibodies were detected by alkaline phosphatase conjugated anti-human IgG or IgE antibodies followed by chemiluminescence. Images of the membranes (**b** and **d**) show the imaged chemiluminescent signal. Relative signal intensities of bound human IgG (**a**) and IgE (**c**) antibodies with 100 % defined by the positive controls and 0 % by the negative control. Only spots with intensities ≥20 % were considered in the overlapping

7. The final result of chemiluminescence and comparison of % shown on the display.

8. Export to Excel and plot the graph (*see* **Note 12**) (Fig. 1a-IgG and 1c-IgE).

4   Notes

1. Free dimethylamine contaminant causes a loss of F-moc protecting groups. For checking contaminants by amines, a 10 μL BPB stock in 1 mL of DMF is added. If the resulting color is yellow color, no further purification is required. If it is contaminated, change the batch or the supplier. This careful attention is relevant because the nascent peptide is in contact for a long time with DMF throughout the synthesis.

2. For purification, treat 100 mL of NMP with 10 g of aluminum oxide acid under constant and vigorous stirring overnight at room temperature. Then, an aliquot of 1 mL in BPB should result in yellow. Filter through a layer of dehydrated silica gel (chromatography, Mallinckrodt Baker BV, Deventer, The Netherlands) on a closed filter funnel. Divide the clear liquid in 100 mL portions in tightly closed jars, storing at –20 °C.

3. The abbreviations correspond to: Fmoc-9-fluorenyl-methoxycarbonyl; mAb—Acetamidomethyl; Trt-Trityl; OtBu-tert butyl ester; Boc-tert-butoxycarbonyl; Pmc—pentamethylchroman-6-sulfonyl; Acm—acetylmethyl; tBu—tert butyl. For the simultaneous preparation of peptides of different size with free amino terminus, couple their terminal amino acid residues as an αN-Boc derivative to prevent acetylation during the normal elongation cycle. Boc is then removed during the final side chain deprotection procedure.

4. The quality of solvents for the required washing steps was at least ACS. Solvents for dissolving reagents for the synthesis were amine- and water-free. Due to the possible decomposition under the influence of light, organic solvents should be stored in the dark with the exception of MeOH and EtOH.

5. A cutting line can be marked using a pencil (H grade) for post-synthesis segmenting the membrane into project specific sections.

6. Add only 1 μL DIC to 100 μL Fmoc-AA mixture stock and repeat spotting four times for the efficient introduction of randomized X positions in the peptide sequences.

7. This harsh treatment is required for complete cleavage of protecting groups.

8. This is for removal of the Boc group from tryptophan.

9. The blocking conditions can be critical to the success of the experiment and, depending on the protein of interest, it may be necessary to test a number of different blocking solutions to optimize the signal and avoid large backgrounds. The TBS blocking solution was the best result in our studies.

10. It is necessary that the solution covers the entire membrane.

11. The machine remains inoperative until the operating temperature has been reached.

12. Signal intensities were quantified with TotalLab Software (Nonlinear Dynamics, USA) using algorithms that compared the intensity between background, spot area and negative control to define the empirical probability that the spot signal intensity was distinct from background signals. Only spots with intensity values above 20 % were considered in the overlapping of reactive peptides.

Acknowledgement

This work received financial assistance from the Conselho Nacional de Desenvolvimento Científico e Tecnológico (CNPq), Coordenação de Aperfeiçoamento de Pessoal de Nível Superior (CAPES) and Fundação Carlos Chagas Filho de Amparo a Pesquisa do Estado do Rio de Janeiro (FAPERJ).

References

1. Winkler DF, Hilpert K (2010) Synthesis of antimicrobial peptides using the SPOT technique. Methods Mol Biol 618:111–124

2. Hilpert K, Behlke J, Scholz C et al (1999) Interaction of the capsid protein p24 (HIV-1) with sequence-derived peptides: influence on p24 dimerization. Virology 254:6–10

3. Reineke U, Sabat R, Kramer A et al (1995) Mapping of protein-protein contact sites using cellulose bound peptide scans. Mol Divers 1:141–148

4. Weiergräber O, Schneider-Mergener J, Grötzinger J et al (1996) Use of immobilized synthetic peptides for the identification of contact sites between human interleukin-6 and its receptor. FEBS Lett 379:122–126

5. Brix J, Rudiger S, Bukau B et al (1999) The mitochondrial import receptors Tom20, Tom22 and Tom70: distribution of binding sequences in a presequence-carrying preprotein and a non-cleavable preprotein. J Biol Chem 274:16522–16530

6. Schneider-Mergener J, Kramer A, Reineke U (1996) Peptide libraries bound to continuous cellulose membranes: tools to study molecular recognition. In: Cortese R (ed) Combinatorial libraries. W. De. Gruyter, Berlin, pp 53–68

7. Sachs EF, Nadler AU (2011) Triostin A derived hybrid for simultaneous DNA binding and metal coordination. Amino Acids 41:449–456

8. Volkmer R, Tapia V, Landgraf C (2012) Synthetic peptide arrays for investigating protein interaction domains. FEBS Lett 586:2780–2786

9. Rudigger S, Germeroth L, Schneider-Mergener J et al (1997) Substrate specificity of the DnaK chaperone determined by screening of cellulose-bound peptide libraries. EMBO J 16:1501–1507

10. Winkler DF, Campbell WD (2008) The spot technique: synthesis and screening of peptide

macroarrays on cellulose membranes. Methods Mol Biol 494:47–70

11. Reineke U, Sabat R, Volk D et al (1998) Mapping of the interleukin-10/interleukin-10 receptor combining site. Protein Sci 7:951–960

12. Bottino CG, Gomes LP, Zauza PL et al (2012) Chagas disease-specific antigens: characterization of CRA/FRA epitopes by synthetic peptide mapping and evaluation on ELISA-peptide assay. BMC Infect Dis 13:568–578

13. De-Simone SG, Napoleão-Pego P, Teixeira-Pinto LAL et al (2013) Linear B-cell epitopes in BthTX-I, BthTX-II and BthA-I, phospholipase $A_{2's}$ from *Bothrops jararacussu* snake venom, recognized by therapeutically neutralizing commercial horse antivenom. Toxicon 72:90–101

14. Cretich M, Damin F, Chiari M (2014) Protein microarray technology: how far of is routine diagnostics? Analyst 139:528–542

15. Uttamchandani M, Wang J, Yao SQ (2006) Protein and small molecule microarrays: powerful tools for high-throughput proteomics. Mol Biosyst 2:58–68

16. Santos-Pinto JRA, Santos LD, Arcuri HA et al (2014) B-cell linear epitopes mapping of antigen-5 allergen from *Polybia paulista* wasp venom. J Allergy Clin Immunol. doi:10.1016/j.jaci.2014.07.006

17. Silva FR, Napoleão-Pego P, De-Simone SG (2014) Identification of linear B epitopes of pertactin of *Bordetella pertussis* induced by immunization with whole and acellular vaccine. Vaccine 32:6251–6258

18. Geysen HM, Meloen RH, Barteling SJ (1984) Use of peptide synthesis to probe viral antigens for epitopes to a resolution of a single amino acid. Proc Natl Acad Sci U S A 81:3998–4002

19. Wang CY, Walfield AM, Fang X et al (2003) Synthetic IgE peptide vaccine for immunotherapy of allergy. Vaccine 21:1580–1590

20. Barteling SJ, Woortmeye R (1984) Formaldehyde inactivation of foot-and-mouth disease virus. Conditions for the preparation of safe vaccine. Arch Virol 80:103–117

21. Kramer A, Schuster A, Reineke U et al (1994) Combinatorial cellulose-bound peptide libraries: screening tool for the identification of peptides that bind ligands with predefined specificity. Methods 6:388–395

22. Houghten RA (1985) General method for the rapid solid-phase synthesis of large numbers of peptides: specificity of antigen-antibody inter-

action at the level of individual amino acids. Proc Natl Acad Sci U S A 82:5131–5135

23. Scott JK, Smith GP (1990) Searching for peptide ligands with an epitope library. Science 249:386–390

24. Furka A, Sebestyen F, Asgedom M et al (1991) General method for rapid synthesis of multicomponent peptide mixture. Int J Pept Protein Res 37:487–493

25. Lam KS, Salmon SE, Hersh EM et al (1991) A new type of synthetic peptide library for identifying ligand-binding activity. Nature 354:82–84

26. Dooley CT, Houghten RA (1993) The use of positional scanning synthetic peptide combinatorial libraries for the rapid determination of opioid receptor ligands. Life Sci 52:1509–1517

27. Fodor SP, Read JL, Pirrung MC et al (1991) Light-directed, spatially addressable parallel chemical synthesis. Science 251:767–773

28. Kramer A, Reineke U, Dong L et al (1999) Spot synthesis: observations and optimizations. J Pept Res 54:319–327

29. Weisser AA, Or-Cuil M, Tapia V et al (2005) SPOT synthesis: reliability of array-based measurement of peptide binding affinity. Anal Biochem 342:300–311

30. Frank R (2002) The SPOT-synthesis technique. Synthetic peptide arrays on membrane supports-principles and applications. J Immunol Methods 267:13–26

31. Frank R (1992) Spot-synthesis: an easy technique for the positionally addressable, parallel chemical synthesis on a membrane support. Tetrahedron 48:9217–9232

32. Frank R, Overwin H (1996) SPOT synthesis. Epitope analysis with arrays of synthetic peptides prepared on cellulose membranes. Methods Mol Biol 66:149–169

33. Kramer A, Schneider-Mergener J (1998) Synthesis and screening of peptide libraries on continuous cellulose membrane supports. Methods Mol Biol 87:25–39

34. Molina F, Laune D, Gougat C et al (1996) Improved performances of spot multiple peptide synthesis. Pept Res 9:151–155

35. Terness P, Kohl I, Hubener G et al (1995) The natural human IgG anti-F(ab′)2 antibody recognizes a conformational IgG1 hinge epitope. J Immunol 154:6446–6452

36. Fernandes I, Lima EX, Takehara HA et al (2000) Horse IgG isotypes and cross-neutralization of two snake antivenoms produced in Brazil and Costa Rica. Toxicon 38:633–644

37. Sheoran AS, Timoney JF, Holmes MA et al (2000) Immunoglobulin isotypes in sera and nasal mucosal secretions and their neonatal transfer and distribution in horses. Am J Vet Res 61:1099–2005

38. Hilpert K, Winkler DFH, Hancock REW (2007) Cellulose-bound peptide arrays: preparation and applications. Biotechnol Genet Eng Rev 24:31–106

39. Hilpert K, Winkler DF, Hancock RE (2007) Peptide arrays on cellulose support: SPOT synthesis, a time and cost efficient method for synthesis of large numbers of peptides in a parallel and addressable fashion. Nat Protoc 2:1333–1349

40. Ledin A, Arnemo JM, Liberg O et al (2008) High plasma IgE levels within the Scandinavian wolf population, and its implications for mammalian IgE homeostasis. Mol Immunol 45:1976–1980

41. De-Simone SG, Napoleão-Pêgo P, Teixeira-Pinto LAL et al (2014) IgE and IgG epitope mapping by microarray peptide-immunoassay reveals the importance and diversity of the immune response to the IgG3 equine immunoglobulin. Toxicon 78:83–93

Chapter 21

Mapping of Epitopes Occurring in Bovine α_{s1}-Casein Variants by Peptide Microarray Immunoassay

Maria Lisson and Georg Erhardt

Abstract

Immunoglobulin E epitope mapping of milk proteins reveals important information about their immunologic properties. Genetic variants of α_{s1}-casein, one of the major allergens in bovine milk, are until now not considered when discussing the allergenic potential. Here we describe the complete procedure to assess the allergenicity of α_{s1}-casein variants B and C, which are frequent in most breeds, starting from milk with identification and purification of casein variants by isoelectric focusing (IEF) and anion-exchange chromatography, followed by in vitro gastrointestinal digestion of the casein variants, identification of the resulting peptides by matrix-assisted laser desorption/ionization time-of-flight mass spectrometry (MALDI-TOF-MS), in silico analysis of the variant-specific peptides as allergenic epitopes, and determination of their IgE-binding properties by microarray immunoassay with cow's milk allergic human sera.

Key words IgE epitope mapping, Casein variants, In vitro gastrointestinal digestion, Peptide microarray, Milk allergy

1 Introduction

Genetic variants of α_{s1}-casein, one of the major allergens in bovine milk, are not considered when discussing the allergenic potential of bovine milk proteins. Within α_{s1}-casein, noticeable genetic variation occurs, which is caused by substitutions or deletions of amino acids [1, 2]. Genetic variants in milk proteins can be identified on protein level by isoelectric focusing (IEF). This is an effective method of electrophoresis that separates proteins according to their isoelectric point. Separation takes place in thin polyacrylamide gels containing carrier ampholytes, which build up a pH gradient during electrophoresis. This allows the proteins to migrate until they reach their isoelectric point where they focus [3]. With IEF numerous genetic variants have been detected in milk proteins from different species [4–7]. Here, we could demonstrate the common α_{s1}-casein variants B and C in homozygous and heterozygous form in milk from several breeds.

Marina Cretich and Marcella Chiari (eds.), *Peptide Microarrays: Methods and Protocols*, Methods in Molecular Biology, vol. 1352, DOI 10.1007/978-1-4939-3037-1_21, © Springer Science+Business Media New York 2016

Ion-exchange chromatography has been widely used in the past for the isolation of the casein fractions from whole casein [8–10]. Anion-exchange chromatography with diethylaminoethyl (DEAE)-cellulose separates proteins on the basis of charge characteristics. Negative charged groups on the surface of the caseins interact with the oppositely positive charged groups that are immobilized on the DEAE-cellulose. The caseins are eluted from the DEAE-cellulose by increasing salt concentration of the buffer, which reduces the casein's affinity for the anion exchanger. The most weakly charged casein elutes first, followed by caseins with gradually stronger charges. Eluted caseins are detected by measuring the absorption at 280 nm at the column outlet [11]. After precipitation and anion-exchange chromatography we were able to recover the α_{S1}-casein variants B and C from the other caseins with high degree of purity.

One main aspect to consider when evaluating the allergenic potential of food allergens is the effect of gastrointestinal digestion as most food allergens are thought to sensitize an individual via the gastrointestinal tract [12]. Digested proteins or fragments thereof must be of sufficient size being able to react with IgE antibodies, which requires the presence of intact IgE-binding epitopes. Several in vitro digestion models have been developed to assess the behavior of food allergens during digestion [13]. Although in vivo methods are of physiological importance and usually provide most accurate results, their implementation is hardly to achieve due to ethical reasons, technical constraints, costs, impracticability of large studies, and high interindividual variability. Therefore, in vitro digestion systems mimicking the physiological process of gastrointestinal digestion are an alternative tool to investigate the digestibility of proteins [13, 14]. Here, we compare the digestibility of α_{S1}-casein variants B and C using an in vitro digestion model according to Moreno et al. [15] that involve a gastric and subsequent duodenal stage with digestive enzymes in physiological concentrations, a good adaption of pH and enzyme/substrate ratio, as well as surfactants like bile salts [13, 15]. Recently, it has been demonstrated that this model of in vitro digestion utilizing commercial non-human enzymes provides a good estimation of the gastrointestinal digestibility and potential allergenicity of proteins as marked similarities in the results obtained with an in vitro digestion system using human oral and gastrointestinal digestive fluids were found [16]. We used matrix-assisted laser desorption/ionization time of flight mass spectrometry (MALDI-TOF-MS) to extensively characterize peptides that resist gastrointestinal digestion of α_{S1}-casein B and C with the aim of identifying the occurrence and differences of resistant regions containing IgE-binding epitopes of these variants. We showed that the amino acid substitutions characterizing the genetic variants affected the arising peptide pattern of α_{S1}-casein B and C and thus variant-specific peptides with modifications in their allergenic epitopes occurred.

To study if these variant-specific peptides retain IgE-binding properties and differ in their allergenic potential, microarray immunoassays were performed. Protein and peptide microarrays are an important tool allowing the determination of allergenic IgE-binding epitopes on milk proteins. The knowledge of IgE-binding epitopes on milk proteins is important for diagnosis and treatment of cow's milk allergy [17–19]. Epitope mapping by peptide microarray immunoassay presents several advantages as it enables the screening of thousands of target peptides simultaneously, requires only a low volume of diluted serum, enables a more robust replication and statistical analysis and gives the possibility to test several immunoglobulin subclasses in parallel [20]. Due to the enormous potential of these tests, studies on epitope mapping of food allergens by microarray immunoassays are increasing [21–24]. Peptides, representing the variant-specific peptides of digested α_{s1}-casein B and C, were commercially synthesized and immobilized on glass slides. Upon incubation with human serum containing IgE antibodies, the specific binding to the peptides was detected using fluorescently labeled secondary anti-human IgE antibodies followed by scanning with a high-resolution microarray scanning system. We were able to show that the amino acid substitutions affect the IgE-binding properties of the variant-specific peptides of α_{s1}-casein B and C indicating that genetic variants of the caseins differ in their allergenicity.

2   Materials

2.1 Isoelectric Focusing

1. Skimmed milk (*see* **Note 1**).

2. Electrophoresis apparatus, with glass plates, spacers, clamps, and cooling unit.

3. Acrylamide/bisacrylamide solution: Weigh 38.5 g acrylamide and 1.5 g bisacrylamide. Add distilled water to a volume of 100 mL and filtrate using a filter with an average retention capacity of 4–12 μm and a diameter of 185 mm (*see* **Note 2**).

4. Stock solution: Weigh 25.5 g urea and add 7.5 mL of the acrylamide/bisacrylamide solution. Make up to 50 mL with distilled water. Stir for about 45 min with 1 g amberlite and filtrate (*see* **Note 2**).

5. Pharmalyte pH 2.5–5 (GE Healthcare Europe GmbH, Freiburg, Germany).

6. Pharmalyte pH 4.2–4.9 (GE Healthcare Europe GmbH, Freiburg, Germany).

7. Servalyt 5–7 (Serva Electrophoresis GmbH, Heidelberg, Germany).

8. *N,N,N,N*-tetramethyl-ethylenediamine (TEMED): Store at 4 °C.

9. Ammonium persulfate (APS): 0.7 % solution in water.

10. Sample preparation: 8 M urea, 3 % β-mercaptoethanol, 0.1 % bromophenol blue. Weigh 24 g urea, 1.5 mL 3 % β-mercaptoethanol, 2.5 mL glycerol, and some bromophenol blue. Make up to 50 mL with distilled water.

11. Anode fluid 3 for IEF (Serva Electrophoresis GmbH, Heidelberg, Germany).

12. Cathode fluid 10 for IEF (Serva Electrophoresis GmbH, Heidelberg, Germany).

13. Electrode strips Whatman chromatography paper no. 17 (Sigma-Aldrich, Steinheim, Germany).

14. Fixation: 20 % (wt/vol) trichloroacetic acid (TCA).

15. Staining solution: 0.1 % (wt/vol) Coomassie brilliant blue R 250, 45 % (vol/vol) ethanol, 10 % (vol/vol) acetic acid, and 45 % (vol/vol) distilled water.

16. Destainig solution: 30 % (vol/vol) ethanol, 10 % (vol/vol) acetic acid, and 60 % (vol/vol) distilled water.

2.2 Casein Precipitation from Milk

1. Skimmed milk.

2. 50 % Acetic acid (vol/vol).

3. 1 M NaOH.

2.3 Purification of Caseins by Anion-Exchange Chromatography

1. Lyophilized casein.

2. DEAE 52-cellulose resin.

3. Equilibration buffer: 0.01 M Tris/imidazole (pH 7.0), 3.3 M urea, 0.01 M β-mercaptoethanol, and 3.1 mM sodium azide.

4. Elution buffer: Prepare elution buffer with three different salt concentrations ranging from 0.075, 0.13 to 0.17 M NaCl by mixing equilibration buffer with 1 M NaCl.

5. Regeneration buffer: Equilibration buffer with 1.5 M NaCl.

6. Peristaltic pump.

7. Absorbance detector set at 280 nm.

8. Dialysis tubing cellulose membrane.

2.4 In Vitro Gastrointestinal Digestion

2.4.1 Gastric Digestion (Phase 1)

1. 12–15 mg lyophilized casein fraction (*see* **Note 3**).

2. Simulated gastric fluid (SGF): 0.15 M NaCl adjusted to pH 2.5 with 1 M HCl.

3. 1 M NaOH.

4. Pepsin solution: 0.32 % (wt/vol) pepsin from porcine gastric mucosa (activity: 3200–4500 U/mg of protein using hemoglobin as substrate) (Sigma-Aldrich, Steinheim, Germany) in SGF.

2.4.2 Duodenal Digestion
(Phase 2)

1. 0.1 M NaOH.

2. 0.125 M bile salt mixture: Weigh 0.167 mg sodium taurocholate and 147 mg sodium glycodeoxycholate and add 2.5 mL water.

3. 1 M $CaCl_2$.

4. 0.25 M Bis–Tris (pH 6.5).

5. Trypsin solution: 0.1 % (wt/vol) trypsin from porcine pancreas (activity: 13,800 U/mg of protein using benzoylarginine ethyl ester as substrate) (Sigma-Aldrich, Steinheim, Germany) in water.

6. α-Chymotrypsin solution: 0.4 % (wt/vol) α-chymotrypsin from bovine pancreas (activity: 40 U/mg of protein using benzoyltyrosine ethyl ether as substrate) (Sigma-Aldrich, Steinheim, Germany) in water.

7. Bowman-Birk trypsin-chymotrypsin inhibitor from soybean (Sigma-Aldrich, Steinheim, Germany): Weigh 0.5 mg inhibitor and dissolve 1 mL in water.

2.5 MALDI-TOF-MS

1. ZipTip$_{18}$ column (Millipore, Eschborn, Germany).

2. 2.5-Dihydroxy benzoic acid and methylene-diphosphonic acid for the matrix solution (Fluka, Neu-Ulm, Germany).

3. Peptide calibration standard mixture.

4. MALDI-TOF-mass spectrometer.

5. Software for data analysis.

6. Protein database Swiss-Prot and TrEMBL and the tools FindPept and Peptide Mass (www.expasy.org).

2.6 Peptide
Microarray
Immunoassay

1. Human sera of cow's milk allergic patients with specific IgE antibodies to cow milk. Their collection and usage required the approval of the local ethical review committees.

2. A set of microarrays consisting of 57 synthetic α_{s1}-casein peptides with up to 20 amino acids in length and 3 control spots (human IgG, mouse IgE, and human IgE), which were produced by JPT Peptide Technologies (Berlin, Germany).

   Peptides were commercially synthesized using PepStar® technique (JPT Peptide Technologies) as described by Funkner et al. [25]. Briefly, peptides were synthesized in a stepwise manner on a cellulose membrane. After coupling a reactive tag on the N-terminus of each peptide, side chains were deprotected and peptides were cleaved from the cellulose membrane by addition of 200 µL of aqueous triethylamine (2.5 % v/v). Peptides were obtained by filtering off the peptide-containing triethylamine solution and by removing the solvent through evaporation under reduced pressure. Peptides (50 nmol) were diluted in printing buffer (70 %

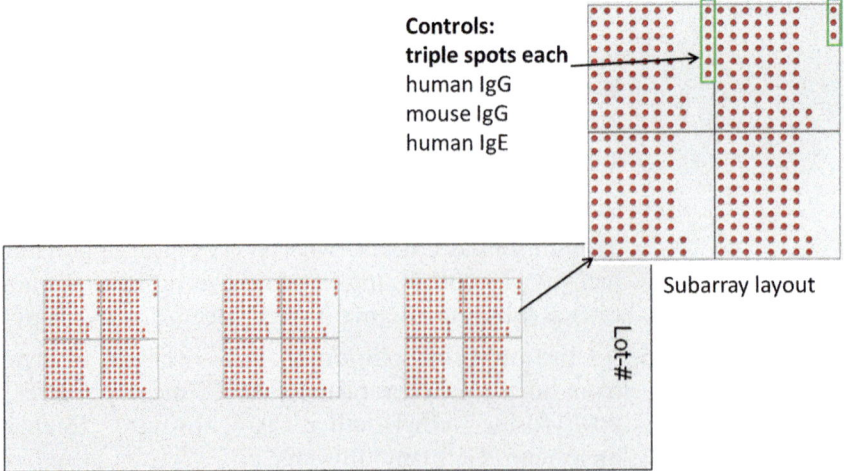

Fig. 1 General microarray layout. The peptide microarrays are printed in three identical subarrays. This enables efficient intra-chip-reproducibility tests. Full-length protein controls (human IgG, mouse IgG, and human IgE) are printed within the peptide array in triplicates and indicated within *green* frames (JPT Peptide Technologies GmbH, Berlin, Germany)

dimethyl sulfoxide, 25 % of 0.2 M sodium acetate pH 4.5, and 5 % v/v glycerol) and 2 droplets of 0.5 nL peptide solution (1 mM) were immobilized chemoselectively onto epoxy-modified glass slides (Corning, Lowell, USA) using a microarray printing system (JPT Peptide Technologies). Peptides and controls were spotted in triplicate onto the glass slides for quality control of the results (Fig. 1). Synthesized peptides covered the variant-specific peptides of α_{S1}-casein B and C, which result from gastrointestinal digestion and correspond to IgE-binding epitopes known from literature.

3. SureLight allophycocyanin-labeled secondary anti-human IgE (Abcam, Cambridge, UK).

4. Blocking buffer (Smartblock; Candor Bioscience GmbH, Wangen, Germany).

5. Diluent buffer [SuperBlock Tris-buffered saline (TBS); Pierce International, Rockford, IL].

6. 50 mM TBS buffer, including 0.1 % Tween 20 (pH 7.2).

7. 3 mM saline-sodium citrate buffer (pH 7.0).

8. Microarray processing station capable of washing and incubating slides in a temperature-controlled environment.

9. High-resolution fluorescence scanner/imager.

10. Spot-recognition software allowing the assignment of signal intensities to spots on the surface of the peptide microarrays.

3 Methods

3.1 Isoelectric Focusing

1. Prepare the electrophoresis apparatus by thoroughly cleaning glass plates. Assemble the gel cassette using 2 glass plates, 1 silanized glass plate, spacers, and clamps and mount in casting stand (*see* **Note 4**).

2. Mix 8.4 mL of stock solution with 0.1 mL of Pharmalyte pH 2.5–5, 0.244 mL of Pharmalyte pH 4.2–4.9, and 0.3 mL of Servalyt 5–7. Add 15 µL TEMED and 1 mL of APS and cast a 0.3 mm thin gel within silanized glass plates without introducing air bubbles by using a syringe connected to a thin tube (*see* **Note 5**).

3. After polymerization of the gel, open the gel cassette with the use of a spatula. The gel remains on the silanized glass plate. Transfer the gel carefully to the electrophoresis chamber with some water as contact agent to the cooling plate and switch on the cooling unit (*see* **Note 6**). Soak the electrode strips with the anode and cathode solutions, cut to gel length and place in the positions provided.

4. Start prefocusing at 3000 V and 20 mA for 14 min.

5. Prepare the samples on a microtiter plate by mixing 7 µL of milk sample with 50 µL of sample preparation and put it on a shaker for 10 min.

6. Place a slotted band 3 mm in front of the anode and apply 10 µL of the sample mix in each gap by using a multiple pipette or individually (*see* **Note 7**). Add milk with known phenotypes as standards at the right and left edges (*see* **Note 8**, Fig. 2).

7. Start final focusing at 3000 V and 40 mA for 60 min.

8. For fixation place the gel in 20 % TCA for 30 min (*see* **Note 9**). Afterwards, transfer it into water for 15 min.

9. After washing place the gel in the staining solution and following in the destaining solution depending on the time needed (*see* **Note 9**).

10. Leaf the gel to air-dry.

3.2 Casein Precipitation from Milk

1. Measure the pH value of skimmed milk samples. Adjust the pH of skimmed milk samples to 4.6 by acidification with 50 % acetic acid.

2. Centrifuge skimmed milk samples at $1620 \times g$ for 10 min at 4 °C.

3. Remove the whey supernatant from the precipitated casein curd and dissolve the casein precipitate in distilled water with 1.0 M NaOH to initial pH value of milk by the use of a mortar and by stirring.

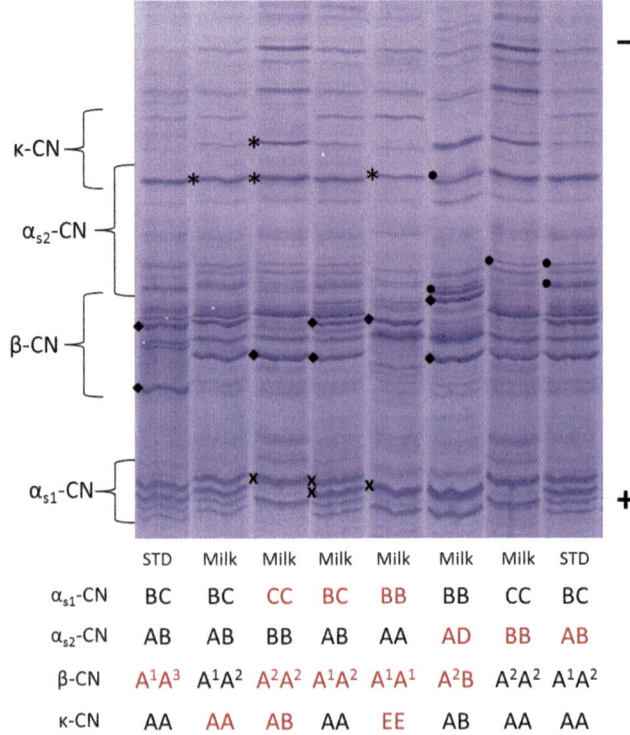

	STD	Milk	Milk	Milk	Milk	Milk	Milk	STD
α_{s1}-CN	BC	BC	CC	BC	BB	BB	CC	BC
α_{s2}-CN	AB	AB	BB	AB	AA	AD	BB	AB
β-CN	A^1A^3	A^1A^2	A^2A^2	A^1A^2	A^1A^1	A^2B	A^2A^2	A^1A^2
κ-CN	AA	AA	AB	AA	EE	AB	AA	AA

Fig. 2 Separation of bovine milk samples with different α_{s1}-, α_{s2}-, β-, and κ-casein (CN) genotypes, indicated in *capitals* at the *bottom* of the illustration, by isoelectric focusing. Reference samples (STD) were used for identification of the casein variants. The positions of the major bands of the individual α_{s1}-, α_{s2}-, β-, and κ-casein variants are indicated by symbols (* ♦ X •)

4. Repeat steps of precipitation and washing twice.

5. Lyophilize whole casein and store at –20 °C.

3.3 Purification of Caseins by Anion-Exchange Chromatography

1. Suspend 30 g DEAE 52-cellulose in 150 mL equilibration buffer and allow the slurry to settle for about 30 min.

2. Siphon off the supernatant using a pipette and bring the slurry back to the initial volume by addition of equilibration buffer.

3. After dispersion and settling repeat **step 2**.

4. Remove supernatant down to the settled slurry plus about 20 % and pack the DEAE 52-cellulose exchanger into a column (*see* **Note 10**). Allow to stand overnight.

5. Equilibrate column with double column volume of equilibration buffer until the baseline absorbance at 280 nm is stable by using a peristaltic pump (*see* **Note 11**).

6. Dissolve an amount of 1.0–1.5 g lyophilized casein in 30–40 mL of equilibration buffer and load onto the DEAE cellulose column (*see* **Note 12**).

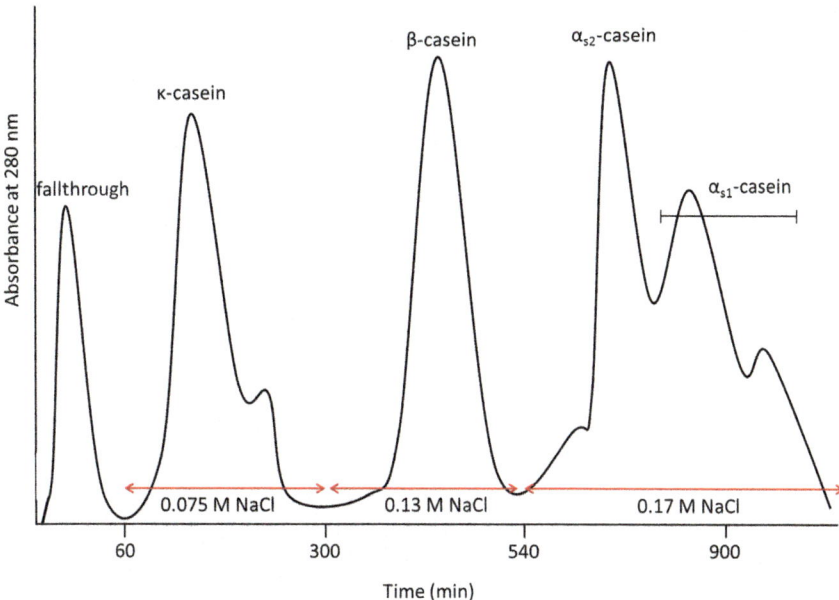

Fig. 3 Elution profile of whole casein on DEAE-cellulose anion-exchange column with a step gradient ranging from 0.075, 0.13 to 0.17 M in 0.01 M Tris/imidazole (pH 7.0), 3.3 M urea, 0.01 M β-mercaptoethanol, and 3.1 mM sodium azide, pH 7.0, which is indicated by *arrow bars*. Elution of each casein fraction is represented at the *top* of the peaks

7. Wash with equilibration buffer until all unbound material has washed through the column and until the baseline absorbance at 280 nm is stable (*see* **Note 13**).

8. Begin elution of caseins using 100–200 mL elution buffer of increasing NaCl concentration starting with 0.075 M NaCl followed by 0.13 M and 0.17 M NaCl (*see* **Notes 14** and **15** and Fig. 3).

9. Choose a flow rate of 0.6 mL/min and monitor the absorbance of the column effluents at 280 nm. Collect fractions of 4–6 mL throughout the column run (*see* **Note 16**).

10. For regeneration of the column wash with around 50 mL of regeneration buffer containing 1.5 M NaCl to elute any remaining ionically bound material.

11. Re-equilibrate with double column volume of equilibration buffer until the baseline absorbance at 280 nm is stable.

12. Determine the main peak fractions of the resulting chromatogram, which correspond to the different purified caseins, by IEF in the way described above (*see* **Note 17**).

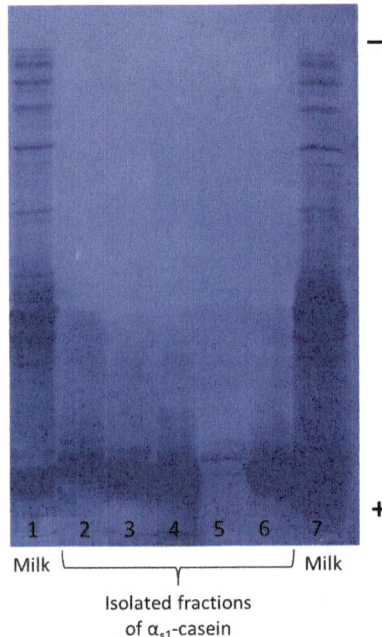

Fig. 4 Determination of the purity and identity of isolated α_{s1}-casein variants obtained after anion-exchange chromatography by isoelectric focusing: Whole skimmed milk (*lanes 1 and 7*), α_{s1}-casein C (*lanes 2 and 5*), α_{s1}-casein B (*lanes 3, 4, and 6*)

13. Pool designated casein fractions, fill them into dialysis tubes, and dialyze three times against 5 L of distilled water for 24 h. Afterwards, lyophilize casein fractions.

14. The identity and purity of the isolated casein fractions are assessed by comparing whole casein with the pooled fractions by IEF (Fig. 4).

3.4 In Vitro Gastrointestinal Digestion

3.4.1 Gastric Digestion (Phase 1)

1. Dissolve 12–15 mg isolated α_{s1}-casein in 1 mL SGF and adjust to pH 2.5 with 1.0 M HCl (*see* **Notes 3** and **18**).

2. Incubate the samples at 37 °C for 15 min.

3. Add the pepsin solution at an approximately physiological ratio of enzyme:casein (1:20 wt/wt) and incubate the samples at 37 °C for 60 min with moderate agitation (*see* **Note 19**).

4. Stop the digestion by increasing the pH to 7.5 using 1.0 M NaOH.

5. Prepare control samples with no enzyme addition.

6. Take aliquots for MALDI-TOF-MS analysis.

3.4.2 Duodenal Digestion (Phase 2)

1. Phase 2 digestion takes place with the gastric casein digests of phase 1 as starting material. At first adjust the pH of the gastric casein digest to 6.5 with 0.1 M NaOH.

2. To simulate a duodenal environment, mix samples with 0.125 M bile salt mixture, 1 M CaCl$_2$, and 0.25 M Bis–Tris and adjust the pH to 6.5 with 1 M HCl (*see* **Note 20**).

3. Add solutions of 0.1 % (wt/vol) trypsin and 0.4 % (wt/vol) α-chymotrypsin at approximately physiological ratios of casein/trypsin/α-chymotrypsin = 1:400:100 (wt/wt/wt), respectively, 1 mg/34.5 U/0.40 U and incubate the samples at 37 °C for 60 min with moderate agitation (*see* **Note 21**).

4. Prepare control samples with no enzyme addition.

5. Stop duodenal digestion by addition of a twofold excess of Bowman-Birk trypsin-chymotrypsin inhibitor calculated to inhibit trypsin and chymotrypsin in the digestion mix.

6. Remove aliquots in triplicate for MALDI-TOF-MS analysis.

3.5 MALDI-TOF-MS

1. Desalt casein digests using ZipTip C$_{18}$ columns.

2. Crystallize the sample by the dried droplet method using 2.5-dihydroxy benzoic acid as matrix.

3. Prepare the matrix solution by dissolving 2.5-dihydroxy benzoic acid and methylene-diphosphonic acid at concentrations of 5 mg/mL in water, respectively.

4. Carry out MALDI-TOF analyses with a mass spectrometer equipped with a UV nitrogen laser ($\lambda = 337$ nm, 3 ns pulse width) to desorb/ionize the matrix/analyte material.

5. Acquire spectra in positive ion reflectron mode over the *m/z* range 400 and 4000 (Fig. 5).

6. Calibrate all mass spectra initially with a peptide calibration standard mixture (Bruker Daltonics).

7. Use software to analyze the data obtained from MALDI-TOF-MS.

8. Based on the known sequences of the different caseins, assign the observed ions to the corresponding amino acid sequences by using the protein database Swiss-Prot and TrEMBL, as well as the tools FindPept and Peptide Mass (www.expasy.org).

9. For peptide identification choose the following search parameters: (a) indicate peptide masses as monoisotopic and $[M+H]^+$; (b) cysteines treated with nothing and methionines oxidized; (c) set mass tolerance to 50 ppm; and (d) select pepsin, trypsin, and chymotrypsin as enzymes (Table 1).

10. To uncover IgE-binding epitopes, compare identified peptides to sequences that overlap with existing IgE-binding epitopes in the literature (Fig. 6).

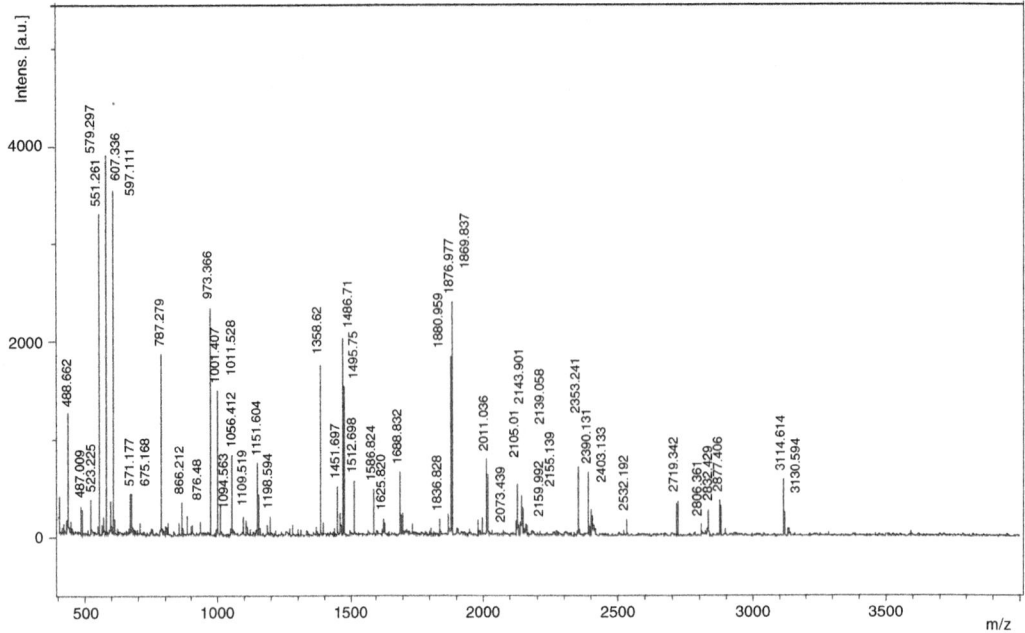

Fig. 5 MALDI-TOF-MS spectrum of peptides survived gastric and duodenal digestion of α_{s1}-casein B. The peptide signals distinctive for the α_{s1}-casein variants are assigned in Table 1

Table 1
Peptides containing amino acid substitutions identified after gastric (phase 1) and duodenal (phase 2) digestion of α_{s1}-CN B and C by MALDI-TOF-MS

	Experimental mass [M + H][a]	Theoretical mass [M + H][a]	Position	Sequence	α_{s1}-Casein
Gastric digestion	876.48	876.47	193–199	(E)/[193]KTTMPLW[199]	B
	1689.79	1689.76	177–192	[177]PSFSDIPNPIGSENSE[192]	B
	1688.78	1688.78	176–192	[176]APSFSDIPNPIGSENSG[192]	C
	1819.83	1819.84	180–196	[180]SDIPNPIGSENSEKTTM[196]	B
	1747.79	1747.83	180–196	[180]SDIPNPIGSENSGKTTM[196]	C
	2216.06	2216.05	180–199	[180]SDIPNPIGSENSEKTTMPLW[199]	B
Duodenal digestion	876.48	876.47	193–199	(E)/[193]KTTMPLW[199]	B
	1358.62	1358.61	180–192	[180]SDIPNPIGSENSE[192]	B
	1407.63	1407.64	184–196	[184]NPIGSENSEKTTM[196]	B
	1407.61	1407.66	187–199	[187]GSENSGKTTMPLW[199]	C
	1495.75	1495.63	187–199	[187]GSENSEKTTMPLW[199]	B
	1486.71	1486.70	180–193	[180]SDIPNPIGSENSEK[193]	B
	1414.67	1414.68	180–193	[180]SDIPNPIGSENSGK[193]	C
	2105.01	2104.97	174–193	[174]TDAPSFSDIPNPIGSENSEK[193]	B
	2105.04	2105.02	179–198	[179]FSDIPNPIGSENSGKTTMPL[198]	C

Bolded letters represent the position of the amino acid substitution differentiating α_{s1}-CN B and C (reproduced from [26] with permission from Elsevier)
[a]Monoisotopic mass values

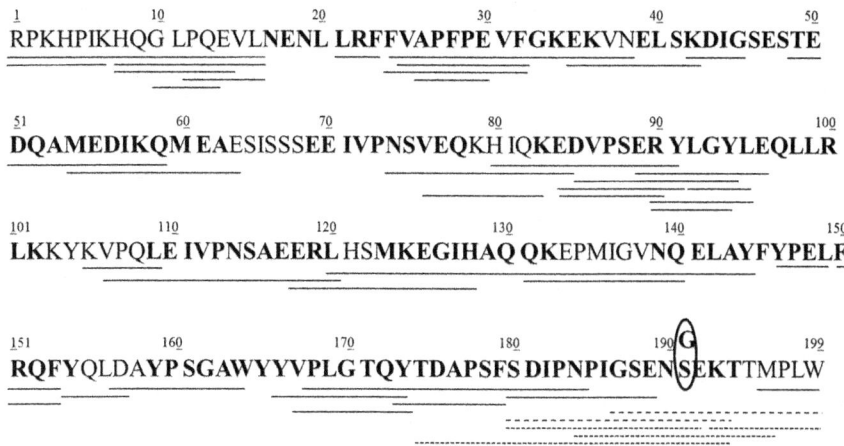

Fig. 6 Peptides (*black continuous lines*) identified after duodenal digestion (phase 2) of α_{s1}-casein variants B and C by MALDI-TOF-MS. Peptides with amino acid substitution occurring in all α_{s1}-casein variants are represented as *small dashed lines* (-------), peptides only present in α_{s1}-casein B are indicated with *small dashed dots* (………) and peptides detected solely in α_{s1}-casein C are shown with *large dashed lines* (_ _ _ _). The position of the amino acid substitution differentiating variant B from C and the IgE-binding epitopes known from literature are highlighted in *bold* (reproduced from Lisson et al. [26] with permission from Elsevier)

3.6 Peptide Microarray Immunoassay

3.6.1 Determination of Peptide Antibody Binding (RepliTope Analysis by JPT Peptide Technologies GmbH)

1. Incubate the array slides with blocking buffer for 60 min (*see* **Note 22**).

2. Place them into a microarray processing station and incubate with 200 μL of the patient's serum, diluted 1:100 in diluent buffer and with diluent buffer only as control for 2 h (*see* **Note 23**).

3. Wash the peptide microarray three times with 50 mM TBS buffer, including 0.1 % Tween 20 (pH 7.2).

4. For detection of specific human IgE-binding incubate the peptide microarray with SureLight allophycocyanin-labeled secondary anti-human IgE at a final concentration of 1 μg/mL in diluent buffer for at least 30 min (*see* **Note 24**).

5. Perform control incubation with secondary antibody only in parallel (*see* **Note 25**).

6. Carry out three washing steps with 50 mM TBS buffer, including 0.1 % Tween 20 (pH 7.2) followed by a washing step with 3 mM saline-sodium citrate buffer (pH 7.0).

7. Dry peptide microarrays using a nitrogen stream.

3.6.2 Signal Detection and Data Analysis

1. Scan peptide microarrays by using a high-resolution fluorescence scanner. Images are saved electronically in tagged image file format (*.tif).

2. Perform image analysis by using the spot-recognition software showing the signal intensity (light units) as single measurements for each peptide.

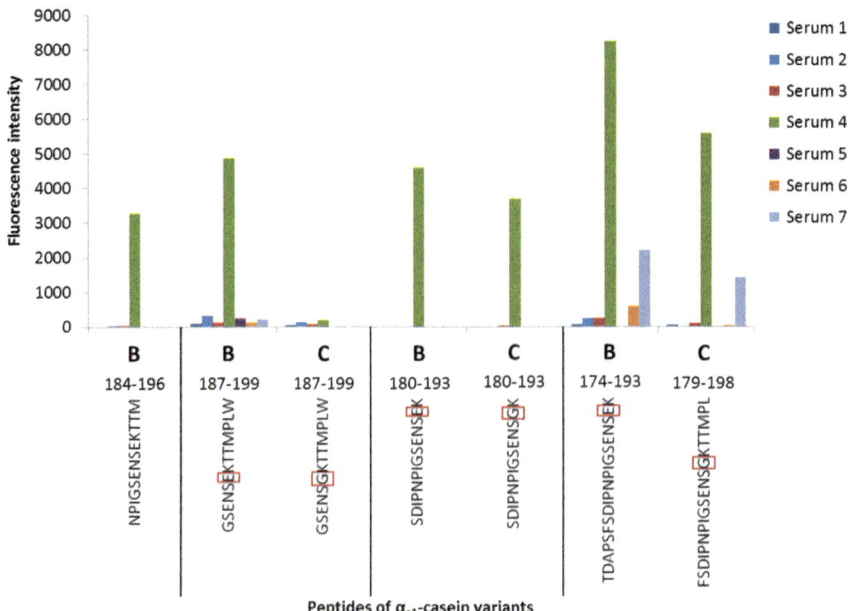

Fig. 7 Interpretation of the data obtained from microarray immunoassays for peptides of α_{s1}-casein variants B and C from cow, using seven sera with cow milk-specific IgE values >100 kU$_A$/L. Differences in amino acid sequence between both variants are framed. The mean fluorescence intensity of the three subarrays on the microarray for each peptide is presented (reproduced from [27] with permission from Elsevier)

3. Analyze each spot feature for total intensity and background intensity.

4. Calculate the mean value for the signal intensities of spots with identical peptides (three identical subarrays) on the microarray. These values were used for creation of data tables and diagrams (Fig. 7).

5. Use negative controls to define a background noise. Determine the final fluorescent intensity of each peptide by subtracting the mean intensity of the corresponding peptide spots from control incubation without serum.

6. Consider IgE binding positive if signal intensities are above the lower limit of detection, meaning that ratios of signal-to-background noise intensities were greater than a minimum threshold.

4 Notes

1. Fresh milk samples should be conserved by addition of sodium azide and stored under freezing conditions (−20 °C).

2. Solution can be stored at 4 °C for some months or frozen as a stock at −20 °C.

3. Regarding the quantity of α_{s1}-casein to the digestion experiments it should reflect the concentrations of this protein found in milk amounting to 12–15 mg/mL.

4. Prior to assembling of the glass plates, clean them with acetone for better detachment of the gel from the plate.

5. The polymerization gets started after addition of APS and TEMED. Thus, work quickly from that moment.

6. Isoelectric focusing requires efficient cooling and constant temperature because isoelectric points are highly dependent on the temperature. This is optimally achieved on the horizontal cooling plate of the electrophoresis chamber, which is connected to a thermostatic circulator.

7. Sometimes it may happen that the bands are too wide. This is due to much sample applied on the gel. In this case, try to reduce the sample volume.

8. Milk with known variants as reference samples are necessary for identification of the protein variants. For evaluation it can be useful to add a further reference sample in the center of the gel.

9. Fixation, staining, and destaining should take place on a shaker with moderate agitation. Duration of the staining depends strongly on the age of the staining solution.

10. Cast the column in one single process without introducing air bubbles as the quality of the cast column material is a decisive factor for a sensible and reproducible separation of the proteins. Also avoid extreme changes in pH value or buffer conditions as they can contribute to a poor resolution of the anion-exchange chromatography.

11. Ensure that the column is always covered with buffer to avoid dry running and damage.

12. To obtain a sufficient amount of isolated casein fraction use at least 1.0–1.5 g of whole casein. Whole casein must be completely dissolved in equilibration buffer before applying onto the column. This ensures that all sample run in the column.

13. After applying of the sample, the first peak appearing is fall through material consisting of components which did not bind to the column. It is important to rinse the column with equilibration buffer until measured absorbance reaches the initial value and remains stable.

14. If not all 4 casein fractions are eluted with 0.075, 0.13 and 0.17 M NaCl, further steps with concentrations of 0.2 M and 0.25 M NaCl in equilibration buffer might be required. The recommended value of the amount of the individual steps is about 100–200 mL and varies depending on the casein fraction, the volume of the column, and the applied amount of casein sample.

15. After the appearance of the first peak within one step gradient, wait until the measured absorbance reaches baseline and remains stable. Then start with the next step gradient.

16. As soon as step gradient elution begins, start collecting fractions. We find that is the best to collect fractions of 4–6 mL for further analysis.

17. All relevant collected fractions of the peaks are analyzed with IEF. Use 20–25 μL of the individual fractions and mix with 10 μL of sample preparation. Apply 11 μL of this sample mixture on the gel. Add whole casein of the isolated sample as a reference at the outside left and right side of the gel.

18. We made the observation that lyophilized casein factions sometimes coagulate in SGF. If this is the case, dissolve them in some μL distilled water before mixing with SGF.

19. It is difficult to choose optimal pepsin:substrate ratio which reflects that existing physiologically, due to the variability of gastric and pancreatic secretions among humans, which also varies with the type of food consumed. The digestion protocol of Moreno et al. [15] kept a ratio of 1:20 (pepsin:substrate).

20. The final duodenal digestion mix contains bile salts, 7.4 mM; $CaCl_2$, 9.2 mM, and Bis–Tris, 24.7 mM.

21. For the addition of the trypsin and chymotrypsin weight ratios of 1:400:100 or 1 mg:34.5 U:0.40 U should be kept. If you take some aliquots in the gastric phase, you have to deduct the removed protein with these aliquots before adding the enzymes for the duodenal digestion.

22. Blocking with solutions containing proteins like bovine serum albumin (BSA) can lead to high background signals, which may impair the final results. JPT do not recommend a blocking step, but if needed, peptide microarray should be scanned after such a blocking step and the image should be as used a basis for following analysis.

23. A dilution series for determining the optimal serum sample concentration was performed. Dilution 1:50, 1:100, and 1:200 of two serum samples were tested on microarray. A final dilution of 1:100 was defined for the experiments.

24. Avoid exposing the fluorescently labelled secondary antibody to light for a longer time as these antibodies are sensitive towards light.

25. Perform control incubations with secondary antibody only in parallel to the epitope mapping. This ensures that detected signals are not a result of unspecific binding of the secondary antibody to the immobilized peptides.

Acknowledgement

This work is supported by the German Research Foundation (DFG ER 122/13-1, 13-2).

References

1. Erhardt G (1993) Allele frequencies of milk proteins in German cattle breeds and demonstration of alphas2-casein variants by isoelectric focusing. Arch Tierz Dummerstorf 36:145–152

2. Caroli AM, Chessa S, Erhardt GJ (2009) Invited review: milk protein polymorphisms in cattle: effect on animal breeding and human nutrition. J Dairy Sci 92:5335–5352

3. Garfin D, Ahuja S (2005) Overview. In: Garfin D, Ahuja S (eds) Handbook of isoelectric focusing and proteomics, 1st edn. Academic, London, pp 1–12

4. Baranyi M, Bosze ZS, Buchberger J et al (1993) Genetic polymorphism of milk proteins in Hungarian spotted and Hungarian grey cattle: a possible new genetic variant of β-lactoglobulin. J Dairy Sci 76:630–636

5. Erhardt G, Jäger S, Budelli E et al (2002) Genetic polymorphism of goat α_{s2}-casein (CSN1S2) and evidence for a further allele. Milchwissenschaft 57:137–140

6. Giambra IJ, Jäger S, Erhardt G (2010) Isoelectric focusing reveals additional casein variants in German sheep breeds. Small Rum Res 90:11–17

7. Shuiep ES, Giambra IJ, El Zubeir I et al (2013) Biochemical and molecular characterization of polymorphisms of α_{s1}-casein in Sudanese camel (Camelus dromedarius) milk. Int Dairy J 28:88–93

8. Hollar CM, Law AJR, Dalgleish DG et al (1991) Separation of major casein fractions using cation-exchange fast protein liquid chromatography. J Dairy Sci 74:2403–2409

9. Turhan KN, Barbano DM, Etzel MR (2003) Fractionation of caseins by anion-exchange chromatography using food-grade buffers. J Food Sci 68:1578–1583

10. Holland B, Rahimi YS, Ion Titapiccolo GI et al (2010) Short communication: separation and quantification of caseins and casein macropeptide using ion-exchange chromatography. J Dairy Sci 93:893–900

11. Amersham Biosciences (2004) Ion exchange chromatography & chromatofocusing. Princ Methods 11-0004-21

12. Moreno FJ (2007) Gastrointestinal digestion of food allergens: effect on their allergenicity. Biomed Pharmacother 61:50–60

13. Hur SJ, Lim BO, Decker EA et al (2011) In vitro human digestion models for food applications. Food Chem 125:1–12

14. Ménard O, Cattenoz T, Guillemin H et al (2014) Validation of a new in vitro dynamic system to simulate infant digestion. Food Chem 145C:1039–1045

15. Moreno FJ, Mellon FA, Wickham MSJ et al (2005) Stability of the major allergen Brazil nut 2S albumin (Ber e 1) to physiologically relevant in vitro gastrointestinal digestion. FEBS J 272:341–352

16. Benedé S, López-Expósito I, Giménez G et al (2014) In vitro digestibility of bovine β-casein with simulated and human oral and gastrointestinal fluids. Identification and IgE-reactivity of the resultant peptides. Food Chem 143:514–521

17. Beyer K, Järvinen KM, Bardina L et al (2005) IgE-binding peptides coupled to a commercial matrix as a diagnostic instrument for persistent cow's milk allergy. J Allergy Clin Immunol 116:704–705

18. Cerecedo I, Zamora J, Shreffler WG et al (2008) Mapping of the IgE and IgG4 sequential epitopes of milk allergens with a peptide microarray-based immunoassay. J Allergy Clin Immunol 122:589–594

19. Hochwallner H, Schulmeister U, Swoboda I et al (2013) Cow's milk allergy: from allergens to new forms of diagnosis, therapy and prevention. Methods 66:22–33

20. Lin J, Bardina L, Shreffler WG et al (2009) Development of a novel peptide microarray for large-scale epitope mapping of food allergens. J Allergy Clin Immunol 124:315–322

21. Ayuso R, Sánchez-Garcia S, Lin J et al (2010) Greater epitope recognition of shrimp allergens by children than by adults suggests that shrimp sensitization decreases with age. J Allergy Clin Immunol 125:1286–1293

22. Rosenfeld L, Shreffler W, Bardina L et al (2012) Walnut allergy in peanut-allergic patients: significance of sequential epitopes of walnut homologous linear epitopes of Ara h 1, 2 and 3 in relation to clinical reactivity. Int Arch Allergy Immunol 157:238–245

23. Perez-Gordo M, Pastor-Vargas C, Lin J et al (2013) Epitope mapping of the major allergen from Atlantic cod in Spanish population reveals different IgE-binding patterns. Mol Nutr Food Res 57:1283–1290

24. Martínez-Botas J, Cerecedo I, Zamora J et al (2013) Mapping of IgE and IgG4 sequential epitopes of ovomucoid with a peptide microarray immunoassay. Int Arch Allergy Immunol 161:11–20

25. Funkner A, Parthier C, Schutkowski M et al (2013) Peptide binding by catalytic domains of the protein disulfide isomerase-related protein ERp46. J Mol Biol 425:1340–1362

26. Lisson M, Lochnit G, Erhardt G (2014) In vitro gastrointestinal digestion of bovine α_{s1}- and α_{s2}-casein variants gives rise to different IgE-binding epitopes. Int Dairy J 34: 47–55

27. Lisson M, Novak N, Erhardt G (2014) Immunoglobulin E epitope mapping by microarray immunoassay reveals differences of immune response to genetic variants of caseins from different ruminant species. J Dairy Sci 97:1939–1954

INDEX

Marina Cretich and Marcella Chiari (eds.), *Peptide Microarrays: Methods and Protocols*, Methods in Molecular Biology, vol. 1352, DOI 10.1007/978-1-4939-3037-1, © Springer Science+Business Media New York 2016

Printed by Printforce, the Netherlands